D1218010

Compound Semiconductor Device Modelling

C.M. Snowden and R.E. Miles (Eds.)

Compound Semiconductor Device Modelling

With 130 Figures

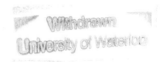

Springer-Verlag
London Berlin Heidelberg New York
Paris Tokyo Hong Kong
Barcelona Budapest

Christopher M. Snowden, BSc, MSc, PhD, CEng, MIEE, Sen. Mem. IEEE

Robert E. Miles, BSc, ARCS, PhD, CEng, MIEE, MIEEE

Microwave Solid State Group, Department of Electronic and Electrical Engineering, University of Leeds, Leeds LS2 9JT, UK

Cover illustration: Plots from a microwave transistor simulation (HEMT)

ISBN 3-540-19827-X Springer-Verlag Berlin Heidelberg New York
ISBN 0-387-19827-X Springer-Verlag New York Berlin Heidelberg

British Library Cataloguing in Publication Data
A catalogue record for this book is available from the British Library

Library of Congress Cataloging-in-Publication Data
A catalog record for this book is available from the Library of Congress

Typesetting: Camera ready by contributors
Printed by Antony Rowe Ltd., Chippenham, Wiltshire
69/3830-543210 Printed on acid-free paper

Contents

Preface

Compound semiconductor devices form the foundation of solid-state microwave and optoelectronic technologies used in many modern communication systems. In common with their low frequency counterparts, these devices are often represented using equivalent circuit models, but it is often necessary to resort to physical models in order to gain insight into the detailed operation of compound semiconductor devices. Many of the earliest physical models were indeed developed to understand the 'unusual' phenomena which occur at high frequencies. Such was the case with the Gunn and IMPATT diodes, which led to an increased interest in using numerical simulation methods. Contemporary devices often have feature sizes so small that they no longer operate within the familiar traditional framework, and hot electron or even quantum-mechanical models are required. The need for accurate and efficient models suitable for computer aided design has increased with the demand for a wider range of integrated devices for operation at microwave, millimetre and optical frequencies. The apparent complexity of equivalent circuit and physics-based models distinguishes high frequency devices from their low frequency counterparts..

Over the past twenty years a wide range of modelling techniques have emerged suitable for describing the operation of compound semiconductor devices. This book brings together for the first time the most popular techniques in everyday use by engineers and scientists. The book specifically addresses the requirements and techniques suitable for modelling GaAs, InP, ternary and quaternary semiconductor devices found in modern technology. Beginning with a review of essential numerical methods and their application to device modelling, the text continues with a chapter on equivalent circuit models and their derivation, for popular microwave and optoelectronic devices. MESFETs, HEMTs, HBTs, Gunn and IMPATT diodes are discussed in detail in separate chapters, with particular emphasis on physical models. Optoelectronic devices are dealt with as a distinct topic, highlighting the modelling of laser diodes. Large-signal and noise-modelling, critical to the development of many high frequency communication circuits, receives special attention. The Monte Carlo technique, which forms the basis of many models for semiconductor material transport parameters is presented together with its application to device simulation. Quantum models, suitable for describing many of the emerging innovative structures is included, with a particularly pragmatic treatment oriented towards engineering solutions. The modern requirements for computer aided design are addressed in chapters on highly efficient physical models (quasi-two-dimensional simulations) and the application of modelling to CAD. The text concludes with a discussion of the relevance of modelling to industry.

The inspiration for this book came from the short course on Compound Semiconductor Device Modelling held at the University of Leeds, in the Spring of 1993. This was the latest in a series of courses held in earlier years, which had previously dealt with the more general topic of semiconductor device modelling. The success of our previous edited text, published in 1989, and the interest expressed by many in industry and universities throughout the world, led us to put together this unique collection of monographs.

We have been fortunate in obtaining contributions to this text from distinguished engineers and scientists, who are internationally recognised for their expertise. We are particularly grateful to them for finding the time, from busy schedules, to compile their chapters. We hope that this text will form a useful foundation in compound semiconductor device modelling for the novice and will also act as an essential reference for the more experienced modeller.

Christopher M. Snowden
Robert E. Miles   University of Leeds, 1992

Contributors

Roel Baets
Laboratory for Electromagnetism and Acoustics, University of Gent-IMEC, Sint-Pietersnieuwsraat 41, B-9000 Gent, Belgium

D. Michael Brookbanks
GEC-Marconi Materials Research Ltd., Caswell, Towcester, NN12 8EQ. UK

Alain Cappy
Département Hyperfréquences et Semiconducteurs, Insitut d'Électronique at de Microélectronique du Nord, Université des Sciences et techniques de Lille, 59655 Villeneuve D'Ascq, France

Eric A.B. Cole
School of Mathematics, University of Leeds, Leeds, LS2 9JT. UK

Tor A. Fjeldly
Division of Physical Electronics, The Norwegian Institute of Technology, University of Trondheim, N 7034 Trondheim-NTH, Norway

Anthony J. Holden
GEC-Marconi Materials Research Ltd., Caswell, Towcester, NN12 8EQ. UK

Stavros Iezekiel
Microwave Solid State Group, Department of Electrical and Electronic Engineering, University of Leeds, Leeds, LS2 9JT. UK

Paulo Lugli
Dip. di Ingengniera Meccanica, Il Universita' di Roma "Tor Vergata", Roma via Carnevale 6. Italy

Robert E. Miles,
Microwave Solid State Group, Department of Electrical and Electronic Engineering, University of Leeds, Leeds, LS2 9JT. UK

Geert Morthier
Laboratory for Electromagnetism and Acoustics, University of Gent-IMEC, Sint-Pietersnieuwsraat 41, B-9000 Gent, Belgium

Michael J. Shur
Department of Electrical Engineering, School of Engineering and Applied Science, University of Virginia, Charlottesville, VA 22901. USA

Christopher M. Snowden

Microwave Solid State Group, Department of Electrical and Electronic Engineering, University of Leeds, Leeds, LS2 9JT. UK

Robert J. Trew

Department of Electrical and Computer Engineering, North Carolina State University, Box 7911, Raleigh, NC 27695-7911. USA

Numerical Methods and their Application to Device Modelling

Eric A.B. Cole
University of Leeds

1. INTRODUCTION

The numerical modelling of devices involves the solution of sets of coupled partial differential equations, which in turn involve the solution of simultaneous nonlinear equations after discretisation has taken place. It is not possible in a single paper to fully describe all the numerical techniques involved in this process, and in this paper we will concentrate on the finite difference approach. Descriptions of the Finite Element method (Selberherr 1984, Mobbs 1989), the Boundary Element method and the Multigrid method (Ingham 1989) can be found elsewhere. Motivation for considering certain techniques will be given by examining the equations involved in the modelling of a two-dimensional MESFET. Finite differences are introduced in section 2, the solution of simultaneous equations discussed in section 3, and the discretisation of the current continuity equations and energy equation discussed in section 4. In section 5 we give details of the implementation for the case of the MESFET , and for a one-dimensional p-n junction. Finally in section 6 we discuss parameter determination.

To motivate the mathematical discussion we introduce the equations governing the simulation of a planar sub-micron gate length GaAs MESFET It is taken as the unipolar device whose cross section in the x-y plane is shown in figure 1, with the ends of the source, gate and drain at $x = s_0$, s_1, g_0, g_1, d_0 and d_1. The electron density n, electron temperature T_e and potential ψ will all be functions of x, y and time t. The equations are:

(i) the Poisson equation

$$\nabla^2 \psi = \frac{q}{\varepsilon}(N_d - n) \tag{1}$$

where ε is the product of the permittivity of the vacuum and the relative permittivity of the semiconductor. This equation is to be solved with the boundary conditions $\psi = V_s$ on the source, $\psi = V_d$ on the drain, $\psi = V_g + \phi_b$ on the gate where ϕ_b is the built-in potential, and $\partial \psi / \partial \underset{\sim}{n} = 0$ at other parts of the

2

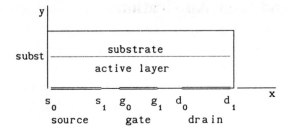

Fig. 1. Cross section of the MESFET.

boundary (In this paper, $\underset{\sim}{n}$ will represent the unit vector normal to the surface). The electric field is

$$\underset{\sim}{E} = -\nabla\psi \ . \tag{2}$$

(ii) The current continuity equation is, neglecting recombination,

$$\frac{\partial n}{\partial t} = \frac{1}{q}\nabla.\underset{\sim}{J} \tag{3}$$

where the current density $\underset{\sim}{J}$ has the form

$$\underset{\sim}{J} = -qn\mu\nabla\psi + k_B\mu T_e\nabla n + ak_B\mu n\nabla T_e + bk_B T_e n\nabla\mu \tag{4}$$

where k_B is Boltzmann's constant, μ is the mobility, and a and b are constants. Tang(1984) has taken $a=b=1$, McAndrew, Singhal and Heasell(1985) and Selberherr(1984) have taken $a=1/2$, $b=0$, while Snowden and Loret(1987) and Feng and Hintz(1988) have taken $a=1$ and $b=0$. The boundary conditions taken here are $n=2.5N_d$ on the source and drain, $n=0$ on the gate, and $\partial n/\partial\underset{\sim}{n}=0$ elsewhere.

(iii) The energy transport equation is

$$\frac{\partial W}{\partial t} = \underset{\sim}{J}.\underset{\sim}{E} - \frac{(W-W_0)}{\tau(\xi)} - \nabla.\underset{\sim}{s} \tag{5}$$

where ξ is the (position and time-dependent) average electron energy, $W=n\xi$, $\tau(\xi)$ is the energy-dependent relaxation time, and

$$\underset{\sim}{s} = -\mu W\underset{\sim}{E} - (k_B/q)\nabla(\mu W T_e) \tag{6}$$

is the energy flux. The electron energy is given by

$$\xi = \frac{3}{2}k_B T_e + \frac{1}{2}m^* v^2 \tag{7}$$

where m^* is the average effective mass and v is the velocity. The kinetic energy term is very often neglected, being typically of an order of magnitude less than the thermal energy. T_e is taken to be the lattice temperature T_0 on the three contacts, with $\partial T_e/\partial \underset{\sim}{n}=0$ elsewhere.

Having obtained a set of modelling equations, they should always be scaled to ensure a good numerical range for the variables (Selberherr 1984, Xue, Howes and Snowden 1991). We assume that this has been done before the following methods are applied.

2. FINITE DIFFERENCES

In getting a simulation going for the first time we very often use a uniform mesh for its simplicity in order to get first-time results quickly. Such a mesh fits very nicely onto devices which have simple one- and two-dimensional geometries. In the one- dimensional case, producing the uniform mesh simply amounts to splitting up the length of the device into M equal segments of length h, while in the two- dimensional device the x-y plane is split up into equal rectangles with sides of lengths h and k in the x- and y-directions respectively. The results of this method usually identify regions, for example of high field, where a finer subdivision is required, and methods are available for doing this (Selberherr 1984, DeMari 1968). Rather than increasing the number of meshpoints, which is wasteful in regions where they are not required, the discretisation can be increased on certain segments of the axes. In the two- dimensional case, this means that the region is split up into strips of varying width, which again is wasteful if there are only certain patches of the x-y plane for which this refinement is necessary. This is cured by use of boxes. The progression from one mesh type to the next involves more complicated code, and is detailed by Barton(1989). In this work we will concentrate on the second, non-uniform mesh discretisation, with the uniform mesh emerging as a special case.

Consider the two-dimensional rectangular mesh shown in figure 2. The one-dimensional discretisation will be a special case of this and the three-dimensional case will be covered by the useful phrase "in a similar manner". Mesh points will be labelled 0,1,2,..M and 0,1,2,..N in the x and y directions respectively. The general mesh point will have coordinate (x_i,y_j). The variable mesh spacings will be $h_i=x_{i+1}-x_i$ $(i=0,..M-1)$ and $k_j=y_{j+1}-y_j$ $(j=1..N-1)$. The case of uniform mesh is given by $h_i=h=$const. and $k_j=k=$const. The value $f(x_i,y_j)$ of any function f will be denoted shortly by $f_{i,j}$ while

4

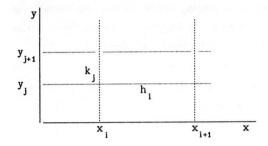

Fig. 2.   The mesh.

its   value   at   the   half-points $(x_i+h_i/2, y_j)$, $(x_i, y_j+k_j/2)$ and $(x_i+h_i/2, y_j+k_j/2)$ will   be   denoted   by $f_{i+1/2,j}$, $f_{i,j+1/2}$ and $f_{i+1/2,j+1/2}$ respectively.

In deriving the following formulae we ease the notation by temporarily resorting to one dimension. On expanding $f(x_{i+1/2})$ and $f(x_{i-1/2})$ about $x=x_i$ and subtracting we find that the half-point formula for the first derivative of f is given by

$$f_i' = \frac{f_{i+1/2} - f_{i-1/2}}{(h_i + h_{i-1})/2}$$

where the next term $-(h_i-h_{i-1})f_i''/4$ has been neglected. For a uniform mesh this   reduces   to $f_i'=(f_{i+1/2}-f_{i-1/2})/h$.   Note   that   this   last   result   is   more accurate   than   its   non-uniform   counterpart   because   the   neglected   term   is identically zero there. An alternative full-point formula is

$$f_i' = \frac{f_{i+1} + (r_i^2-1)f_i - r_i^2 f_{i-1}}{(1+r_i)h_i}$$

where $r_i = h_i/h_{i-1}$ and the next term $-h_i h_{i-1} f_i'''/6$ has been neglected. This result reduces to $f_i'=(f_{i+1}-f_{i-1})/(2h)$ in the case of a uniform mesh.

The second order derivative at x_i can be written

$$f_i'' = \frac{f'_{i+1/2} - f'_{i-1/2}}{(h_i + h_{i-1})/2} = \frac{(f_{i+1}-f_i)/h_i - (f_i-f_{i-1})/h_{i-1}}{(h_i + h_{i-1})/2} .$$

The neglected term here is $(h_i-h_{i-1})f_i'''/3$,   so again this approximation can be of an order worse than its uniform mesh counterpart.

We often need an expression for the normal derivative at a boundary, in which case it is very often zero. False nodes can be introduced outside the boundary but this increases storage requirements. Alternatively, this derivative may be expressed at x=0 in terms of values at internal nodes by

$$
f_0' = -\frac{h_1(h_1+2h_0)f_0 - (h_0+h_1)^2 f_1 + h_0^2 f_2}{h_0 h_1 (h_0+h_1)} \tag{8a}
$$

and at $x=x_M$ by

$$
f_M' = \frac{h_{M-2}(h_{M-2}+2h_{M-1})f_M - (h_{M-2}+h_{M-1})^2 f_{M-1} + h_{M-1}^2 f_{M-2}}{h_{M-1} h_{M-2} (h_{M-1}+h_{M-2})} . \tag{8b}
$$

In the case of uniform mesh, these reduce to the forms

$$
f_0' = -(3f_0 - 4f_1 + f_2)/(2h) \qquad \text{and} \qquad f_M' = (3f_M - 4f_{M-1} + f_{M-2})/(2h).
$$

Formulae involving values at more internal nodes are available (Hildebrand 1956).

Returning to the case of functions of two variables, the Poisson equation

$$
\nabla^2 \psi = g
$$

can be discretised in the form

$$
\frac{\dfrac{\psi_{i+1,j}-\psi_{i,j}}{h_i} - \dfrac{\psi_{i,j}-\psi_{i-1,j}}{h_{i-1}}}{(h_i+h_{i-1})/2} + \frac{\dfrac{\psi_{i,j+1}-\psi_{i,j}}{k_j} - \dfrac{\psi_{i,j}-\psi_{i,j-1}}{k_{j-1}}}{(k_j+k_{j-1})/2} - g_{i,j} = 0. \tag{9}
$$

In the case of uniform mesh this reduces to the standard 5-point formula

$$
p^2 \psi_{i-1,j} + p^2 \psi_{i+1,j} + \psi_{i,j-1} + \psi_{i,j+1} - 2(1+p^2)\psi_{i,j} = p^2 h^2 g_{i,j} \tag{10}
$$

where $p=k/h$. This equation has an error of order h^4, while the more accurate 9-point formula

$$
\begin{aligned}
(10p^2-2)(\psi_{i-1,j}+\psi_{i+1,j}) &+ (10-2p^2)(\psi_{i,j-1}+\psi_{i,j+1}) \\
+ (1+p^2)(\psi_{i-1,j-1}+\psi_{i+1,j-1}+\psi_{i-1,j+1}&+\psi_{i+1,j+1}-20\psi_{i,j}) \\
&= p^2 h^2 (g_{i-1,j}+g_{i+1,j}+g_{i,j-1}+g_{i,j+1}+8g_{i,j})
\end{aligned} \tag{11}
$$

has an error of order h^6 (Scraton 1987).

Very often a quantity has to be integrated across a device. The trapezium rule for this is

$$\int_{x_0}^{x_M} f(x)dx = h_0 f_0/2 + (h_0 + h_1)f_1/2 + (h_1 + h_2)f_2/2 + \ldots + h_{M-1} f_M/2$$

and has error of order $\langle h_i^2 \rangle$ (the average value of h_i^2). This should be replaced by Simpson's rule

$$\int_{x_0}^{x_M} f(x)dx = \sum_{i=0}^{M} A_i f_i \qquad \text{(M even)}$$

where $A_0 = (h_0 + h_1)(2 - r_1)/6$

$$A_M = (h_{M-2} + h_{M-1})(2 - r_{M-1}^{-1})/6$$

$$A_i = \begin{cases} (h_i + h_{i-1})(2 + r_i + r_i^{-1})/6 & \text{(i odd)} \\ (h_i + h_{i-1})(2 - r_i^{-1})/6 + (h_{i+1} + h_{i+2})(2 - r_{i+2})/6 & \text{(i even)} \end{cases}$$

This scheme has error of order $\langle h_i^4 \rangle$.

3. SOLUTION OF SIMULTANEOUS EQUATIONS

The discretisation of the differential equations of the model generally gives rise to the problem of solving a set of analytical equations of the form

$$f_i(X_0, X_1, \ldots, X_M) = 0 \qquad (i=0,1,\ldots,M), \qquad (12)$$

where the X_i will represent typical physical quantities at each point i. If there are ν physical quantities described at each point, then the top index is replaced by $\nu(M+1)-1$. The method of solution will depend on whether all the equations are linear or at least one equation is nonlinear.

A. Linear Equations

When equations (12) are linear they take the form

$$A\underset{\sim}{X} = \underset{\sim}{b} \qquad (13)$$

where A is an $(M+1) \times (M+1)$ matrix and $\underset{\sim}{b}$ is an $(M+1)$ component vector, both with elements which may depend on $\underset{\sim}{X}$. These can be solved by direct methods or by indirect methods using iteration.

(a) Direct Methods. Very often the matrix is tridiagonal and a simple solution is available. Suppose that the equations have the form

$$\alpha_i X_{i-1} + \beta_i X_i + \gamma_i X_{i+1} = \delta_i \qquad (i=0,1,..,M)$$

and the terms involving α_0 and γ_M do not appear. The standard method of eliminating down from the first equation and solving the resulting final two equations for X_{M-1} and X_M gives the solution

$$X_M = \left((\beta'_{M-1})^{-1} \gamma_{M-1} - (\alpha_M)^{-1} \beta_M \right)^{-1} \left((\beta'_{M-1})^{-1} \delta'_{M-1} - (\alpha_M)^{-1} \delta_M \right)$$

$$X_i = (\beta'_i)^{-1} \left(\delta'_i - \gamma_i X_{i+1} \right) \qquad (\ i=M-1,..,0)$$

(14)

where the β'_i and δ'_i are generated by

$$\beta'_0 = \beta_0 , \qquad \delta'_0 = \delta_0$$

$$\left. \begin{array}{l} \beta'_i = \beta_i - \alpha_i (\beta'_{i-1})^{-1} \gamma_{i-1} \\[2mm] \delta'_i = \delta_i - \alpha_i (\beta'_{i-1})^{-1} \delta'_{i-1} \end{array} \right\} \qquad (i=1,..,M-1).$$

Efficient coding of this routine is available (Varga 1962, Press et al. 1986). We can usually get away without pivoting in this case because the simulation will probably not cause problems with the sizes of the elements of the tridiagonal system. In one-dimensional simulations which involve ν physical variables at each node (for example, $\nu=3$ when the variables are n, p and ψ), the equations at each node can be grouped so that equations (14) still hold. In that case, however, the X_i, δ_i and δ'_i are ν-component vectors and the quantities α_i, β_i, γ_i and β'_i are $\nu \times \nu$ matrices. The multiplications in equations (14) are in the correct order for matrix multiplication (Kurata 1982).

Lower-upper decomposition may be used when the matrix **A** is not simply tridiagonal. In this method we decompose **A** in the form $\mathbf{A} = \mathbf{LU}$ where **L** and **U** are upper and lower triangular respectively. The equation $(\mathbf{LU})\underset{\sim}{X} = \underset{\sim}{b}$ is solved successively by solving $\mathbf{U}\underset{\sim}{X} = \underset{\sim}{Y}$ when $\underset{\sim}{Y}$ has first been solved from $\mathbf{L}\underset{\sim}{Y} = \underset{\sim}{b}$. Since **L** and **U** are triangular then the two solutions are straightforward using simple back substitution in each. It is usually necessary to use partial pivoting with scaling to control the growth of rounding errors (Smith 1985). Efficient methods for this process can be found in Press et al.(1986) and Wilkinson and Reinsch(1971).

One advantage of having a tridiagonal matrix **A** is that we need to store at most $4(M+1)$ elements α_i, β_i, γ_i and δ_i. When the matrix does not have this simple structure then these direct methods are wasteful in that storage of **A** requires $(M+1) \times (M+1)$ elements, most of which are zero. Further, methods such as LU decomposition replace many of the original zeros by nonzero values. In such a case, methods for handling large sparse matrices are necessary, and many are now available (for example, Press et al.1986).

Rounding errors can accumulate in direct elimination methods involving large matrices, and it is often said that this is a drawback of these methods compared with the iteration methods described below. However, the direct solution may itself be iterated to give a more accurate solution in the following way: let $\underset{\sim}{X}_c$ be a computed approximate solution to $A\underset{\sim}{X}=\underset{\sim}{b}$ and let $\underset{\sim}{e}$ be a correction to this, so that $A(\underset{\sim}{X}_c+\underset{\sim}{e})=\underset{\sim}{b}$. Then $\underset{\sim}{e}$ is found by solving the equation $A\underset{\sim}{e}=\underset{\sim}{b}-A\underset{\sim}{X}_c$. This equation can easily be solved for $\underset{\sim}{e}$ because it has the same structure as the original equation except for a different right hand side, and most of the work has gone into the original solution (obtaining the LU decomposition, say). This process may be performed once or iterated as many times as necessary

(b) Relaxation Methods.   Why relaxation? First, consider the iterative process

$$\underset{\sim}{X}^{k+1} \;=\; B\underset{\sim}{X}^k + \underset{\sim}{c} \tag{15}$$

where the *iteration matrix* **B** is an $(M+1) \times (M+1)$ matrix with constant coefficients. If the solution converges to $\underset{\sim}{X}$ then equation (15) is equivalent to the equation $(I-B)\underset{\sim}{X}=\underset{\sim}{c}$. For simplicity, let **B** have distinct eigenvalues λ_0, λ_1,..,λ_M with eigenfunctions $\underset{\sim}{y}_i$:

$$B\underset{\sim}{y}_i \;=\; \lambda_i\underset{\sim}{y}_i \qquad (i=0,..,M).$$

On expanding

$$\underset{\sim}{X}^0 \;=\; \sum_{i=0}^{M} \alpha_i\underset{\sim}{y}_i \qquad \text{and} \qquad \underset{\sim}{c} \;=\; \sum_{i=0}^{M} \gamma_i\underset{\sim}{y}_i$$

and substituting into equation (15), it follows that

$$\underset{\sim}{X}^k \;=\; \sum_{i=0}^{M} \left(\alpha_i\lambda_i^{\,k} + \gamma_i(1+\lambda_i+\lambda_i^{\,2}+...+\lambda_i^{\,k-1}) \right)\underset{\sim}{y}_i \;. \tag{16}$$

As iteration number $k\to\infty$, this diverges if *any* of the $|\lambda_i|>1$, but if all the $|\lambda_i|<1$ then

$$\underset{\sim}{X}^k \rightarrow \sum_{i=0}^{M} \frac{\gamma_i}{1-\lambda_i} \underset{\sim}{\chi}_i .$$

However, this convergence is slow if the *spectral radius* $\rho_s \equiv \max_i |\lambda_i|$ is close to unity. So instead take a modified iteration process

$$\underset{\sim}{X}^{k+1} = \underset{\sim}{X}^k + w \left(\underset{\sim}{c} - (I-B) \underset{\sim}{X}^k \right) = B' \underset{\sim}{X}^k + w \underset{\sim}{c} \tag{17}$$

where w is some parameter and $B' = I - w(I-B)$. The eigenvalues of B' are $1 - w(1-\lambda_i)$, and so the trick is to adjust the value of w so that the spectral radius of B' is as small as possible. This is the basis of the relaxation method. To find ρ_s, perform the iteration (15) with $\underset{\sim}{c} = \underset{\sim}{0}$. If the eigenvalue λ_{max} corresponds to the spectral radius then equation (16) gives

$$\underset{\sim}{X}^k \rightarrow (\lambda_{max})^k \sum_{i=0}^{M} \alpha_i (\lambda_i / \lambda_{max})^k \underset{\sim}{\chi}_i \rightarrow (\lambda_{max})^k \alpha_{max} \underset{\sim}{\chi}_{max} \rightarrow \lambda_{max} \underset{\sim}{X}^{k-1}$$

so that $\lambda_{max} = \| \underset{\sim}{X}^k \| / \| \underset{\sim}{X}^{k-1} \|$ for large k and some suitably defined norm $\| \ \|$. Care must be taken when implementing this routine to avoid overflows and underflows by re-scaling during each iteration. Writing the error at the k'th iteration as $\underset{\sim}{e}^k \equiv \underset{\sim}{X} - \underset{\sim}{X}^k$ so that $\underset{\sim}{e}^{k+1} = B \underset{\sim}{e}^k \rightarrow \lambda_{max} \underset{\sim}{e}^k$, then $|\underset{\sim}{e}^{k+1}| = \rho_s |\underset{\sim}{e}^k|$ and so $|\log_{10} \rho_s|$ is a measure of the number of decimal digits by which the error is asymptotically decreased by each iteration.

In our original problem given by equation (13), we may write $A = L_0 + D + U_0$ where D is diagonal and L_0 and U_0 are lower and upper triangular with zeros on the diagonals (not the L and U defined previously). This decomposition suggests the Jacobi iteration

$$D \underset{\sim}{X}^{k+1} = \underset{\sim}{b} - (L_0 + U_0) \underset{\sim}{X}^k \tag{18}$$

for which the iteration matrix is $B = -D^{-1}(L_0 + U_0)$. In Gauss-Seidel iteration we use new values as soon as they become available, and this leads to the iteration scheme

$$(L_0 + D) \underset{\sim}{X}^{k+1} = -U_0 \underset{\sim}{X}^k + \underset{\sim}{b}$$

for which the iteration matrix is $B = -(L_0 + D)^{-1} U_0$. It can be shown (Varga 1962, Young 1971, Stoer and Bulirsch 1980) that the relaxation scheme (17) is convergent only for $0 < w < 2$. If $0 < w < 1$ then we have Successive Under-Relaxation (SUR) while if $1 < w < 2$ we have Successive Over-Relaxation (SOR). Further, the Jacobi spectral radius for equation (10) is

$$\rho_J = \frac{p^2 \cos \frac{\pi}{M+1} + \cos \frac{\pi}{N+1}}{1 + p^2}$$

and the optimal choice for w is

$$w = \frac{2}{1 + (1-\rho_J^2)^{1/2}} \quad . \tag{19}$$

This is not the exact value for the equation (11) but it very often works well enough.

It is common to change the value of w as the iteration progresses. The problem is that although the value of w in equation (19) is a good value at later stages in the iteration, it is not good in the earlier stages. The ordering of the indices in equation (9) shows that the mesh can be divided into even and odd meshes, and that a half-iteration may be done on one sub-mesh followed be a half-iteration on the other. The value of w on each half sweep may be changed using *Chebyshev Acceleration* in the form

$$w^{(0)} = 1 \quad , \quad w^{(1/2)} = \frac{1}{1 - \rho_J^2/2}$$

$$w^{(k+1/2)} = \frac{1}{1 - \rho_J^2 w^{(k)}/4} \qquad (k = \tfrac{1}{2}, 1, \tfrac{3}{2}, ...)$$

As k→∞ then w becomes its optimal value. Again, Stern(1970) uses the scheme

$$\underset{\sim}{X}^{k+1} = (1-w^{k+1})\underset{\sim}{X}^k + w^{k+1}(B\underset{\sim}{X}^k + \underset{\sim}{c})$$

where

$$w^{k+1} = \frac{w^k}{1 - q_k/q_{k-1}} \quad \text{and} \quad q_k \equiv \max | B\underset{\sim}{X}^k + \underset{\sim}{c} - \underset{\sim}{X}^k |$$

on the k'th sweep. If convergence is slow then q_k/q_{k-1} is close to unity and the new value of w will be substantially larger, while w remains almost unchanged if convergence is rapid. When the solution is diverging the values of w become extreme, and in practice they are limited to between 0.1 and 0.8.

Stone's strongly implicit iteration method (Stone 1968, Jesshope 1979) solves the equation (13) when **A** is tridiagonal with outlying sub-diagonals. Standard LU decomposition fills in the zeros between the diagonals with nonzero values. Instead, in Stone's method we choose a matrix **N** and use the iteration scheme

$$(A+N)\underset{\sim}{X}^{k+1} = (A+N)\underset{\sim}{X}^k + (\underset{\sim}{b}-A\underset{\sim}{X}^k) .$$

This is solved exactly using LU decomposition on A+N. The matrix N is chosen so that the factorisation of A+N into $L_1 U_1$ involves much less arithmetic than the original factorisation A=LU, the elements of L_1 and U_1 are easily found, and $\|N\| \ll \|A\|$. This last condition ensures rapid convergence.

Reiser(1972a) uses a non-symmetric matrix whose convergence properties are affected by the complex eigenvalues of the iteration matrix. An upper limit of $w_{up} = \dfrac{2(1-a^2)^{1/2}}{b^2+(1-a^2)^{1/2}}$ was found for w, where $|\text{Real}(\lambda)| \le a \le 1$ and $|\text{Im}(\lambda)| \le b$ is a rectangle enclosing the eigenvalues of $-D^{-1}(L+U)$. He found the optimal value of w=0.825 for the case in which w_{up} =1.3.

B. Non-linear Equations

The standard method in this case is the Newton-Raphson method. For a given set $\underset{\sim}{X}^k$ let $\underset{\sim}{X}^k+\delta\underset{\sim}{X}^k$ be the true solution of equations (12). Then since

$$f_i(\underset{\sim}{X}^k+\delta\underset{\sim}{X}^k) = f_i(\underset{\sim}{X}^k) + \sum_{j=0}^{M} \frac{\partial f_i}{\partial X^k_j} \delta X^k_j + O(\delta\underset{\sim}{X}^2)$$

and the left hand side is zero, we can neglect the higher order terms and get the set of M+1 linear equations

$$\sum_{j=0}^{M} \frac{\partial f_i}{\partial X^k_j} \delta X^k_j = -f_i(\underset{\sim}{X}^k) , \qquad (i=0,..,M) . \qquad (20)$$

This set of linear equations is solved for the $\delta\underset{\sim}{X}^k$ to give a corrected solution $\underset{\sim}{X}^{k+1}=\underset{\sim}{X}^k+\delta\underset{\sim}{X}^k$ and the whole process iterated to convergence. Bad results can be found if the initial guess is too far from the true solution, but convergence is rapid when close to the limit. Hence hybrid methods can be employed to get the root close enough for the Newton-Raphson process to take over.

The partial derivatives in equation (20) must be evaluated in advance. In simulations in which the functions f_i are complicated, or in which flexibility of the model is to be built in, it is possible to differentiate the functions f_i numerically using

$$\frac{\partial f_i}{\partial X_j} \approx \frac{f(X_0,..,X_j+\Delta X_j,..,X_M)-f(X_0,..X_j,..X_M)}{\Delta X_j}$$

where the ΔX_j are suitably chosen increments. If the ΔX_j are too small then the roundoff errors can swamp the calculation, while if the ΔX_j are too large the convergence will be linear (as opposed to quadratic convergence if the partial derivatives are installed as functions). For details of how to choose the ΔX_j automatically, see Curtis and Reid(1974). Hybrid methods could be used, in which quantities which do not change between iterations (for example, the Bernoulli function of the next section) are differentiated explicitly while, for example, mobility curves fed in from other simulations could be differentiated numerically.

4. DISCRETISING THE CURRENT CONTINUITY AND ENERGY EQUATIONS

To discuss the time discretisation of the current continuity and energy equations, first consider the simplified case of constant electron temperature, only one type of carrier and no recombination-generation. The current continuity for electrons will be

$$\frac{\partial n}{\partial t} = \frac{1}{q} \nabla.\underset{\sim}{J} \tag{21}$$

where the current density $\underset{\sim}{J}(n,\mu,\underset{\sim}{E})$ is a function of n, mobility μ and electric field $\underset{\sim}{E}=-\nabla\psi$. Let Δt be the chosen timestep and let superscript denote the timepoint. Discretising equation (21) using

$$n_{i,j}^{k+1} - n_{i,j}^{k} = \frac{\Delta t}{q} \nabla.\underset{\sim}{J}_{i,j}^{k}$$

although very simple, just will not work unless Δt is excessively small (Reiser 1972b). Instead, we often use an implicit Crank-Nicholson scheme (Smith 1985) in the form

$$n_{i,j}^{k+1} - n_{i,j}^{k} = \frac{\Delta t}{2q}\left(\nabla.\underset{\sim}{J}_{i,j}^{k+1} + \nabla.\underset{\sim}{J}_{i,j}^{k}\right)$$

which is fully implicit in n and all other variables. Although this scheme is stable for all values of the time and space intervals, it is very difficult to solve. Instead, a semi-implicit scheme is often used in which $\underset{\sim}{J}_{i,j}^{k+1}$ is taken to depend on the n^{k+1} at timepoint k+1 but on all other quantities at timepoint k. This linearised semi-implicit scheme has the form

$$n_{i,j}^{k+1} - n_{i,j}^k = \frac{\Delta t}{2q}\left(\nabla.\underset{\sim}{J}(n^{k+1},\mu^k,\underset{\sim}{E}^k) + \nabla.\underset{\sim}{J}(n^k,\mu^k,\underset{\sim}{E}^k)\right) \ ,$$

and a fully written version of this equation is given by Snowden (1988). The iterative scheme needed to solve this is normally SUR. However, substantial errors and instability arise if Δt is taken too large. This limitation can be avoided if we use the Scharfetter – Gummel method in which we take an exponential variation in the carrier deviations between nodes (Scharfetter and Gummel 1969, McAndrew, Singhal and Heasell 1985).

We now return to the full description given by equations (3)-(7) and in the following produce a full generalisation of the Scharfetter – Gummel method. The constants a and b in equation (4) depend on the particular model chosen. Results will be derived for the current continuity equation and then applied to the energy equation.

Multiply the x-component of equation (4) by μ^{b-1}, assume that $\mu^{b-1}J_x$ is a constant A over the segment ($x\in[x_i,x_{i+1}]$, y_j), and that ∇n and ∇T_e are constants over the segment. The equation can be written

$$\frac{\partial}{\partial x}(n\mu^b) + \left[\frac{a}{T_e}\frac{\partial T_e}{\partial x} - \frac{q}{k_B}\frac{\partial \psi}{\partial x}\right](n\mu^b) = \frac{A}{k_B T_e}$$

which has integrating factor

$$\exp\left[\int\int\left(\frac{a}{T_e}\frac{\partial T}{\partial x} - \frac{q}{k_B}\frac{\partial \psi}{\partial x}\right)dx\right] = T_e^{\ r} \tag{22}$$

where $r \equiv a - \frac{q}{k_B}\frac{\partial \psi}{\partial x}\left(\frac{\partial T}{\partial x}\right)^{-1}$. The resulting equation is

$$\frac{\partial}{\partial x}(n\mu^b T_e^{\ r}) = A T_e^{\ r-1}/k_B$$

which can be integrated between $x=x_i$ and $x=x_{i+1}$ and re-arranged to give

$$(J_x)_{i+1/2,j} = \frac{k_B}{h_i}(\mu_{i+1/2,j})^{1-b}\left[C_a\left(\frac{q}{k_B}(\psi_{i+1,j}-\psi_{i,j}), \ T_{i+1,j}, \ T_{i,j}\right)(n\mu^b)_{i+1,j}\right.$$

$$\left. -C_a\left(\frac{q}{k_B}(\psi_{i,j}-\psi_{i+1,j}), \ T_{i,j}, \ T_{i+1,j}\right)(n\mu^b)_{i,j}\right]$$

with a similar expression for $(J_y)_{i,j+1/2}$. Here, C_a is the function defined by

$$C_a(x,y,z) \equiv \frac{x-a(y-z)}{\exp\left[\dfrac{x-a(y-z)}{z-y}\ln\left(\dfrac{z}{y}\right)\right] - 1} \qquad (y\neq z)$$

$$= p(y,z)B\left(\frac{x-a(y-z)}{p(x,y)}\right) \qquad (23)$$

where $p(y,z) \equiv \dfrac{(z-y)}{\ln(z/y)}$ and $B(t) \equiv \dfrac{t}{e^t - 1}$ is the Bernoulli function.

One important property of this C-function, namely

$$\lim_{T \to T_0} C_a(x,T_0,T) = T_0 B(x/T_0)$$

is useful for applying to models in which the electron temperature is a constant T_0. Having obtained expressions for $(J_x)_{i+1/2,j}$ etc, equation (3) is discretised as (with b=1)

$$
\begin{aligned}
n_{i,j}^{k+1} = \frac{\Delta t}{q}\Bigg\{ & \frac{2k_B}{h_i(h_{i-1}+h_i)} C_a\left(\frac{q}{k_B}(\psi_{i+1,j}^{k+1}-\psi_{i,j}^{k+1}),T_{i+1,j}^{k+1},T_{i,j}^{k+1}\right)\mu_{i+1,j}^{k+1} n_{i+1,j}^{k+1} \\[2mm]
& \frac{-2k_B}{h_i(h_{i-1}+h_i)} C_a\left(\frac{q}{k_B}(\psi_{i,j}^{k+1}-\psi_{i+1,j}^{k+1}),T_{i,j}^{k+1},T_{i+1,j}^{k+1}\right)\mu_{i,j}^{k+1} n_{i,j}^{k+1} \\[2mm]
& \frac{-2k_B}{h_{i-1}(h_{i-1}+h_i)} C_a\left(\frac{q}{k_B}(\psi_{i,j}^{k+1}-\psi_{i-1,j}^{k+1}),T_{i,j}^{k+1},T_{i-1,j}^{k+1}\right)\mu_{i,j}^{k+1} n_{i,j}^{k+1} \\[2mm]
& \frac{+2k_B}{h_{i-1}(h_{i-1}+h_i)} C_a\left(\frac{q}{k_B}(\psi_{i-1,j}^{k+1}-\psi_{i,j}^{k+1}),T_{i-1,j}^{k+1},T_{i,j}^{k+1}\right)\mu_{i-1,j}^{k+1} n_{i-1,j}^{k+1} \\[2mm]
& \frac{+2k_B}{k_j(k_{j-1}+k_j)} C_a\left(\frac{q}{k_B}(\psi_{i,j+1}^{k+1}-\psi_{i,j}^{k+1}),T_{i,j+1}^{k+1},T_{i,j}^{k+1}\right)\mu_{i,j+1}^{k+1} n_{i,j+1}^{k+1} \qquad (24) \\[2mm]
& \frac{-2k_B}{k_j(k_{j-1}+k_j)} C_a\left(\frac{q}{k_B}(\psi_{i,j}^{k+1}-\psi_{i,j+1}^{k+1}),T_{i,j}^{k+1},T_{i,j+1}^{k+1}\right)\mu_{i,j}^{k+1} n_{i,j}^{k+1} \\[2mm]
& \frac{-2k_B}{k_{j-1}(k_{j-1}+k_j)} C_a\left(\frac{q}{k_B}(\psi_{i,j}^{k+1}-\psi_{i,j-1}^{k+1}),T_{i,j}^{k+1},T_{i,j-1}^{k+1}\right)\mu_{i,j}^{k+1} n_{i,j}^{k+1} \\[2mm]
& \frac{+2k_B}{k_{j-1}(k_{j-1}+k_j)} C_a\left(\frac{q}{k_B}(\psi_{i,j-1}^{k+1}-\psi_{i,j}^{k+1}),T_{i,j-1}^{k+1},T_{i,j}^{k+1}\right)\mu_{i,j-1}^{k+1} n_{i,j-1}^{k+1}\Bigg\} \\[2mm]
& + n_{i,j}^{k}
\end{aligned}
$$

which arises from

$$\frac{\partial n_{i,j}}{\partial t} = \frac{(J_x)_{i+1/2,j} - (J_x)_{i-1/2,j}}{q(h_{i-1}+h_i)/2} + \frac{(J_y)_{i,j+1/2} - (J_y)_{i,j-1/2}}{q(k_{j-1}+k_j)/2} .$$

The discretisation of the energy equation follows in exactly the same way. Notice that equation (6) has the same form as equation (4) if we make the substitutions $a=b=1$, $n \rightarrow W$ and $\underset{\sim}{J} \rightarrow \underset{\sim}{qs}$ to get

$$(s_x)_{i+1/2,j} = -\frac{k_B}{h_i}\left[C_1\left(\frac{q}{k_B}(\psi_{i+1,j}-\psi_{i,j}), T_{i+1,j}, T_{i,j}\right)(W\mu)_{i+1,j}\right.$$

$$\left. - C_1\left(\frac{q}{k_B}(\psi_{i,j}-\psi_{i+1,j}), T_{i,j}, T_{i+1,j}\right)(W\mu)_{i,j}\right]$$

with a similar expression for $(s_y)_{i,j+1/2}$. The energy equation (5) is then discretised as

$$\frac{\partial W_{i,j}}{\partial t} =$$

$$\frac{(E_x)_{i,j}}{2}\left[(J_x)_{i+1/2,j}+(J_x)_{i-1/2,j}\right] + \frac{(E_y)_{i,j}}{2}\left[(J_y)_{i,j+1/2}+(J_y)_{i,j-1/2}\right]$$

$$- \frac{1}{h'_i}\left[(s_x)_{i+1/2,j}-(s_x)_{i-1/2,j}\right] - \frac{1}{k'_j}\left[(s_y)_{i,j+1/2}-(s_y)_{i,j-1/2}\right] \qquad (25)$$

$$- \frac{1}{\tau_{i,j}}\left[W_{i,j} - W_{0i,j}\right]$$

where $h'_i \equiv (h_{i-1}+h_i)/2$ and $k'_j \equiv (k_{j-1}+k_j)/2$. The simpler and less accurate linear approximation would enter the analysis at the integrating factor stage equation (22) by assuming that ψ and T_e are constant over the interval.

It is necessary to avoid overflows and underflows when evaluating the C-functions. The usual method (Selberherr 1984) is to approximate $B(t)$ by

$$B(t) = \begin{cases} -t & \text{for } t \le t_1 \\ t/(e^t-1) & t_1 < t \le t_2 (<0) \\ 1-t/2 & t_2 < t \le t_3 (>0) \\ te^{-t}/(1-e^{-t}) & t_3 < t \le t_4 \\ te^{-t} & t_4 < t \le t_5 \\ 0 & t_5 < t \end{cases} \qquad (26)$$

where $t_1,..,t_5$ are machine-dependent values which are calculated at the start of the routine. For example, using Turbo Pascal on a PC produced the values $t_1=-t_4=-22.8738565$, $t_2=-t_3=-2.5266464601\times10^{-3}$, and $t_5=89.4159861$. The C-function of equation (23) is evaluated by taking $p=z$ if $|z-y|<10^{-4}$. Use of the Newton–Raphson scheme requires the differentiation of the C- function with respect to x, y and z. The derivative of the Bernoulli function can be approximated by

$$B'(t) = \begin{cases} (t-2)/4 & \text{for } t_2 \le t \le t_3 \\ \dfrac{B(t)}{t}\left(1-B(t)-t\right) & \text{otherwise.} \end{cases} \qquad (27)$$

5. IMPLEMENTATION FOR THE MESFET AND P-N JUNCTION

We now revisit the MESFET model introduced in section 1 and also consider a simple p-n junction simulation. One input into these models is the doping profile. The simplest profile to model is the abrupt junction, but a useful graded junction can be written using the function $G(x)$ of Stern and Das Sarma(1984). For parameters c, ℓ and m, write

$$\eta(x) = \frac{|x-c|+\ell+2m+\left(\dfrac{\ell-m}{\pi}\right)\cos\left(\dfrac{\pi(|x-c|-(\ell+m)/2)}{\ell-m}\right)}{2(\ell+m)}$$

and take

$$G(x) = \begin{cases} 0 & \text{for } x \le x-\ell \\ \eta & c-\ell < x < c-m \\ (x-c)/(\ell+m)+0.5 & c-m \le x \le c+m \\ 1-\eta & c+m < x < c+\ell \\ 1 & c+\ell \le x \end{cases} \qquad (28)$$

The plot of $G(x)$ is shown in figure 3. It consists of a straight line between $x=c-m$ and $x=c+m$ joined to smooth curves involving the cosine function. The values of the function lie between 0 and 1, and it and its derivative are continuous. Its position can be adjusted using the value c and the sharpness of the junction can be adjusted using the values of ℓ and m.

A. Implementation for the MESFET

The implementation chosen for the MESFET described in Section 1 was that for a non-uniform mesh. Referring to figure 1, the dimensions taken were

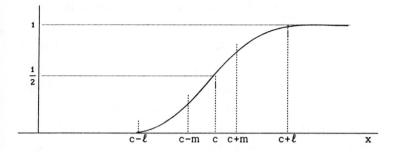

Fig.3.   The G-function used in fitting the doping profile.

s_0=0.0µm, s_1=0.2µm, g_0=0.6µm, g_1=1.1µm, d_0=1.6µm and d_1=1.8µm. The total thickness including substrate was 0.45µm. An abrupt junction at y=0.35µm was taken, although a graded junction using equation (28) is easy to program. The doping profile taken was

$$N_d = \begin{cases} 10^{23}\ m^{-3} & \text{for} \quad 0\mu m \leq y \leq 0.35\mu m \\ 10^{19}\ m^{-3} & 0.35\mu m < y \leq 0.45\mu m. \end{cases}$$

Monte Carlo simulations and experimental data on the steady state transport characteristics provide curves of ξ and τ in terms of the static electric field E_{ss} which is used as an intermediate parameter. This enables τ to be found in terms of ξ. The mobility is given by

$$\mu = \frac{300\mu_0}{T_0}\left[1 + \frac{\dfrac{8.5\times10^4 E^3}{\mu_0(1-5.3\times10^{-4}T_0)E_0^{\ 4}}}{1+(E/E_0)^4}\right]$$

where $E = 4\times10^5 Vm^{-1}$ and

$$\mu_0 = \frac{0.8}{1 + (N_d/10^{23})^{1/2}} \ .$$

The iterations for determining ψ, n and W had their own relaxation factors w_{psi}, w_n and w_T respectively. The Poisson equation (1) was solved by iterating the equation

$$\psi_{i,j} = (1-w_{psi})\psi_{i,j} + w_{psi}A_{i,j}$$

where $A_{i,j}$ is the left hand side of equation (9). Taking w_{psi}=1.4 (SOR) seemed to give the most efficient value for the iteration. The boundary condition $\partial\psi/\partial\underset{\sim}{n}$=0 was implemented using equations (8) with the derivatives zero. The current continuity equation was solved by iterating the equation

$$n_{i,j} = (1-w_n)n_{i,j} + w_n B_{i,j}$$

where $B_{i,j}$ is the right hand side of equation (24). The boundary condition $\partial n/\partial \underset{\sim}{n}=0$ was again implemented using equations (8). It was found that a value of $w_n=0.8$ (SUR) gave the fastest iteration. The energy equation (25) was similarly iterated. A value of $w_T=0.5$ seemed to give the fastest iteration.

A mixture of convergence tests was used. In the case of the ψ iteration, we required convergence at each point, that is, we required that $|\psi_{i,j}-\psi 1_{i,j}|$ had to be less than some prescribed value at every point, where $\psi 1$ was the value at the previous iteration. The same test was applied to the electron temperature T_e. In the case of the $n_{i,j}$ we applied a weaker test by requiring that only the average relative difference $\sum |n_{i,j}-n1_{i,j}|/n_{i,j}$ be smaller than some prescribed value. It was found that this average condition gave a smoother time plot of the total current

$$J_{tot} = \int_0^{0.45} \left(J_x + \varepsilon \partial E_x/\partial t\right)dy \quad .$$

Three levels of iteration were used at each time point: (a) separate iterations around ψ, n and T_e inside a total iteration; (b) no individual iterations around ψ, n and T_e but an iteration around all three, and (c) no iterations at all. Level (b) gave the more stable results for the transient solution. Level (c) gave the fastest iteration to the steady state without obtaining the proper transient solution. This faster iteration can be seen to work in the following way: because we are not iterating at each timestep then the method becomes explicit and requires a small timestep Δt. The scheme is therefore

$$n_{i,j}^{k+1} = n_{i,j}^k + \frac{\Delta t}{q}(\nabla.\underset{\sim}{J}^k) = (1-w_{i,j})n_{i,j}^k + w_{i,j}\left(n_{i,j}^k + \frac{\Delta t}{qw_{i,j}}\nabla.\underset{\sim}{J}^k\right)$$

where $w_{i,j}$ is chosen from the factors in equation (24), and then the value of Δt is adjusted accordingly. This is just an iteration scheme (17) with a position-dependent relaxation factor. This method of iterating to a steady state is cheap and cheerful, easy to program, but slow and should not be ultimately used when direct methods can do the job more quickly. For the full transient solution it is possible to vary Δt at each step in order to give the most efficient time progression (Kurata 1982).

B. The p-n Junction

As the second example we consider the one dimensional p-n junction with constant electron temperature T with the contacts on the p-type and n-type material being at $x=x_0$ and $x=x_M$ respectively. The two carrier densities are n and p, N_d and N_a are the donor and acceptor concentrations, μ_n and μ_p the mobilities, τ_n and τ_p the carrier lifetimes, ϕ_n and ϕ_p the quasi Fermi levels, $J_{\sim n}$ and $J_{\sim p}$ the current densities, R the recombination rate, ψ the potential, and E_g the bandgap energy. The equations describing the system are taken to be (Snowden 1988) the Poisson equation

$$\nabla^2 \psi = -\frac{q}{\varepsilon}(N_d - N_a - n + p), \tag{29}$$

the current continuity equations

$$\frac{dn}{dt} = \frac{1}{q}\nabla \cdot J_{\sim n} - R \quad , \qquad\qquad \frac{dp}{dt} = -\frac{1}{q}\nabla \cdot J_{\sim p} - R$$

where

$$J_{\sim n} = -q\mu_n n\nabla\psi + k_B T\mu_n \nabla n \quad , \qquad\qquad J_{\sim p} = -q\mu_p P\nabla\psi - k_B T\mu_p \nabla p \quad ,$$

and

$$n = N_c \exp\left(\frac{q}{k_B T}(\psi + \phi_n)\right) \quad , \qquad\qquad p = N_v \exp\left(-\frac{q}{k_B T}(\psi + \phi_p + E_g)\right)$$

$$R = \left(\nu_i^2 - np\right)\Big/\left(\tau_p(n+\nu_i) + \tau_n(p+\nu_i)\right) \quad , \qquad\qquad \nu_i^2 = N_c N_v \exp\left(-\frac{q}{k_B T}E_g\right)$$

The boundary conditions on the solutions are (where $C \equiv N_d - N_a$)

$$n = \frac{1}{2}\left((C^2 + 4\nu_i^2)^{1/2} + C\right) \quad , \qquad\qquad p = \frac{1}{2}\left((C^2 + 4\nu_i^2)^{1/2} - C\right) \quad ,$$

$\phi_n = \phi_p = 0$ at $x=x_0$, $\phi_n = \phi_p = $ bias at $x=x_M$, $\psi = y - \phi_p$ at $x=x_0$ and $x=x_M$ where y is the solution at each end of the equation

$$N_v \exp\left(-\frac{q}{k_B T}(y + E_g)\right) - N_c \exp\left(\frac{q}{k_B T}y\right) + N_d - N_a = 0 \quad .$$

The implementation chosen for GaAs was that of a uniform mesh with a total device length of 2.0μm, and an abrupt junction with $N_a = 10^{23}$m^{-3} and $N_d = 10^5$m^{-3} for $0 \leq x < 1.0\mu$m, and $N_a = 10^5$m^{-3} and $N_d = 10^{23}$m^{-3} for $1.0\mu m \leq x \leq 2.0\mu$m. Values $N_c = 3.9 \times 10^{23}$, $N_v = 8.3 \times 10^{24}$, $\varepsilon = 1.151 \times 10^{-10}$Fm^{-1}, $\tau_n = \tau_p = 1.0 \times 10^{-9}$s, $E_g = 1.41$eV, and

$T=300°K$ were used. The mobilities were taken as constants $\mu_e=0.4m^2V^{-1}s^{-1}$ and $\mu_p=0.026m^2V^{-1}s^{-1}$.

The equations may be solved for the steady state using the Newton–Raphson method (Kurata 1982). Alternatively, we can again use the simplified method of iterating the transient case to the steady state. At timepoint k+1 we solve the Poisson equation

$$\psi_{i-1}^{k+1} - 2\psi_i^{k+1} + \psi_{i+1}^{k+1} = h^2 g_i^k$$

where g_i^k is the right hand side of equation (29) – this is solved using the direct method based on equation (14). Then writing $a_i^{k+1}\equiv q(\psi_{i+1}^{k+1}-\psi_i^{k+1})/k_B T$, the one dimensional constant-T equivalents of equation (24), namely

$$n_i^{k+1} = \frac{\Delta t T}{qh^2}k_B\mu_n\left(B(a_i^{k+1})n_{i+1}^k-\left[B(-a_i^{k+1})+B(a_{i-1}^{k+1})\right]n_i^k+B(-a_{i-1}^{k+1})n_{i-1}^{k+1}\right) - \Delta tR_i^k + n_i^k$$

$$p_i^{k+1} = \frac{\Delta t T}{qh^2}k_B\mu_p\left(B(-a_i^{k+1})p_{i+1}^k-\left[B(a_i^{k+1})+B(-a_{i-1}^{k+1})\right]p_i^k+B(a_{i-1}^{k+1})p_{i-1}^{k+1}\right) - \Delta tR_i^k + p_i^k$$

are iterated. Note the superscripts k and k+1 on the n and p – these equations are swept *only once* from $x=x_1$ to $x=x_M$. We then pass on to the next timepoint, whereas a true transient solution would require iteration of these equations before passing on. After approximately 5000 such timesteps, ψ has settled to a sensible solution ψ^∞ but the solutions for n and p are not quite there.. We then finish off with a further loop of 100 iterations, not involving the Poisson equation or the timestep, by simultaneously solving the equations

$$B(-a_{i-1}^\infty)n_{i-1}^\infty - \left[B(a_{i-1}^\infty)+B(-a_i^\infty)\right]n_i^\infty + B(a_i^\infty)n_{i+1}^\infty = \frac{qh^2}{k_B\mu_n T}R_i^\infty$$

$$B(a_{i-1}^\infty)p_{i-1}^\infty - \left[B(-a_{i-1}^\infty)+B(a_i^\infty)\right]p_i^\infty + B(-a_i^\infty)p_{i+1}^\infty = \frac{qh^2}{k_B\mu_p T}R_i^\infty \quad ,$$

where $R_i^\infty\equiv R(n_i^\infty,p_i^\infty)$, using the direct method based on equation (14).

6. PARAMETER DETERMINATION

An important problem for circuit designers is the definition and extraction of the values of a complete set of device model parameters to guarantee correct simulation results (Clarke 1989). General-purpose constrained optimisation techniques are available for determining the parameters for many arbitrary models (Ward and Doganis 1982, Doganis and Scharfetter 1983, Wang, Lee and Chang 1986, de Graaff and Klaassen 1990). This approach uses the *Levenberg–Marquardt* method which is modified to incorporate linear constraints in order to keep the parameter values within reasonable bounds.

The model equation for the quantity I will take the form

$$I = g(V; P_1, P_2, \ldots, P_q) \tag{30}$$

where the V represents a set of measurable quantities and $P_1, P_2, \ldots, P_q$ is a set of q parameters which must be adjusted. For example, for a MOSFET, the quantity I may represent the drain current and V may represent the set $\{L, W, V_d, V_g, V_b\}$ where L and W are the channel length and width, while V_d, V_g and V_b are the drain, gate and substrate bias respectively. A set of Q measurements is made resulting in measured values (I_1^*, V_1^*), $(I_2^*, V_2^*), \ldots (I_Q^*, V_Q^*)$. The problem is to adjust the values of P_1, $P_2, \ldots, P_q$ so that a best fit of the set of predicted values $\{g_k(P_1, \ldots, P_q) \equiv g(V_k; P_1, \ldots, P_q) : k = 1, \ldots Q\}$ with the set of observed values $\{I_k^* : k = 1, \ldots Q\}$ is obtained.

Let $\underset{\sim}{P}$ be the q-component column vector $(P_1, P_2, \ldots, P_q)^T$ where the superscript T denotes the transpose of a matrix, and define the *objective function*

$$F(\underset{\sim}{P}) \equiv \sum_{k=1}^{Q} \left(\frac{I_k^* - g_k(\underset{\sim}{P})}{a_k} \right)^2 = \sum_{k=1}^{Q} R_k^2 \tag{31}$$

where

$$R_k \equiv \frac{I_k^* - g_k(\underset{\sim}{P})}{a_k} .$$

We then try to find the parameter set which minimises this function. Normally we would think of the value of the number a_k as being I_k^* so that R_k has the form $R_k = 1 - g_k(\underset{\sim}{P})/I_k^*$ which is the relative error at the k'th measurement. However, in practice, we take $a_k = \max(I_k^*, I_{min})$ for some suitably chosen value I_{min}: this ensures that no undue weight is contributed to the summation in equation (31) by very small measured values of I. Alternative forms for the objective function are

$$\sum_{k=1}^{Q} \left| \frac{I_k^* - g_k(\underset{\sim}{P})}{a_k} \right| \qquad \text{and} \qquad \max_k \left| \frac{I_k^* - g_k(\underset{\sim}{P})}{a_k} \right|$$

but form (31) is the easiest to handle analytically and will be used in this work. Writing $\underset{\sim}{R}$ as the Q-component column vector $(R_1, R_2, \ldots, R_Q)^T$, then the minimum value of the objective function F is given by

$$f_i(\underset{\sim}{P}) \equiv \frac{\partial F}{\partial P_i} = \sum_{k=1}^{Q} \frac{\partial R_k}{\partial P_i} R_k = 0 \qquad (i=1,2,\ldots,q) \qquad (32)$$

or

$$J^T \underset{\sim}{R} = \underset{\sim}{0}$$

where

$$J_{ij} \equiv \frac{\partial R_i}{\partial P_j}$$

is the Jacobian of $\underset{\sim}{R}$. Thus equations (32) are a set of q generally nonlinear equations of the form of equations (12), and the Newton-Raphson result (20) may be used in the form

$$\sum_{j=0}^{q} \frac{\partial f_i}{\partial P_j} \delta P_j = -f_i(\underset{\sim}{P}) \qquad (i=1,2,\ldots,q). \qquad (33)$$

The coefficients of the δP_j in the summation are

$$\frac{\partial f_i}{\partial P_j} = \frac{\partial}{\partial P_j} \sum_{k=1}^{Q} \frac{\partial R_k}{\partial P_i} R_k = \sum_{k=1}^{Q} \left(\frac{\partial R_k}{\partial P_i} \frac{\partial R_k}{\partial P_j} + R_k \frac{\partial^2 R_k}{\partial P_j \partial P_i} \right)$$

$$= \left(J^T J + G \right)_{ij}$$

where

$$G_{ij} \equiv \sum_{k=1}^{Q} R_k \frac{\partial^2 R_k}{\partial P_i \partial P_j} \quad .$$

Equation (33) can then be written in matrix form as

$$\left(J^T J + G \right) \delta \underset{\sim}{P} = -J^T \underset{\sim}{R} \quad . \qquad (34)$$

The Levenberg – Marquardt method approximates **G** with the diagonal matrix $\lambda \mathbf{D}$ where

$$D_{ij} = \begin{cases} (\mathbf{J}^T\mathbf{J})_{ii} & \text{for } i=j \\ 0 & \text{otherwise} \end{cases}$$

and λ is some suitably chosen non–negative constant which may be altered from one iteration to another. Using superscripts to denote the iteration number, then equation (34) gives

$$\underset{\sim}{P}^{k+1} = \underset{\sim}{P}^k - \left(\mathbf{J}^T\mathbf{J} + \lambda^k\mathbf{D} \right)^{-1} \mathbf{J}^T\underset{\sim}{R} \tag{35}$$

where, of course, $\mathbf{J} \equiv J(\underset{\sim}{P}^k)$, $\mathbf{D} \equiv D(\underset{\sim}{P}^k)$ and $\underset{\sim}{R} \equiv R(\underset{\sim}{P}^k)$. When λ^k is large, the algorithm is equivalent to the *steepest descent* method which is based on the algorithm

$$\underset{\sim}{P}^{k+1} = \underset{\sim}{P}^k - \alpha^k\nabla F(\underset{\sim}{P}^k)$$

where the scalar $\alpha^k \geq 0$ is chosen to minimise the quantity on the right hand side. When λ^k is small (or the elements of **G** are small), the method is equivalent to Gauss-Newton iteration. The Marquardt algorithm is therefore based on the idea that the optimal strategy lies somewhere between these two methods. The value of λ^k is chosen to be large for a poor initial guess.

In implementing the routine based on equation (35), the following main observations may be made:

(i) Evaluation of the Jacobian **J** is performed numerically but the process is very time consuming, and requires at least q+1 function evaluations at each iteration k. Instead, it is common to evaluate **J** this way at one iteration and then for the next q iterations to use *Broyden's rank-one correction* (Broyden 1972)

$$\mathbf{J}^{k+1} \approx \mathbf{J}^k + \frac{\{\underset{\sim}{R}^k - \underset{\sim}{R}^{k-1} - \mathbf{J}^k(\underset{\sim}{P}^k - \underset{\sim}{P}^{k-1})\}(\underset{\sim}{P}^k - \underset{\sim}{P}^{k-1})^T}{(\underset{\sim}{P}^k - \underset{\sim}{P}^{k-1})^T(\underset{\sim}{P}^k - \underset{\sim}{P}^{k-1})} \ .$$

This process eliminates the need to perform q^2 function evaluations in each cycle of q iterations.

(ii) Calculation of the objective function F, and any sum with the upper limit of Q, can be time-consuming if the measured data set is large. However, due to the fact that the I-V characteristics of most semiconductors are smooth, these summations can be performed over a carefully selected subset of the data.

(iii) Values used in the calculations can vary enormously. For example, in MOSFET parameter extraction, values of the P_i can range between 10^{-5} and 10^{17} while Jacobian entries can range between 10^{-9} and 10^3. Consequently, the matrices need to be scaled to avoid serious roundoff error: each column of J is multiplied by a normalising factor while each row of $\delta \underset{\sim}{P}^k$ is divided by the same factor. In this way, entries of J and $\delta \underset{\sim}{P}^k$ are centred on 1.

(iv) It is possible that at least one of the parameters obtained at the end of an iteration may have a value which is non-physical. Consequently, the parameters $\underset{\sim}{P}^k$ are subjected to constraint equations

$$\begin{pmatrix} I \\ -I \end{pmatrix} \underset{\sim}{P} \leq B$$

where I is the $q \times q$ identity matrix and B is a $2q \times 1$ matrix whose rows consist of upper and lower bounds on the set of parameters. This condition is applied before the next iteration takes place, and parameters are reset to these upper or lower bounds if their values are found to lie outside.

REFERENCES

Barton T (1989) Computer simulations. Semiconductor device modelling (Snowden CM ed, Springer-Verlag, London, Berlin):227-247

Broyden CG (1972) in Numerical methods for unconstrained optimisation (Murray W ed., Academic Press, New York)

Clarke ME (1989) Equivalent circuit models for silicon devices. Semiconductor device modelling (Snowden CM ed.,Springer-Verlag, London, Berlin):128-142

Curtis A and Reid JK (1974) The choice of step-length when using differences to approximate Jacobian matrices. J Inst Math Appl 13:121-126

de Graaff HC and Klaassen FM (1990) Compact transistor modelling for circuit design. Springer-Verlag, Wien, New York

De Mari A (1968) An accurate numerical steady-state one-dimensional solution of the p-n junction. Solid State Electronics 11:33-58

Doganis K and Scharfetter DL (1983) General optimisation and extraction of IC device model parameters. IEEE Trans ED-30:1219-1228

Feng YK and Hintz A (1988) Simulation of submicrometer GaAs MESFETs using a full dynamic transport model. IEEE Trans ED-35:1419-1431

Hildebrand FB (1956) Introduction to numerical analysis. McGraw-Hill, New York, Toronto, London

Ingham DB (1989) Numerical techniques - finite difference and boundary element method. Semiconductor device modelling (Snowden CM ed., Springer-Verlag, London, Berlin):34-48

Jesshope CR (1979) Comp Phys Comm 17:383-391

Kurata M (1982) Numerical analysis for semiconductor devices. Lexington Books, Lexington, Toronto

McAndrew CC, Singhal K and Heasell EL (1985) A consistent nonisothermal extension of the Scharfetter-Gummel stable difference approximation. IEEE ED Lett EDL6:446-447

Mobbs SD (1989) Numerical techniques – the finite element method. Semiconductor device modelling (Snowden CM ed., Springer-Verlag, London, Berlin):49–59

Press WH, Flannery BP, Teukolsky SA and Vetterling WT (1986) Numerical Recipes. Cambridge University Press, Cambridge, London, New York

Reiser M (1972a) Large-scale numerical simulation in semiconductor device modelling. Computer Methods in App Mech and Eng 1:17–38

Reiser M (1972b) A two-dimensional numerical FET model for dc- ac- and large signal analysis. IBM Research J, April, RZ499

Scharfetter DL and Gummel HK (1969) Large signal analysis of a silicon Read diode oscillator. IEEE Trans ED-16:64–77

Scraton RE (1987) Further numerical methods. Edward Arnold, London, Baltimore

Selberherr S (1984) Analysis and simulation of semiconductor devices. Springer-Verlag, Vienna, New York

Smith GD (1985) Numerical solution of partial differential equations: finite difference methods. Clarendon Press, Oxford

Snowden CM(ed) (1988) Semiconductor device modelling. Peter Peregrinus,London

Snowden CM and Loret D (1987) Two-dimensional hot electron models for short-gate-length GaAs MESFETs. IEEE Trans ED-34:212–223

Stern F (1970) Iteration methods for calculating self-consistent fields in semiconductor inversion layers. J Comp Phys 6:56–67

Stern F and Das Sarma S (1984) Electron energy levels in GaAs-Ga$_{1-x}$Al$_x$As heterojunctions. Phys Rev B30:840–848

Stoer J and Bulirsch R (1980) Introduction to numerical analysis. Springer-Verlag, New York

Stone HL (1968) Iterative solution of implicit approximations of multi-dimensional partial differential equations. SIAM J Num Anal 5:530–558

Tang T-W (1984) Extension of the Scharfetter-Gummel algorithm to the energy balance equation. IEEE Trans ED-31:1912–1914

Varga RS (1962) Matrix iterative algebra. Prentice Hall Inc, New Jersey

Wang S-J, Lee J-Y and Chang C-Y (1986) An efficient and reliable approach for semiconductor device parameter extraction. IEEE Trans CAD-5:170–179

Ward DE and Doganis K (1982) Optimised extraction of MOS model parameters. IEEE Trans CAD-1:163–168

Wilkinson JH and Reinsch C (1971) Handbook for automatic computation, vol 2. Springer-Verlag, Berlin

Xue H, Howes MJ and Snowden CM (1991) The modified semiconductor equations and associated algorithms for physical simulation. Int J Num Modelling: Electronic Networks, Devices and Fields 4:107–122

Young DM (1971) Iterative solution of large linear systems. Academic Press, New York

MESFET Modelling

Christopher M. Snowden
University of Leeds

INTRODUCTION

Metal semiconductor field effect transistors (MESFETs) play a central role in microwave technology over the frequency range 1 to 30GHz. They are incorporated in to monolithic microwave integrated circuits as well as being used as discrete elements. They can be used as small-signal and power amplifiers, oscillators, switches, mixers, attenuators and in digital circuits as the fundamental active device. This wide range of applications leads to the requirement for models capable of accurately representing the operation of MESFETs at DC, and microwave frequencies and in both small- and large-signal operation.

Traditionally, microwave transistors have been modelled using equivalent circuit models, where the electrical behaviour of the device at the connecting terminals is represented by a circuit consisting of linear and non-linear circuit elements. However, this type of model can only give limited insight into the physical behaviour of the device and is often difficult apply to wideband large-signal requirements. In contrast, physics-based models which describe the device in terms of the carrier transport properties and the material and geometrical attributes of the transistor allow both a physical and electrical description of the device. Physical models are by their nature more complex than equivalent circuit models and normally require powerful numerical techniques to obtain solutions to the set of partial differential equations which constitute a major part of the model. However, dramatic advances in computer technology and improvements in numerical methods have allowed physical models to address complex device structures, whilst being suitable for use on relatively modest computer systems. Today, many two- and three-dimensional models are available that can be used in conjunction with workstations, and the need for powerful mainframes (supercomputers) is diminishing.

The MESFET is a surface-oriented type of semiconductor device which may or may not have a recessed gate structure, Figure 1. The gate length of microwave MESFETs plays a fundamental role in determining the cut-off frequency of the transistor and its inherent current gain. In practice this gate length usually lies in the range 0.1 to 1.0 microns, the shorter gate lengths being used at the higher frequencies of operation. The gate width determines the overall current and power handling capacity of the FET and can vary from a few tens of microns to several millimetres. The electron current flow in MESFETs is from the source to the drain, through a narrow conducting channel which is modulated by the gate depletion region under the control of the Schottky gate contact. In most practical MESFETs, with short gate lengths and small aspect-ratios (ratio of gate length to channel thickness), substrate injection plays a significant role in determining the current flow through the device. These factors lead to a two-dimensional distribution of the physical variables, potential, carrier densities, electric field and carrier energy.

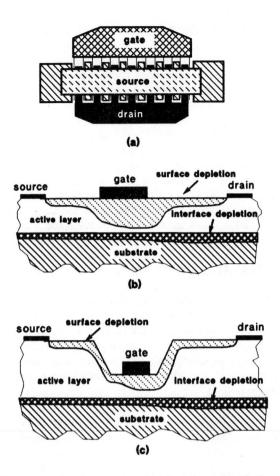

Figure 1 MESFET cross-sectional structures (a) plan view of typical power MESFET (b) planar MESFET cross-section (c) recessed-gate MESFET cross-section.

MESFETs are essentially unipolar devices and are fabricated from n-type materials (usually gallium arsenide GaAs), taking advantage of the higher mobility of electrons. Many devices have non-uniform doping profiles, which vary over many orders of magnitude within a few hundred nanometres of the surface. In some devices p⁻ layers or superlattices are included immediately below the active layer to improve carrier confinement (reducing substrate injection), and isolate the active layer from the semi-insulating bulk substrate.

The preceding outline of MESFET operation clearly reveals that in all FETs with small aspect-ratios, one-dimensional models can only provide a very simplistic view of device operation. In practice, at least a two-dimensional model is required, and in some (digital) FETs with very small gate widths a three-dimensional model may be required to fully account for fringing effects. The majority of MESFETs are satisfactorily modelled using a two-dimensional (cross-sectional) model. Physical modelling techniques applied to two-dimensional simulations can be broadly classified as either being based on carrier transport equation solutions or as Monte Carlo particle simulations. Two-dimensional FET models based on solutions to the carrier transport equations began to appear in the early 1970's, with the pioneering work of Kennedy and O'Brien (1970). In 1973 Reiser published detailed descriptions of rigorous two-dimensional silicon MESFET simulation (Reiser, 1973), which was followed by papers describing two-dimensional GaAs MESFET simulations (for example Yamaguchi, Asai and Kodera 1976). Later, papers on short channel GaAs MESFET modelling appeared (for example Wada and Frey, 1979). The role of hot electron transport in short gate length MESFETs was addressed in a number of papers published in the early 1980's (for example Carnez et al, 1980, Curtice and Yun, 1981, Cook and Frey, 1982, Snowden 1984). More complete hot electron models began to appear later, solving both energy and momentum conservation equations (for example Snowden and Loret 1987, Feng and Hintz 1988). Monte Carlo simulations for MESFETs have been reported since the mid 1970's (for example Warriner 1977, Moglestue 1982, Yoshii et al 1983, Al-Mudares 1984). In this chapter we will concentrate on models based on solutions to the carrier transport equations and the Monte Carlo technique will be presented in a later chapter.

CARRIER TRANSPORT EQUATIONS

All the early MESFET simulations were based on the drift-diffusion approximation, obtained from the time-independent Boltzmann transport equation. This is a reasonable approximation for FETs with large aspect ratios and long gate lengths (greater than 1 micron). The drift-diffusion approximation assumes that there is no electron temperature gradient ($\nabla T_e = 0$) and that consequently the average electron temperature throughout the device is fixed at the lattice temperature. The basic transport equations consist of the current continuity equations,

$$\frac{\partial n}{\partial t} = \frac{1}{q}\nabla.J_n - qG \qquad \text{for electrons} \qquad (1)$$

$$\frac{\partial p}{\partial t} = -\frac{1}{q}\nabla.J_p - qG \qquad \text{for holes} \qquad (2)$$

and the current density equations,

$$J_n = qn\mu_n E + qD_n\nabla.n \qquad \text{for electrons} \qquad (3)$$

$$J_p = qp\mu_p E - qD_p\nabla p \qquad \text{for holes} \qquad (4)$$

where n and p are the electron and hole densities, μ_n and μ_p are the electron and hole mobilities and q is the magnitude of the charge on the electron, G is the generation-recombination rate, E is the electric field, D_n and D_p are the electron and hole diffusion coefficients. The diffusion coefficients are often obtained from the well known Einstein relationships,

$$D_n = \frac{kT\mu_n}{q} \qquad\qquad D_p = \frac{kT\mu_p}{q} \qquad (5)$$

However, it has been demonstrated that the anisotropic nature of compound semiconductor materials leads to diffusion coefficients which depart significantly from the Einstein relations (Bauhann et al 1973). Some simulations have taken account of anisotropic diffusion (Snowden, 1982, Feng and Hintz 1988).

The electron and hole mobilities μ_n and μ_p are usually determined from the steady-state velocity-field characteristic. The mobilities and diffusion coefficients are a function of electron temperature and doping level (Snowden 1983) as well as electric field. The velocity-field characteristics for GaAs, which allow the mobility to be determined ($v = \mu E$), and which were obtained from contemporary Monte-Carlo models are shown in Figure 2.

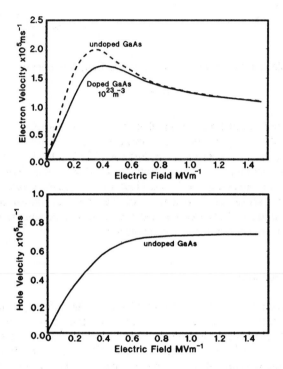

Figure 2. Velocity-field characteristics for GaAs obtained from Monte-Carlo simulations for electrons and holes in GaAs.

The generation-recombination mechanism is often represented using a Shockley-Read-Hall model for thermal effects,

$$G_{thermal} = \frac{n_i^2 - pn}{\tau_n(p + p_t) + \tau_p(n + n_t)} \tag{6}$$

where τ_n and τ_p are the electron and hole lifetimes, n_t and p_t are carrier densities dependent on the position and occupancy of the traps. n_i is the intrinsic carrier density. In some circumstances it is necessary to account for other generation-recombination mechanisms, such as impact ionisation, Auger recombination, surface recombination, and optical generation (see Selberherr 1984 and Snowden 1988 for models).

The electric field and electrostatic potential are related to the charge by the Poisson equation,

$$\nabla.(\epsilon_0 \epsilon_r E) = -q\left(n - p - N_D^+ + N_A^- - N_T^i\right) \tag{7}$$

where N_D^+ and N_A^- are the ionized donor and acceptor doping densities, N_T^i is the net ionized trap density ϵ_0 is the permittivity of free space and ϵ_r is the permittivity of the semiconductor. In the case of homogeneous structures, the Poisson equation simplifies to,

$$\nabla.E = -\nabla^2\psi = -\frac{q}{\epsilon_0 \epsilon_r}\left(n - p - N_D^+ + N_A^- - N_T^i\right) \tag{8}$$

where ψ is the electrostatic potential. The electric field E is obtained as,

$$E = -\nabla\psi \tag{9}$$

Many MESFET simulations assume that the device is entirely unipolar and utilise a reduced set of the transport equations, considering only the electron continuity and current density equations (eg Reiser 1973, Snowden 1983). This is usually adequate for normal operating conditions, but is inappropriate for devices with lightly doped p-type buffer layers or for operating conditions close to breakdown (Wada and Tomizawa 1988). The reduced set of equations is relatively fast to solve. Bipolar devices naturally require a full solution of both electron and hole transport equations, together with a suitable treatment of generation and recombination.

Hot Electron Modelling

The preceding description of carrier transport based on the drift-diffusion approximation assumes that the carrier energy distribution remains close to its equilibrium form. In short gate length MESFETs the electric field, carrier density gradient and current densities are often very large in magnitude. The high electric fields present in short channel FETs lead to substantial electron heating, and the carriers attain very high energies relative to the equilibrium levels. In these circumstances the carriers experience non-equilibrium transport conditions and their velocity may transiently exceed the equilibrium value (associated with the 'steady-state velocity field characteristics).

Non-stationary transport models may be derived from the Boltzmann Transport Equation to obtain a set of essentially hydrodynamic equations which include the current continuity (carrier conservation), energy and momentum conservation equations. These equations are available in several degrees of approximation, varying from a full dynamic transport model to a simplified energy-transport model. A useful assessment of the relative merits of the various hydrodynamic approximations is given by Feng and Hintz (1988). A comprehensive transport model, which accounts for degeneracy, multiple conduction sub-bands and spatially inhomogeneous effective masses and band edges has been described by McAndrew et al (1987).

A full set of hydrodynamic equations which describe non-stationary transport for electrons in two dimensions is, *particle (current) conservation,*

$$\frac{\partial n}{\partial t} + \nabla.(nv) = 0 \tag{10}$$

momentum conservation,

$$\frac{\partial v}{\partial t} = -\frac{qE}{m*(w)} - \frac{2}{3m*(w)n}\nabla(nw) - v.\nabla v + \frac{1}{3n}\nabla(m*(w)v^2) - \frac{v}{\tau_m(w)} \tag{11}$$

energy conservation

$$\frac{\partial w}{\partial t} = -qv.E - v.\nabla w - \frac{2}{3n}\nabla\left[nv\left(w - \frac{m*(w)}{2}v^2\right)\right] - \frac{w - w_0}{\tau_w(w)} \tag{12}$$

where the electron energy is given by,

$$w = \frac{1}{2}m*(w)v^2 + \frac{3}{2}kT_e \tag{13}$$

$$w_0 = \frac{3}{2}kT_0 \tag{14}$$

and $m*(w)$, $\tau_w(w)$ and $\tau_m(w)$ are the average effective mass and the effective energy and momentum relaxation times, T_e is the electron temperature. The parameters $m*(w)$, $\tau_w(w)$, $\tau_m(w)$, are determined from their relationship with the steady-state electric field E_{ss} obtained from Monte Carlo simulations (Carnez et al 1980, Snowden and Loret 1987). The mobility, average electron energy, energy relaxation time and upper valley occupancy for GaAs obtained from Monte Carlo simulations are shown in Figure 3.

The complete set of hydrodynamic equations is often simplified by neglecting the terms $v.\mathcal{W}$ and the kinetic energy term $\frac{1}{2}m*v^2$. Here momentum and energy conservation equations reduce to the following forms,

$$v = \frac{\tau_m(w)}{m*(w)}\left(-qE - \frac{2}{3}\nabla w - \frac{2w}{3n}\nabla n\right) \tag{15}$$

$$\frac{\partial w}{\partial t} = -qvE - v\nabla w - \frac{2}{3n}\nabla(nwv) - \frac{w - w_0}{\tau_w(w)} \tag{16}$$

A single-electron-gas approximation is often used in compound semiconductor models, to avoid the requirement for solving a set of coupled transport equations for each valley (Blotekjaer 1970). The single-electron-gas approach accounts for inter-valley transfer effects and determines the transport properties in terms of average electron energy. Simulation results for a single-electron-gas model of a 0.3 micron gate length MESFET are shown in Figure 4.

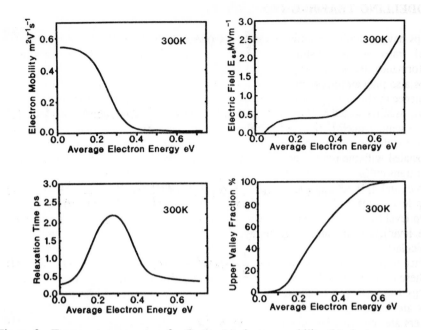

Figure 3. Transport parameters for GaAs. (a) electron mobility (b) electron energy (c) relaxation time (d) upper valley occupancy

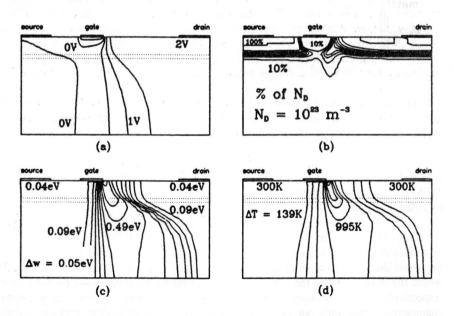

Figure 4 Simulation results for a 0.3 micron gate length MESFET using a hydrodynamic model for V_{DS}=2V, V_{DS}=0V. (a) potential, (b) electron density (c) electron energy (d) electron temperature (Santos 1991)

MODELLING TRAPPING PHENOMENA

Traps have a significant effect on compound semiconductor MESFET operation. Deep level traps in the substrate play an important role in determining the transient performance and substrate current of MESFETs (Barton and Snowden 1990). Surface traps also play an important role in determining the DC and RF operation and strongly influence the breakdown process (Barton and Ladbrooke 1985, Mizuta et al 1987). These traps, usually treated as deep level traps reflecting their position in the energy-band gap of the semiconductor, can act as either electron or hole traps. The total response of the device is due to a complex interaction between carriers in the active channel and those associated with the traps. The observed differences between steady-state DC and pulsed (fast transient) *I-V* characteristics of MESFETs and low frequency dispersion associated with g_m and g_d (Ladbrooke and Blight 1987) are attributable to the dynamic behaviour of deep level traps in the bulk material and at the surface of GaAs MESFETs, Figure 5. Deep level traps are due to both unwanted impurities such as carbon (associated with the EL2 level), and to intentional doping with chromium. The impact of bulk traps is particularly significant for ion-implanted FETs, but the time-dependent behaviour of surface traps remains important for both epitaxial and ion-implanted devices. The filling of these traps is dependent on their concentration, capture cross-section, position in the energy band-gap and on the carrier distribution in the channel and substrate. Acceptor sites are ionized when the trap energy level is below the Fermi level, whilst donor centers are ionized when the trap level is above the energy level.

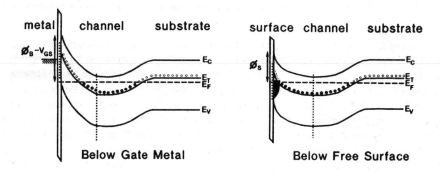

Figure 5. Energy-band diagrams below the gate metal and free-surface of a GaAs MESFET, under equilibrium conditions. Empty traps are donated by circles o, full traps by filled circles ●.

Traps play a key role in producing semi-insulating behaviour in GaAs substrates. Semi-insulating GaAs has intrinsic behaviour in addition to high resistivity properties. Semi-insulating substrates are characterized by donor- or acceptor-type deep level traps, where the donors and acceptor traps compensate residual shallow acceptors or donors respectively. Semi-insulating substrates (undoped and lightly Cr-doped) are often characterised using only one or two deep level traps - the EL2 midgap trap (dominant for LEC undoped substrates) and a shallower trap (with an activation energy of approximately 0.39eV). A second level, possibly a donor level due to Cr or the donor level EL2, occurs at E_c -0.75eV in GaAs. Alternatively, Cr levels at 0.89 and 0.45eV below the conduction band edge may fix the position of the Fermi level in Cr doped SI

GaAs. The EL2 level has been found to have an emission time constant that has a strong dependence on electric field (probably due to a shallow-dopant concentration in implanted channels). A wide range of capture cross-sections and densities have been reported for GaAs samples, Table 1.

Table 1. Characteristics of EL2 Trap in GaAs

Material Type	σ_T 10^{-18} m^{-2}	E_T eV	Density of Traps m^{-3}	Mobility m^{2}V^{-1}s^{-1} at 300K
LEC Undoped	0.04	0.745	2.3×10^{18}	0.408
LEC Cr-Doped	0.04	0.756	1.6×10^{18}	0.041
LEC (undoped)	1.5	0.75	4×10^{20}	0.450
LEC (Cr low Doped)	1.5	0.75	3.2×10^{20}	0.410
FET Material	0.1	0.76	6×10^{20}	-
LEC S.I.(donor compensated)	2.0	0.69	2×10^{21}	-
LEC (Cr low doped)	1.0	0.72	1.9×10^{21}	-
LEC Undoped	1.4	0.72	5.1×10^{20}	-

Active layers formed using epitaxial technology have lower traps densities than those produced using annealed ion-implantations. MOVPE layers grown on high quality S.I. substrates, with EL2 and E01 substrate traps with activation energies of 0.81eV and densities of 10^{20}m^{-3}, have demonstrated very low electron and hole trap densities below 10^{17} m^{-3} and 2×10^{18} m^{-3} respectively in the active layer. The dominant trap level found in epitaxial layers grown on SI GaAs substrates is the EL2 level (at 0.81 eV), which is often the only one significant enough to be identified (typically at levels of 10^{20} m^{-3}). It can be deduced that the EL2 trap level is present in the active channel under the gate as well as in substrate and buffer regions. The EL3 level is also found in epitaxial layers, usually at relatively low densities.

Table 2. Traps Located in FET Epitaxial Active Layers

Description	Doping Density m^{-3}	Type	Trap Energy eV	Trap Density m^{-3}
MOCVD	1.0×10^{23}	EL2	0.81	1.0×10^{17}
VPE	1.4×10^{23}	EL3	0.575	$3 - 6 \times 10^{20}$

The trap density in FET structures appears to be non-uniform with clustering of traps around the interface between the active channel and the substrate (and also close to the surface). It is likely that although the EL2 level (0.83 eV) is not substantially affected by ion implantation and annealing it undergoes re-distribution, accumulating near the surface. Deep level traps in the transition region at the active channel-substrate interface can cause effects such as hysteresis in drain-source characteristics and affect the mobility and carrier concentration profiles. It is interesting to note that to observe hysteresis the FET must be biased beyond the pinch-off voltage V_{po}, and that trapping effects are not observed before pinch-off in carrier concentration or mobility profiles for Cr-doped substrates. There is little evidence from C-V or transconductance data of significant concentrations of deep traps in the active channel, although FET characteristics show hysteresis when biased beyond pinch-off and then returned to $V_{GS}=0V$ (the opposite pulse sequence to DLTS). This suggests that electrons in the channel are driven into the substrate region when $|V_{GS}| > |V_{Po}|$, and that they are released back into the channel when V_{GS} returns to 0V, producing a transient peak in the current. The impact of deep-level traps in the substrate of short gate-length MESFETs on substrate current has been investigated using two-dimensional numerical simulations (Horio et al 1988, Barton and Snowden 1990). This work has shown that the substrate current becomes significant for FETs fabricated with substrates with low acceptor and trap densities.

Spectra associated with hole-type trap behaviour have been observed in the active layer of FETs as well as in substrates. Whilst it is clear that bulk GaAs substrates may have a number of hole-trap levels, it now appears likely that the hole-type spectra observed in GaAs and associated with anomalously high trap densities (typically above 10^{22} m^{-3}) may be attributed to the surface of FETs (Blight et al 1986). This is substantiated by the fact that comparison of short gate length FETs fabricated on the same wafer, but with different surface passivation, drastically modifies the 'hole-like' DLTS spectra, even eliminating them completely in some cases. Zhao has modelled the effects of surface states on DLTS spectra of GaAs MESFETs, showing that surface states do indeed exhibit hole-like DLTS signatures (Zhao 1990). This comprehensive model includes deep-level traps in the active channel under the gated region and surface states on the ungated surfaces.

Trap Dynamics - Basic Equations

A straightforward approach to determining trap filling can be based on a Shockley-Read-Hall model, with recombination through a single level. The steady-state trap filling f of a particular level is given by,

$$f = \frac{\tau_{cp}n_0 + \tau_{cn}p_0}{\tau_{cp}(n + n_0) + \tau_{cn}(p + p_0)} \tag{17}$$

where τ_{cn} and τ_{cp} are the time constants for capture and emission of electrons and holes respectively (Son and Tang, 1989). n_0 and p_0 are the electron and hole concentrations when the Fermi level is at the same energy as the trap,

$$n_0 = n_i \exp\left(\frac{E_T - E_i}{kT}\right) \qquad\qquad p_0 = \frac{n_i^2}{n_0} \tag{18}$$

In circumstances where the electron concentration is much higher then the hole concentration, which occurs in most regions of interest in MESFETs (including the substrate close to the active channel), the trap filling is modified predominantly by changes in the electron concentration and hence,

$$f \approx \frac{n}{n + n_0} \tag{19}$$

This equation can then be used to calculate the charge density due to each trap. The sign of the charge resulting from each trap is determined by its type. Donor traps are neutral when filled and positive when empty, and acceptor traps are negative when filled and neutral when empty. A four level trap model such as that used by Snowden and Barton (1990) leads to a Poisson equation of the form

$$\nabla^2 \psi = -\frac{q}{\epsilon_o \epsilon_r} \left[N_D^+ - n + p - N_A^- + N_{Dd}(1 - f_d) - N_{Ad}f_a + N_{Ds} - N_{As} \right] \tag{20}$$

where N_{Dd} and N_{Ad} are deep donors and acceptors, N_{Ds} and N_{As} are shallow donors and acceptors. The filling factors f_a and f_d refer to the filling of acceptor and donor traps. The shallow donors are assumed to be completely empty and the shallow acceptors are assumed to be completely filled.

Steady-state unipolar simulations which neglect the contribution from holes and consider only the EL2 trap level (for example) can utilize the following relationship to determine the ionized deep-donor density,

$$N_T^+ = N_T \frac{N_c \exp[-(E_c - E_T)/kT]}{n + N_c \exp[-(E_c - E_T)/kT]} \tag{21}$$

where E_c is the conduction band energy level, E_T is the deep-donor trap energy level and N_c is the effective density of states of the conduction band.

The transient stimulation of traps leads to changes in the trap filling which have time constant in the range 100ns to 50ms, where,

$$\tau_n = \tau_{cn} = \frac{1}{\sigma_n v_{th} N_T} \tag{22}$$

$$\tau_p = \tau_{cp} = \frac{1}{\sigma_p v_{th} N_T} \tag{23}$$

where τ_n and τ_p are the minority carrier lifetimes and σ_n and σ_p are the capture cross-sections for electrons and holes respectively. N_T is the trap density and v_{th} is the thermal velocity. In the case of transient simulations it is necessary to solve the rate equation for traps in addition to the other transport equations, where,

$$\frac{\partial(N_T - N_T^i)}{\partial t} = C_n n N_T^i - e_n(N_T - N_T^i) - C_p p(N_T - N_T^i) + e_p N_T^i \tag{24}$$

where the capture coefficients C_n and C_p are given by,

$$C_n = \sigma_n v_{nth} \qquad\qquad C_p = \sigma_p v_{pth} \qquad (25)$$

and e_n, e_p are the emission rates for electrons and holes respectively. The emission coefficients are related to the capture coefficient by,

$$e_n = C_n N_c g^{-1} \exp[(E_T - E_c)/kT] \qquad (26)$$

$$e_p = C_p N_v g \exp[(E_v - E_T)/kT] \qquad (27)$$

g is the degeneracy factor of the deep donor. The emission rate and trap energy levels are all temperature dependent. The EL2 level can be characterized semi-empirically using (Lo and Lee 1991),

$$e_n = 3.42 \times 10^7 T^2 \exp(-0.825.q/kT) \qquad s^{-1} \qquad (28)$$

$$E_{T EL2} = 0.75 - 3.36 \times 10^{-4} \frac{T^2}{T + 204} \qquad eV \qquad (29)$$

The behaviour of MESFETs when bias-transients occur can be explained by considering the combined influence of deep levels in the substrate, active layer and surface-states. The energy-band diagrams for regions below the gate metallisation and below the free surface some distance from the gate are shown in Figure 6 for equilibrium conditions with zero bias on all contacts. It is assumed here that the Fermi-level below the gate is pinned by the surface-states. As a consequence of this the surface-states play little role in the transient behaviour of the channel below the gate. The built-in potential of the gate ϕ_B leads to a gate depletion region of depth w_{dg}. The negative charge associated with the filled surface states between the gate and drain (or gate and source) causes band-bending near the surface creating a surface potential ϕ_{so} and an associated surface depletion region of depth w_{dso}.

If the gate bias is now stepped to a negative value at time $t=0$, the depletion region below the gate is instantly extended into the active channel towards the substrate and the fringing regions increase towards the source and drain, Figure 7. Note that here $V_{DS}=0$ hence the depletion region will be symmetrical in the active channel. At this early stage in the transient, the deep levels below the gate have not responded to the change and electrons remain trapped above the Fermi level (denoted by filled circles at the trap level E_{TA} in Figure 7). Similarly, the surface-states have not responded at this time and there is no change in the surface depletion depth at $t=0$.

At some time later, where $0<t<\tau$, where τ is the time constant of the traps, electrons will be emitted from the deep levels which are above the Fermi-level below the gate, Figure 8. This will in turn start to reduce the depth of the gate depletion region w_{dg} slightly. Gate leakage caused by the increase in reverse-bias and an increase in tunnelling due to higher electric fields at the edges of the gate, starts to increase the negative charge captured by the surface-states on both the drain and source sides of the gate. This increases the local band-bending, increasing the surface depletion depth and effective surface-potential. This effect decreases with increasing distance from the gate metal.

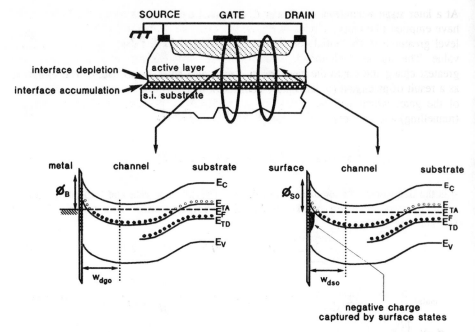

Figure 6. Depletion regions and energy-band diagrams for zero-bias equilibrium in a GaAs MESFET.

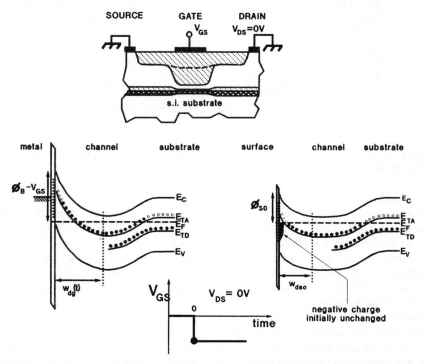

Figure 7. Depletion regions and energy-band diagrams immediately following the transient change in V_{GS}.

At a later stage where $t > \tau$ equilibrium is restored and all traps above the Fermi level have emptied (the empty circles in Figure 9). The gate depletion depth has settled at a level greater than the initial zero-bias value, but slightly less than the initial transient value. The surface-depletion depth has increased from its zero-bias value, with the greatest change closer to the gate, where the surface-states have filled to higher levels as a result of gate-leakage. This results in a variation in surface-potential on either side of the gate, which will be a function of surface-state density, gate-leakage current (tunnelling) and gate-bias.

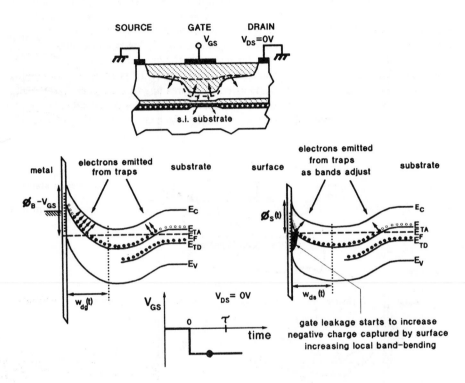

Figure 8. Depletion regions and energy-band diagrams following the transient change in V_{GS} for $t < \tau$

The presence of a SI substrate below the active channel has two main effects on the characteristics of the interface. Firstly, the potential barrier is much more abrupt than that found in FETs with epitaxial buffer layers below the active channel, where the barrier does not reach its full height for several microns below the interface. This leads to greater confinement of the electrons, reducing the steady-state substrate current in comparison with FETs with buffer layers. Secondly, the effective active channel is narrowed by the formation of a significant depletion region on the active channel side of the interface. Barton and Snowden (1990) have investigated the influence of trapping phenomena in the substrate of GaAs MESFETs, using a two-dimensional simulation, and have shown that the presence of deep levels at the substrate interface results in an increase in the output conductance of the device at higher frequencies compared with DC and low frequencies. Their simulation reveals that during the early stages of a transient the barrier height at the substrate interface is much reduced leading to injection of electrons into the substrate, transiently increasing the drain-source current. In the steady-state, following a transient, the substrate current is again reduced as a result of the increase in interface barrier height, which accompanies the increase in the negative trapped charge in deep levels in the substrate near the interface.

The above phenomena have been incorporated into a MESFET simulation, which utilizes a physical device model incorporating an energy-transport model (Snowden and Pantoja 1992). Simulation results for a 0.5 micron recessed gate MESFET fabricated using an MBE doping profile are shown in Figure 10, for pulsed I-V ($\tau < 1$ ns) and true DC characteristics. Two key features of the trap-dependent I-V characteristics emerge. Firstly, the pulsed I-V characteristics reveal a higher output conductance, with associated higher current levels at any given value of V_{DS}, and a slightly lower transconductance g_m. Secondly, in this particular FET the breakdown voltage is increased for the pulsed conditions. This latter effect is attributable to a lowering of the peak electric field in the region of the drain edge of the gate and associated spreading of the space charge region towards the drain as a consequence of the increased penetration of the substrate by hot carriers due to the reduced substrate barrier height which occurs during the transient.

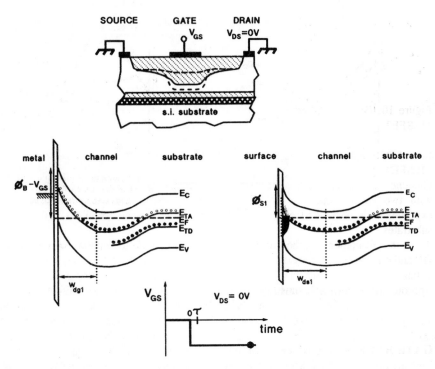

Figure 9. Depletion regions and energy-band diagrams in equilibrium for V_{GS}, for $t \gg \tau$

Look et al (1991) have shown that the presence of even a relatively thin buffer layer between the active layer and the substrate can dramatically lower the free-carrier loss to substrate interface states. They utilized a solution of Poisson's equation for the depletion approximation, to obtain an expression for the sheet free-carrier charge transferred from a conductive active layer to acceptor or donor states at interfaces or in bulk material.

approximation, to obtain an expression for the sheet free-carrier charge transferred from a conductive active layer to acceptor or donor states at interfaces or in bulk material.

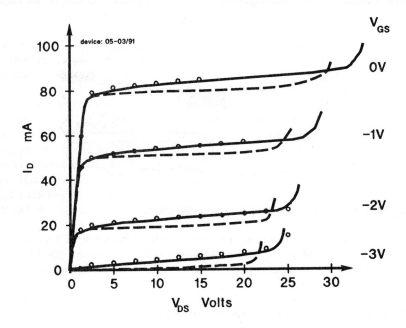

Figure 10. DC and Pulsed I-V Simulated Characteristics for a 0.5 micron gate length MESFET _____ Pulsed I-V, ------- DC. o pulsed I-V measured data.

Li and Dutton (1991) have used a two-dimensional numerical model of a recessed gate MESFET with a deep-level-trap model, to investigate the influence of trapping phenonema on small-signal a.c. operation. Implemented using PISCES-IIB their model, with two traps levels at 0.69eV (EL2) and 0.39eV, was able to reproduce the experimentally observed frequency-dependent output conductance of MESFETs with SI substrates.

It is well known that the process of surface passivation can substantially modify the DC and microwave characteristics of MESFETs, significantly influencing the breakdown voltage. At the present time very little progress has been reported in developing a rigorous non-empirically based model for passivated surfaces of MESFETs.

GATE RECESSES AND PLANAR DEVICES

All early two-dimensional FET simulations were restricted to planar structures, where the gate, source and drain contacts are located at the same level. Although many devices are still fabricated in planar form (often self-aligned digital structures), modern MESFETs (and HEMTs) are frequently fabricated with recessed gates. Single gate recesses are widely used, with recess depths of up to 0.3 microns (and commonly 0.1 microns). Recent power FET designs have utilised double-recess structures with tailored doping profiles to achieve higher breakdown voltages, Figure 11.

The presence of a recessed gate greatly modifies the behaviour of a MESFET. Recessed gate FETs should be modelled using a two-dimensional simulation capable of accurately reflecting the surface geometry, including the nature of the gate recess side-walls.

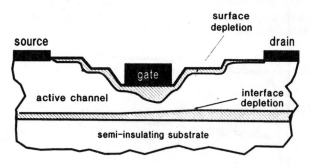

Figure 11. Double-recessed gate power FET structure.

The highly non-planar nature of single- and double-recessed gate FET surfaces leads to some difficulties in obtaining a model which can faithfully represent gate recesses. In many respects the finite-element method is more attractive for modelling the non-planar FETs.

CONTACTS AND BOUNDARY CONDITIONS

The semiconductor equations require a set of consistent boundary conditions to obtain a solution for a given simulated structure (the 'domain'). This is usually achieved by assuming that the potential and carrier concentrations are fixed at a particular value (Dirichlet conditions) or have zero derivatives (Neumann boundary conditions). Surfaces which exist between contacts or define the limit of the simulation inside the device are usually assumed to have Neumann boundary conditions, equivalent to establishing zero current flow across these boundaries. The source, drain and gate contacts are usually defined using Dirichlet boundary conditions. Hence, assuming charge neutrality and equilibrium conditions at the ohmic contacts, the boundary conditions for the source and drain contacts are given by,

$$n = \frac{1}{2}\left(\sqrt{(N_A - N_D)^2 + 4n_i^2} + (N_A - N_D)\right) \tag{30}$$

$$p = \frac{1}{2}\left(\sqrt{(N_A - N_D) + 4n_i^2} - (N_A - N_D)\right) \tag{31}$$

$$\psi_{source} = V_s - \phi_b \tag{32}$$

$$\psi_{drain} = V_D - \phi_b \tag{33}$$

where the built-in potential of the contacts ϕ_b is given by,

$$\phi_b \approx \frac{kT}{q} \ln\left(\frac{N_D^*}{n_i}\right) \qquad for \ N_D^* > N_A^-$$

and the Schottky gate boundary conditions are often determined using thermionic emission-diffusion theory, where the normal component of the current density at the gate J_n is related to the electron density at the gate n_s, by,

$$J_n.\vec{n} = qv_r(n_s - n_0) \qquad (35)$$

where the equilibrium electron density n_s is given by,

$$n_s = N_c \exp\left(-q\frac{\phi_{bi}}{kT}\right) \qquad (36)$$

where ϕ_{bi} is the built-in potential of the Schottky barrier (usually 0.7 to 0.8 V), v_r is the recombination velocity. In simple models the reverse-bias value of n_s, is often assumed to be negligible.

Alternative boundary conditions for the free surfaces may be formulated to account for surface and interface trapping effects, temperature gradients and passivation (change in dielectric constant). These types of boundary condition are discussed elsewhere (for example Selberherr 1984, Snowden 1988).

CHARGE CONTROL MODELS

The comprehensive two-dimensional solution of the electron transport equations is computationally intensive and there have been a number of more efficient approximations introduced to overcome these restraints. Many models resolve the analysis into a quasi-two-dimensional problem with an essentially one-dimensional electric field in the undepleted active channel, parallel to the surface, and a description of the two-dimensional active channel cross-section (for example Carnez et al 1980, Cook and Frey 1982, Snowden and Pantoja 1992).

In practice, the active channel is defined by the location of the depletion regions, associated with the gate, surface and substrate interface. These boundaries are often obtained using a charge-control model. A charge-control model relates the net sheet carrier concentration to the potential difference between the surface/gate and the conducting channel, V_{CG}. It may also determine the depletion depth. The model is usually obtained by solving the Poisson equation, assuming an abrupt depletion region interface, to obtain a look-up table for the relationship between the depleted and undepleted sheet charge concentrations and the voltage V_{CG}.

This process typically requires the progressive numerical solution of:

$$V_{CG}(y_d) = \frac{q}{\epsilon} \int_0^{y_d} N(y) y \, dy \qquad (37)$$

where $N(y)$ is the net ionized donor density at depth y in the profile and y_d is the depth of the depletion edge. Typical results are shown in Figure 12. The details of quasi-two-dimensional models will be discussed in a later chapter.

44

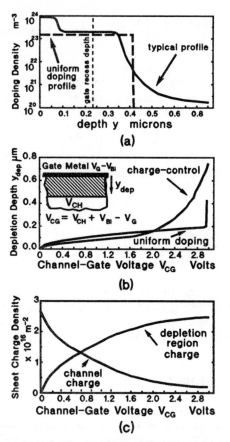

(a)

(b)

(c)

Figure 12. Charge control model for a typical MESFET. (a) doping profile (b) depletion depth as a unction of V_{CG} (c) sheet charge density as a function of V_{CG}.

PULSE-DOPED AND ULTRA-SHORT-GATE LENGTH MESFETS

Pulse-doped FETs, also known as δ-doped FETs, have active layers with narrow regions of very highly doped material, Figure 13. These structures can exhibit quantization of energy levels (as in HEMTs). In these devices it is sometimes necessary to obtain a quantum mechanical solution for the electron density and potential distribution, requiring a self-consistent solution of the Poisson and Schrödinger equations. This is usually attempted for a one-dimensional cross-section of the transistor, solving,

$$\frac{\partial}{\partial x}\left(\epsilon_r \frac{\partial V}{\partial x}\right) + \frac{q}{\epsilon_0}\left(N_D^+ - n + p - N_a^-\right) = 0 \qquad \textit{Poisson's Equation} \qquad (38)$$

$$- \frac{h^2}{8\pi^2} \frac{\partial}{\partial x} \left(\frac{1}{m*} \frac{\partial \psi_n}{\partial x} \right) + (V_{tot} - E_n)\psi_n = 0 \qquad \textit{Schrödinger's Equation} \quad (39)$$

$$n_{2d} = \frac{4\pi m^* kT}{h^2} \sum_n |\psi_n|^2 \ln\left[1 + \exp\left(\frac{E_F - E_n}{kT} \right) \right] \qquad (40)$$

where ψ_n is the wave function corresponding to sub-band n, E_n is the energy at the bottom of sub-band n, and V_{tot} is the sum of the electrostatic potential and exchange correlation potential.

The conduction band profile for a δ-doped FET is shown in Figure 14. A doping pulse of 3×10^{24} m^{-3}, 20nm thick is situated 30nm below the gate, with a background doping of 10^{20} m^{-3}. The first four sub-band eigen-energies are shown for a temperature of 300K.

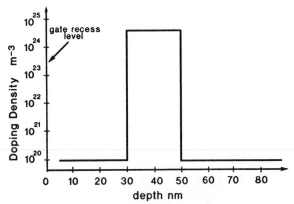

Figure 13 . Doping profile for δ-doped MESFETs

Ultra-small scale GaAs MESFETs, with gate lengths of less than 50nm have been simulated using quantum moment methods (Zhou and Ferry 1992). A set of the lowest three quantum-moment equations derived from the Wigner distribution function are analogous to the classical moment equations obtained from the Boltzmann transport equations, except that the former can be derived to preserve quantum corrections. The particle conservation equation is identical to the classical continuity equation. Expressing the quantum-moment equations in electron temperature representation, the momentum and energy conservation equations may be written as (Grubin and Kreskovsky 1989, Zhou and Ferry 1992),

momentum conservation,

$$\frac{\partial v}{\partial t} = - \frac{qE}{m*} - \frac{1}{nm*}\nabla(nkT_q) - v.\nabla v - \frac{v}{\tau_m} \qquad (41)$$

energy conservation

$$\frac{\partial T_e}{\partial t} = -\frac{1}{3}v.\nabla T_q - \frac{2}{3}\nabla(vT_q) + \frac{m*v^2}{3k}\left(\frac{2}{\tau_m} - \frac{1}{\tau_w}\right) - \frac{T_e - T_0}{\tau_w} \qquad (42)$$

where the electron energy is given by,

$$w = \frac{1}{2}m*v^2 + \frac{3}{2}kT_e + U_q \qquad (43)$$

where

$$U_q = -\frac{\hbar^2}{8m*}\nabla^2\ln(n) \qquad (44)$$

and

$$T_q = T_e + \frac{3}{2k}U_q \qquad (45)$$

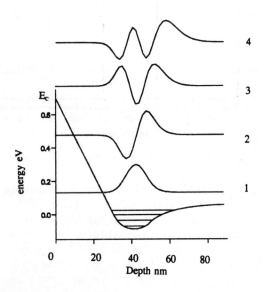

Figure 14. Quantum-mechanical solution for the conduction band profile in the δ-doped FET structure shown in Figure 13. Note the quantization in the vicinity of the doping pulse. The wavefunctions for the first four eigen-energies are shown (Drury 1992).

The quantum correction U_q tends to smooth the electron distribution, particularly in regions of rapid change. The quantum correction is most significant at low thermal energies. If $\hbar$ is set to zero the classical hydrodynamic equations are recovered. It should be noted that the above set of quantum-moment transport equations does not include heat flow.

The numerical simulations of Zhou and Ferry (1992) have shown that MESFETs with gate lengths of 24nm can exhibit very high frequency oscillations (> 500 GHz). Their work has also indicated that the effect of nonlocal quantum potential on the transient response of FETs is to soften the potential barriers in time and space, diminishing the charge-control by the gate.

NUMERICAL SOLUTION METHODS FOR MESFET SIMULATIONS

The set of semiconductor equations chosen to represent the operation of MESFETs requires discretizing over the domain of the model. Two-dimensional models are most commonly used to represent MESFET operation. Hence, the domain of the numerical model is defined by the cross-section of the device normal to the gate length. In practice a limited region of this domain is chosen for analysis, encompassing the active region of the device bounded by the edges of the ohmic source and drain contacts, the surface (and gate), and a suitable section of the substrate, Figure 15. This region is sub-divided into smaller regions to allow the semiconductor equations to be discretized across the domain of the model. The most popular numerical methods used to discretize and solve these equations are the finite-difference and finite-element techniques.

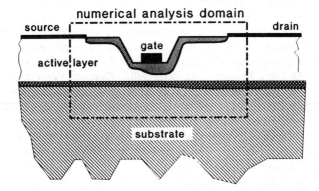

Figure 15. Typical domain of the numerical analysis chosen for MESFET simulation.

There are a wide variety of numerical methods available for solving the set of non-linear partial differential equations formed by the transport equations. The broad approach to solving the system of equations can be considered in terms of either a coupled (simultaneous) or decoupled (successive) solutions. The latter approach is more commonly termed the Gummel Algorithm (Gummel 1964), whilst the former is essentially based around a Newton scheme. The relative merits of each approach depend on the computing resources available and the nature of the simulation. Schemes based on the successive method decouple the equations and solve each equation sequentially, iterating until the solution converges to the required accuracy. This is a relatively simple method to program, but can be slow to converge, particularly in the case of bipolar devices where minority carrier levels require many iterations to reach the required solution. This approach is particularly attractive for computers with relatively small memory resources, and full time-domain two-dimensional unipolar simulations have been successfully written for use on personal computers using this method. The decoupled approach requires less memory storage than the coupled Newton when using direct solution methods. In circumstances where the carrier densities are less than or equal to the fixed charge densities, decoupled solutions are more efficient. However, when the mobile carrier densities exceed the fixed charge densities, the problem becomes more non-linear and a coupled algorithm is more efficient.

Classical models, which solve the Poisson and continuity equations, are readily implemented using coupled schemes, based on the Newton method. However, models which incorporate energy and momentum conservation equations are not as amenable to fast direct solution methods and sequential decoupled methods are more popular (Santos 1991). Newton algorithms are well known for their quadratic convergence properties, which allow a rapid simultaneous solution of the transport equations. However, the computer memory storage requirements and complex coding are considerably more demanding than the Gummel algorithm. The simultaneous solution method is particularly useful for bipolar simulations.

There are a number of difficulties which are often encountered in the numerical solution of the semiconductor equations. Numerical instabilities may arise from the solution of the continuity equation. Typically, oscillations in the carrier densities may occur when using standard approaches such as the Galerkin finite-element method and central-difference schemes (linearized Crank-Nicholson). These oscillations can be suppressed in multi-dimensional simulations using modified techniques, such as upwind methods. However, this may in turn result in numerical diffusion (Huang 1985). These problems can be overcome by careful choice of algorithm (for example Shigyo et al 1989), allowing accurate solutions to be obtained using an appropriate mesh (Gresho and Lee 1981). Discretization techniques suitable for the hydrodynamic models, incorporating energy and momentum relaxation equations, have been described by several authors (Rudan and Odeh 1986, Snowden and Loret 1987, Zhou and Ferry 1992).

The following comments apply mainly to finite-difference schemes. The Poisson equation is readily solved using either LU decomposition methods or successive over-relaxation, with a five-point central-difference discretization. In the case of Gauss-Siedel methods an SOR factor in the region of 1.8 is typically found to be close to optimum for most rectangular domains (there is no analytic method of determining the optimum relaxation factor for arbitrary non-rectangular domains). The current continuity equation normally requires a half-point finite-difference expansion for the current density terms. The $\nabla.(nv)$ term in the continuity equation may be discretized using a second-upwind method, ensuring conservation in the discretized form (see Roache 1982 for example). Forward-difference schemes are used for the time-dependent term of the continuity equation. Successive-iterative continuity schemes require under-relaxation because of the presence of complex eigenvalues in the Jacobi iteration matrix. This is often implemented using a Scharfetter-Gummel method in schemes based on electron and hole densities (for both finite-difference and finite-element methods). Crank-Nicholson schemes are frequently used in semi-implicit formulations, although fully implicit schemes are preferred allowing larger time-steps. Many bipolar simulations utilise quasi-Fermi level representation. This can lead to difficulties in obtaining accurate descriptions of n and p because of the exponential dependence of any error associated with the solutions for the quasi-Fermi levels.

The energy conservation and momentum conservation equations require careful discretization to achieve stable convergent solutions. In the energy conservation equation the $v.\nabla w$ (or $v.\nabla T_e$) term can be discretized using a simple first-upwind scheme, whereas the $\nabla(nvw)$ or the $\nabla.(vT_e)$ term is conservative in nature and requires a second-upwind method or similar treatment. The energy-conservation equation is often solved using a modified Scharfetter-Gummel scheme (Tang 1984, Feng and Hintz). In the momentum conservation equation, central-difference discretization is used for the $\triangledown(nw)$ (or $\triangledown.(nkT_e)$) term using half-point schemes, and to avoid numerical instability a first-upwind discretization is employed for the $v.\triangledown v$ term. The time-dependent nature of the energy and continuity equations can be treated using backward time differences or better still an integration and expansion method (for example Bosch and Thim 1974). A rational Runge-Kutta method is suitable for solving the momentum conservation equation.

MODELLING OF THERMAL EFFECTS IN MESFETS

The DC and rf characteristics of most semiconductor devices are strongly temperature dependent, yet the majority of simulations assume that the device is at a constant lattice temperature, usually room temperature (300K). A rigorous thermal model requires the solution of the heat flow equation,

$$c_L \rho_L \frac{\partial T_L}{\partial t} = \nabla.\left(\kappa_L \nabla T_L\right) + H_S \tag{46}$$

where c_L and ρ_L are the specific heat and density of the lattice (approximating the total heat capacity), κ_L is the lattice thermal conductivity, H_s is the heat generation and T_L is the lattice temerature. The thermal generation is often obtained from the scalar product of the electric field E and total conduction current density J,

$$H_S = J.E + qE_g G \tag{47}$$

G is the net generation-recombination rate, E_g is the energy band-gap of the semiconductor (GaAs). The first term on the right-hand side of this equation is the Joule heating and the second-term represents the energy exchange with the lattice through generation-recombination. It should be noted that band-gap narrowing effects have been neglected here. In GaAs, the thermal conductivity above 300K is given by,

$$\kappa_L = 108(T_L - 273.2)^{-0.26} \quad Wm^{-1}K^{-1} \tag{48}$$

The temperature dependence of the carrier mobilities play a fundamental role in determining the terminal current. A convenient expression for the temperature dependent low-field electron mobility is,

$$\mu_{n0} = \mu_{n0}(300)\left(\frac{300}{T_L}\right)^{2.3} \tag{49}$$

where $\mu_{no}(300)$ is the low-field electron mobility at 300K. The temperature- dependent saturation velocity is given by,

$$v_{sat} = (1.28 - 0.0015T_L)\times10^5 \quad m.s^{-1} \tag{50}$$

The net electron mobility is then obtained using the expression,

$$\mu_n(E,T) = \frac{\mu_{no}(T) + v_{sat}(T)\left(\dfrac{E^3}{E_0^4}\right)}{1 + \left(\dfrac{E}{E_0}\right)^4} \qquad (51)$$

where E_0 is the characteristic field (2.69×10^5 Vm^{-1}).

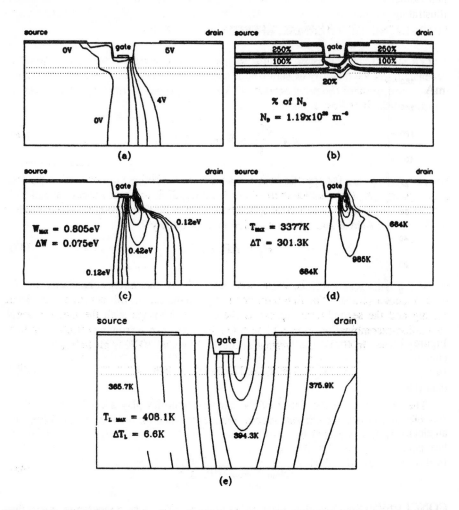

(a)   (b)

(c)   (d)

(e)

Figure 16. Simulation results for a 0.5 micron gate length MESFET obtained using a coupled thermal model ($V_{DS}=5$V, $V_{GS}=0$V). (a) potential (b) electron concentration (c) average electron energy (d) electron temperature (e) lattice temperature. (Santos 1991).

The solution of the thermal model requires careful consideration of the thermal boundary conditions. If the thermal boundary is restricted to the same domain as the carrier transport equations, equivalent third-order boundary conditions are required (Ghione et al 1987). The thermal model couples with the carrier transport model to form a coupled electro-thermal model. Ghione et al (1988) have suggested that to obtain accurate results, the domain for analysis should be extended horizontally for two to three times the source-drain spacing and to a depth of five to ten times the active region of the device.

Simulation results for a 0.5 micron gate length MESFET obtained using this approach are shown in Figure 16 ($V_{DS}=5V$, $V_{GS}=0V$). The simulation includes a full treatment of energy conservation (Santos 1991). A comparison between isothermal and non-isothermal simulated I_D-V_{DS} characteristics are shown in Figure 17 for this FET, illustrating the importance of including thermal modelling in MESFET simulations.

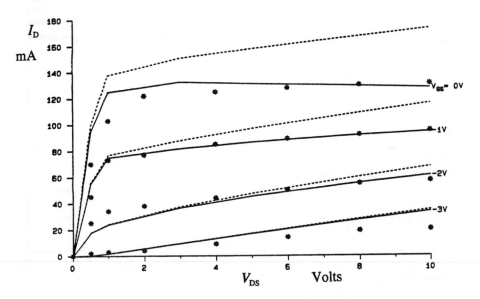

Figure 17. Comparison between isothermal and non-isothermal simulated I_D-V_{DS} characteristics for the 0.5 micron gate length FET simulated in Figure 16. - - - Isothermal model results, ___ non-isothermal model results (as in Figure 16), * measured data (DC test).

The thermal model can alternatively be implemented by dividing the analysis into an electro-thermal component over the original domain of the transport model and a non-heat generative model over a much larger region outside the transport simulation domain. In this latter case Dirichlet or Neumann boundary conditions are appropriate.

CONCLUSIONS

Contemporary MESFET modelling using physical simulation techniques has evolved to allow a wide range of key physical process to be fully accounted for. A wide variety of two-dimensional models which include hot electron effects, thermal properties and trapping phenomena are now available for describing planar and recessed gate MESFET structures. This chapter has provided an outline of many essential steps in microwave FET simulation using physical models.

REFERENCES

Al-Mudares, M., (1984) Computer Simulation Studies of Microwave FETs, PhD Thesis, University of Surrey

Bauhann, P.E., Haddad, G.I., and Masnari, N.A., (1973) Comparison of the hot electron diffusion rates for GaAs and InP, Electronic Letters, 9, 19, pp.460-461

Barton, T.M., and Ladbrooke, P.H., (1985) Dependence of maximum gate-drain potential in GaAs MESFET's upon localized surface charge, IEEE Electron Device Letters, Vol. EDL-6, No.3, pp.117-119

Barton, T.M., and Snowden, C.M., (1990) Two-dimensional numerical simulation of trapping phenomena in the substrate of GaAs MESFETs, IEEE Trans. Electron Devices, Vol. ED-37, No.6, pp.1409-1415

Blight, S.R., Wallis, R.H. and Thomas, H., (1986) Surface influence of the conductance DLTS spectra of GaAs MESFET's, IEEE Trans. Electron Devices, Vol. ED-33, No.10, pp.1447-1453

Blotekjaer, K., (1970) Transport equations for electrons in two-valley semiconductors, IEEE Trans. Electron Devices, Vol. ED-17, No.1, pp.38-47

Bosch, R., and Thim, H.W., (1974) Computer simulation of transferred electron devices using the displaced Maxwellian approach, IEEE Trans. ED-21, 1, pp.16-35

Carnez, B., Cappy, A., Kaszynski, A., Constant, E., and G. Salmer, (1980) Modeling of submicrometer gate field-effect transistor including the effects of nonstationary electron dynamics, J. Appl. Physics, 51, 1, pp.784-790

Cook, R.K. and Frey, J. (1982) An efficient technique for two-dimensional simulation of velocity overshoot in Si and GaAs devices, COMPEL, 1, 2, pp.65-87

Curtice, W, and Yun, Y-H, (1981) A temperature model for the GaAs MESFET, IEEE Trans. Electron Devices, ED-28, 8, pp.954-962

Drury, R. (1992) (diagram), University of Leeds, Leeds.

Feng, Y-K., and Hintz, A., (1988) Simulation of submicrometer GaAs MESFET's using a full dynamic transport model, IEEE Trans. Electron Devices, ED-35, pp.1419-1431

Ghione, G., Golzio P., and Naldi, C., (1987) Thermal analysis of power GaAs MESFETs, Proc. NASECODE V Conf., Ed. J.J.H. Miller, (Boole Press, Dublin), pp.195-200

Ghione, G., Golzio P., and Naldi, C., (1988) Self-consistent thermal modelling of GaAs MESFETs: a comparative analysis of power device mountings, Alta Frequenza, Vol. LVII, No.7, pp.311-319

Gresho, P.M. and Lee, R.L., (1981) Don't suppress the wiggles - they're telling you something!, Comp. and Fluids, Vol. 9, pp.223-254

Grubin, H.L., and Kreskovsky, J.P., (1989) Quantum moment balance equations and resonant tunneling substrates, Solid-State Electron., Vol. 32, No.12, pp.1071-1075

Gummel, H.K., (1964) A self-consistent interactive scheme for one-dimensional steady-state transistor calculations, IEEE Trans. Electron Devices, Vol. ED-11, No. 10, pp.455-465

Horio, K., Yanai, H. and Ikoma, T. (1988) Numerical simulation of GaAs MESFET's on the semi-insulating substrate compensated by deep traps, IEEE Trans. Electron Devices, ED-35, pp.1778-1785

Huang, M.D.D., (1985) The constant-flow patch test - a unique guideline for the evaluation of discretization schemes for the current continuity equations, IEEE Trans. Electron Devices, ED-32, No.10, pp.2139-2164

Kennedy, D.P and O'Brien, R.R, (1970) Computer aided two-dimensional analysis of the junction field effect transistor, IBM J. Res. Dev., 14, pp. 95-116

Ladbrooke, P.H. and Blight, S.R., (1987) Low-field low frequency dispersion of transconductance in GaAs MESFETs, GEC Journal of Research, Vol 5, No.4, pp.217-225

Li, Q, and Dutton, R.W. (1991) Numerical small-signal AC modeling of deep-Level-trap related frequency-dependent output conductance and capacitance for GaAs MESFET's on semi-insulating substrates, IEEE Trans. Electron devices, ED-38, No.6, pp.1285-1288

Lo, S-H, and Lee, C-P, (1991) Numerical analysis of frequency-dependent output conductance of GaAs MESFETs, IEEE Trans. Electron Devices, Vol. ED-38, No.8, pp.1693-1700

Look, D.C., Evans, K.R. and Stutz, C.E., (1991) Effects of a buffer layer on free-carrier depletion in n-type GaAs, IEEE Trans. Electron Devices, ED-38, No.6, pp.1280-1284

McAndrew, C.C., Heasell, E.L., and Singhal, (1987) A comprehensive transport model for semiconductor device simulation, IoP Semicond. Sci. Technol., 2, pp.643-648

Mizuta, H., Yamaguchi, K. and Takahashi, (1987) Surface potential effect on gate-drain avalanche breakdown in GaAs MESFET's, IEEE Trans. Electron Devices, Vol. ED-34, No. 10, pp.2027-2033

Moglestue, C., (1982) Comp. Meth. App. Mechanics and Eng., 30, pp.173-208

Pone, J.F., Castagnè, R.C., Courat, J.P. and Arnodo, C., (1982) Two-dimensional particle modeling of submicrometer gate GaAs FET's near pinch off, IEEE Trans. Electron Devices, ED-29, pp.1244-1255

Reiser, M. (1973) A two-dimensional numerical FET model for DC, AC, and large-signal analysis, IEEE Trans. Electron Devices, ED-20, 1, pp.35-45

Roache, P.J., (1982) Computational Fluid Dynamics, (Albuquerque, NM) Hermosa.

Rudan, M. and Odeh, F., (1986) Multi-dimensional discretization scheme for the hydrodynamic model of semiconductor devices, COMPEL, Vol. 5, No.3, pp.149-183

Santos, J.C.A.D., (1991) Modelling of short-gate-length metal semiconductor field-effect transistors for power amplifiers, PhD Thesis, University of Leeds

Selberherr, S., (1984) Analysis and Simulation of Semiconductor Devices (Springer-Verlag, Vienna)

Shigyo, N., Wada, T. and Yasuda S., (1989) Discretization problem for multidimensional current flow, IEEE Trans. Computer-Aided Design, Vol. CAD-8, No. 10, pp.1046-1050

Snowden, C.M. (1982) Microwave FET oscillator development based on large-signal characterisation, PhD Thesis, University of Leeds

Snowden, C.M., Howes, M.J. and Morgan D.V., (1983) Large signal modeling of GaAs MESFET operation, IEEE Trans. Electron Devices, ED-30, pp.1817-1824

Snowden, C.M., (1984) Numerical Simulation of Microwave GaAs MESFETs, Proc. Int. Conf. on Simulation of Semiconductor Devices and Processes, (Pineridge, Swansea, UK) pp.406-425

Snowden, C.M., and Loret, D. (1987) Two-dimensional hot electron models for short-gate length GaAs MESFETs, IEEE Trans. Electron Devices, ED-34, pp.212-223

Snowden, C.M., (1988) Semiconductor Device Modelling, (Peter Peregrinus, London)

Snowden, C.M., and Pantoja, R.R., (1992) GaAs MESFET Physical Models for Process-Oriented Design, IEEE Trans. MTT-40

Son, I, and Tang, T-W, (1989) Modeling of deep-level trap effects in GaAs MESFET's, IEEE Trans. Electron Devices, Vol. ED-36, No.4, pp.632-640

Tang, T.W., (1984) Extension of the Scharfetter-Gummel algorithm to the energy-balance equation, IEEE Trans. Electron Devices, Vol. ED-31, No.12, pp.1912-1914

Wada, T. and Frey, J. (1979) Physical basis of short channel MESFET operation, IEEE J. Solid-State Circuits, SC-14, 2, pp.398-412

Wada, Y. and Tomizawa, M., (1988) Drain avalanche breakdown in gallium arsenide MESFET's, IEEE Trans. Electron Devices, Vol. ED-35, No.11, pp.1765-1770

Warriner, R.A., (1977) Computer simulation of gallium arsenide field effect transistor using Monte Carlo methods, Solid State Electron. Dev., 1, pp105-109

Yamaguchi, K., Asai, S. and Kodera, H. (1975) Two-dimensional numerical analysis of stability criteria of GaAs FETs, IEEE Trans. Electron Devices, ED-23, pp.1283-1290

Yoshii, A., Tomizawa, M. and Yokoyama, K., (1983) Accurate modeling for submicrometer-gate Si and GaAs MESFET's using two-dimensional particle simulation, IEEE Trans. Electron Devices, ED-30, pp.1376-1379

Zhao, J.H.,(1990) Modeling the effects of surface states on DLTS spectra of GaAs MESFETs, IEEE Trans. Electron Devices, Vol.37, No.5, pp.1235-1244

Zhou J-R, and Ferry, D.K., (1992) Simulation of ultra-small GaAs MESFET using quantum moment methods, IEEE Trans. Electron Devices, ED-39, pp.473-478

HEMT Modelling

Michael Shur and Tor A. Fjeldly*
University of Virginia and University of Trondheim*

The High Electron Mobility Transistor (HEMT) is a contender for the coveted place as the fastest solid state device, even though this claim is being challenged from time to time by MESFET technology (see Feng et al. 1990). This prominence has caused a great deal of activity in the area of HEMT modeling. HEMTs (which are also called HFETs – for Heterostructure Field Effect Transistors or MODFETs – for Modulation Doped Field Effect Transistors, or even TEGFETs – for Two-dimensional Electron Gas Field Effect Transistors) have been modeled at different levels – from advanced (relying primarily on self-consistent Monte Carlo simulations), to intermediate (utilizing the numerical solutions of phenomenological semiconductor equations), to analytical or semi-analytical (based on the calculation of the drift current and the total sheet charge in the HEMT channel.) These models of varying degrees of sophistication have different applications. Self-consistent Monte Carlo models are indispensable for revealing the device physics and verifying novel device concepts and ideas (see Hess and Kizilyalli 1986, and Jensen et al. 1991). Two-dimensional HEMT simulators can be used to optimize the device design and fine tune the HEMT fabrication process (e.g., PRIZM 1991). The analytical or semi-analytical HEMT models help understand the device operation, are suitable for engineering design, and are used in modern circuit simulator such as AIM-Spice (see Lee at al. 1993). In this Chapter, we will primarily consider analytical models suitable for circuit simulators, but we will also discuss more sophisticated techniques, such as full two-dimesional simulation and Monte Carlo technique. Quasi two-dimensional models are considered in other Chapters of this book.

The first basic HEMT charge control model was developed by Delagebeaudeuf and Linh 1982 and Lee et al. 1983, and has been reviewed, for example, by Shur 1987, Shur 1990 and Morkoç et al. 1991. The analytical model considered in this Chapter was first reported by Fjeldly and Shur 1991 and by Lee et al. 1993.

We will start from a brief review of HEMT device structures. We will then consider band diagrams and threshold voltages followed by models for HEMT current-voltage and capacitance-voltage characteristics, and models for the HEMT gate leakage current. Finally we will briefly review some relevant aspects of two-dimensional HEMT simulation.

HFET DEVICE STRUCTURES

Fig. 1 shows typical examples of heterostructures used for HEMTs (or HFETs). In all HFET structures, a two-dimensional electron gas forms in the undoped semiconductor channel of the narrow gap semiconductor (the GaAs buffer layer in Fig. 1). Electrons are capacitively induced into this channel by the gate voltage and are supplied by the ohmic source and drain contacts (not shown in the figure). The dopants in the wide gap semiconductor layer control the device threshold voltage. Usually, these donors are separated from the two-dimensional electron gas by a thin spacer layer. Such a layer

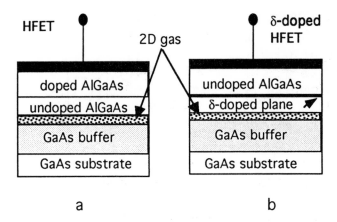

Fig. 1. Examples of modulation doped layers for HFETs:
a) conventional modulation doped layer, b) delta-doped layer.

reduces the impurity scattering of channel electrons by the remote donors.
In the structure shown in Fig. 1a, the threshold voltage is controlled by the dopants in the doped AlGaAs layer. However, deep impurities (DX centers) in AlGaAs lead to a variety of problems such as the time dependence of the device current-voltage characteristics (see Lee et al. 1983 and Subramanian 1988). These problems are somewhat reduced in the δ-doped structure shown in Fig. 1b. Other HFET structures include quantum well devices which can have a doped or undoped channel (see Fig. 2a), Inverted HFETs (see Fig. 2b), and n-channel and p-channel complementary Heterostructure Insulated Gate Field Effect Transistors (HIGFETs) (see Fig. 3), π-HFETs (see Fig. 4a), and Dipole HFETs (see Fig. 4b). Quantum well HFETs have the advantage of a better localization of the 2D electron gas in the channel, leading to a higher output resistance and smaller leakage current. In Inverted HFETs, the electron 2D gas is induced close to the gate which allows an increase of the effective gate capacitance and, hence, of the device transconductance.

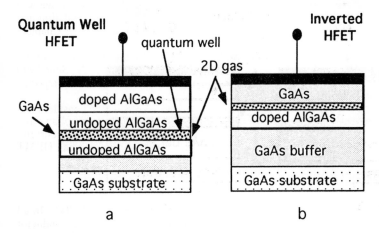

Fig. 2. Quantum well HFET (a) and Inverted HFET (b).
Note that in the quantum well HFET, the 2D electron gas is localized in a potential well created by two AlGaAs layers. In the Inverted HFET, the 2D gas is induced into the GaAs layer under the gate at the GaAs/AlGaAs heterointerface.

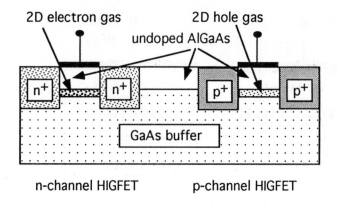

Fig. 3. Complementary *n*-channel and *p*-channel HIGFETs.

Complementary HIGFETs (C-HIGFETs) represent an analog to Si CMOS technology. Both π-HFETs and Dipole HFETs have a larger barrier separating the gate from the channel which leads to a smaller gate leakage current. All these HFET structures can be implemented in a large variety of heterostructure systems, such as AlGaAs/GaAs, AlGaAs/InGaAs/GaAs, AlInAs/InGaAs/InP, etc. This variety of structures and material systems makes analytical device models especially valuable for device and circuit designers both for optimizing the device design and predicting the device and circuit performance, and for establishing a feedback loop between the device design and the fabrication process based on a meaningful parameter extraction utilizing analytical models.

The model reviewed in this Chapter is fairly universal. It applies to *n*-channel HFETs and HIGFETs as well as to π-HFETs and Dipole FETs. With a trivial change in signs, it can equally well apply to *p*-channel devices. A similar approach (with modifications) may also apply to Inverted HFETs. The model has been recently implemented in a new circuit simulator – AIM-Spice – suitable for simulations of large scale HFET circuits, as well as for statistical analysis and quality control.

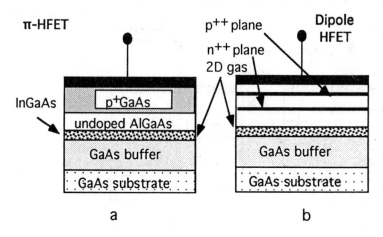

Fig. 4. π-HFET (a) and Dipole HFET (b).

HFET THRESHOLD VOLTAGE

Fig. 5 (from Lee et al. 1993) shows the HFET band diagram at flat band conditions. This figure assumes that the GaAs buffer layer is doped *p*-type with Fermi energy E_{Fp}.

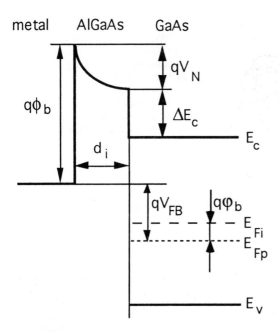

Fig. 5 Band diagram of a conventional HFET structure at flat band condition. (From Lee et al. 1993.)

As can be seen from the figure, the flat-band voltage, V_{FB}, is given by

$$V_{FB} = \phi_b - V_N - \left(\Delta E_c - \Delta E_F\right)/q \qquad (1)$$

where $q\phi_b$ is the metal-semiconductor energy barrier, ΔE_c is the conduction band discontinuity, $\Delta E_F = E_c - E_{Fp}$,

$$V_N = q\int_0^d \frac{N_d(x)}{\varepsilon_i(x)} x\, dx \qquad (2)$$

is the voltage drop across the AlGaAs layer at flat band, and x is the distance from the gate. In Eq. (2), $N_d(x)$ is the donor doping profile in the AlGaAs layer and $\varepsilon_i(x)$ is the position dependent dielectric permittivity in this layer.

In MOSFETs, the threshold voltage is usually defined as that of strong inversion, i.e., when the band bending at the semiconductor surface is such that the electron concentration at the surface is equal to the hole concentration in the bulk (see, for example, Shur 1990 for details). For the band diagram in Fig. 5, this threshold voltage is given by :

$$V_T \approx \phi_b - V_N - \left(\Delta E_c + \Delta E_{F1}\right)/q \qquad (3)$$

where $\Delta E_{F1} = E_c - E_{Fi} - q\phi_b$,

$$\phi_b = V_{th} \ln\left(\frac{N_a}{n_i}\right) \qquad (4)$$

is the potential difference between the position of the Fermi level in the GaAs layer, E_{Fp}, and the intrinsic Fermi level, E_{Fi}, n_i is the intrinsic carrier concentration in the GaAs layer, and V_{th} is the thermal voltage. (Here we assume a very low doping level in the GaAs buffer, so that we can neglect the depletion charge in the buffer.) However, strong inversion does not have much meaning in an HFET where the doping density in the GaAs buffer is very low since this condition corresponds to very low densities of the 2D electron gas. Hence, the definition of the threshold voltage in the basic HFET charge control model must differ from eq. (3). A better definition is based on a crude linear approximation of the dependence of the position of the Fermi level in the GaAs layer on the sheet density of the two-dimensional electron gas, n_s :

$$E_F \approx E_{Fo} + a_F n_s \tag{5}$$

where a_F is a constant and E_{Fo} is the intercept of this linear approximation with $n_s = 0$. Using this approximation, we obtain

$$V_{To} \approx \phi_b - V_N - (\Delta E_c + \Delta E_{Fo})/q \tag{6}$$

where $\Delta E_{Fo} = E_c - E_{Fo}$. V_{To} can be interpreted as the threshold voltage at which the carrier density of the two-dimensional gas is extrapolated to zero. This allows us to write the equation of the charge control model as follows:

$$n_s = \frac{\varepsilon_i V_{GT}}{q(d_i + \Delta d)} \tag{7}$$

where

$$\Delta d = \frac{\varepsilon_i a_F}{q} \tag{8}$$

(see Lee et al. 1983 and Byun et al. 1990 for more details).
Performing the integration in eq. (2) for uniform doping in the AlGaAs layer, we find

$$V_{To} \approx \phi_b - \frac{qN_d d_d^2}{2\varepsilon_i} - \Delta E_{ceff}/q \tag{9}$$

where $\Delta E_{ceff} = \Delta E_c + \Delta E_{Fo}$ (ΔE_{Fo} is approximately zero for GaAs). For a delta-doped structure, we obtain

$$V_{To} \approx \phi_b - \frac{qn_\delta d_\delta}{\varepsilon_s} - \Delta E_{ceff}/q \tag{10}$$

where d_δ is the distance between the metal gate and the doped plane, and n_δ is the sheet concentration of donors in the δ-doped plane. The calculated dependencies of V_{To} are shown in Fig. 6.
The dependence of the concentration of the two-dimensional gas, n_s, on the gate-source voltage can be crudely approximated by a straight line only in the above-threshold regime. This dependence also becomes invalid at large gate-source voltages because the electron quasi-Fermi level, E_{Fn}, in the AlGaAs layer may approach the bottom of the conduction band so that the carriers are induced into the conduction band minimum. Most of these carriers (with concentration n_t per unit area) are trapped in AlGaAs and do not contribute much to the device current.

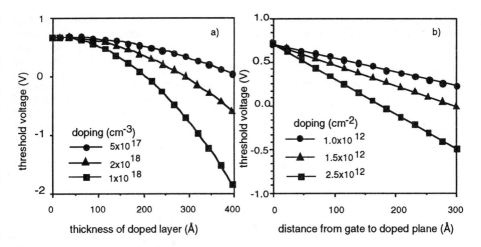

Fig. 6 Threshold voltage of conventional (a) and delta-doped (b) HFETs. (From Shur 1990.)

In Fig. 7, we show the computed dependencies of n_s and n_t on the gate-source voltage. This calculation is based on a self-consistent solution of Poisson's equation and Schrödinger's equation as described by Stern 1972. As can be seen from this figure, the concentration of electrons in the AlGaAs layer increases sharply and the slope of the n_s versus V_{GS} dependence decreases at high gate-source voltages. At such gate voltages, the dependence of n_s on V_{GS} becomes sublinear and gradually saturates.

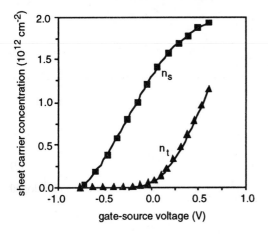

Fig. 7. Computed surface carrier densities of electrons in the two-dimensional electron gas, n_s, and in the AlGaAs layer, n_t, versus gate-source voltage. At V_{GS} close to zero, real space transfer of electrons into the AlGaAs layer becomes important, limiting the concentration of electrons induced into the quantum well. (After Chao et al. 1989, © 1989 IEEE.)

HFET CHARGE CONTROL MODEL

We will first consider a basic HFET model without taking the sublinear dependence of the electron sheet density on the gate-source voltage into account. In this case, the resulting model is identical to that of the MOSFET discussed, for example, by Shur 1990. This model leads to the following expression for the drain saturation current:

$$I'_{sat} = \frac{g'_{chi} V_{gte}}{1 + g'_{chi} R_s + \sqrt{1 + 2 g'_{chi} R_s + \left(V_{gte}/V_L\right)^2}} \tag{11}$$

Here, $V_L = F_s L$ where F_s is the saturation field and L is the gate length, R_s is the source series resistance, $g'_{chi} = q n_s' W \mu_n/L$ is the intrinsic linear channel conductance where W is the gate width and μ_n is the low-field electron mobility, and the electron sheet density in the channel can be expressed as

$$n'_s = 2n_o \ln\left[1 + \frac{1}{2} \exp\left(\frac{V_{gt}}{\eta V_{th}}\right)\right] \tag{12}$$

where n_o is the electron sheet density at threshold and η is the subthreshold ideality factor (see Fjeldly and Shur 1991 and Ruden 1989). We note that, in order for eq. (11) to reach the correct asymptotic value in the subthreshold regime, the effective gate voltage swing, V_{gte}, to be used in eq. (11) can be written as

$$V_{gte} = V_{th}\left[1 + \frac{V_{gt}}{2V_{th}} + \sqrt{\delta^2 + \left(\frac{V_{gt}}{2V_{th}} - 1\right)^2}\right] \tag{13}$$

where $V_{gt} = V_{gs} - V_T$ is the extrinsic gate voltage swing, and the parameter δ (typically, 3 to 5) determines the width of the transition region.

The value of the electron sheet density to be used in the expression for the intrinsic linear channel conductance can be approximated by

$$n_s = \frac{n'_s}{\left[1 + (n'_s/n_{max})^\gamma\right]^{-1/\gamma}} \tag{14}$$

where n_s' is given by eq. (12) and γ is a characteristic parameter for the transition to saturation in n_s. The drain current can be expressed as

$$I_{sat} = \frac{I'_{sat}}{\left[1 + (I'_{sat}/I_{max})^\gamma\right]^{-1/\gamma}} \tag{15}$$

where I_{sat}' is given by eq. (11), and

$$I_{max} = q n_{max} v_s W \tag{16}$$

is an upper limit for the channel current determined by n_{max} and the saturation velocity, v_s. Eqs. (15) and (16) account for real space carrier transfer into the wide band gap semiconductor layer (as shown in Fig. 7).

As was discussed by Lee et al. 1993, the *I-V* characteristics for FETs can be described by one unified expression which combines the linear and the saturation regions, and which includes the subthreshold regime:

$$I_{ds} \approx \frac{g_{ch} V_{ds}}{\left[1 + \left(g_{ch} V_{ds} / I_{sat}\right)^m\right]^{1/m}} \tag{17}$$

Here, m is a parameter that determines the shape of the characteristics in the knee region, $g_{ch} = g_{chi}/(1 + g_{chi}R_t)$ is the extrinsic linear channel conductance and $R_t = R_s + R_d$ is the sum of the source and the drain series resistances.

CAPACITANCE MODEL

As was discussed by Fjeldly and Shur (1991), the charge control in FETs can be described in terms of a unified expression for the channel capacitance, valid for all bias voltages, with the differential gate-channel capacitance per unit area written as

$$c_{gc} = \frac{c_a c_b}{c_a + c_b} \tag{18}$$

where c_a is the above-threshold contribution and c_b is the subthreshold contribution. For HFETs, c_{gc} (through c_a) has to reflect the limitations imposed on the maximum carrier concentration in the channel, n_s, caused by the finite energy band discontinuity between the GaAs channel and the AlGaAs layer. Eqs. (14) and (15) already incorporate both the above-threshold and the subthreshold behavior of the channel charge density, including the saturation of n_s at large gate voltage swings. Hence, at small drain-source voltages, c_{gc} can be expressed as

$$c_{gc} = q \frac{dn_s}{dV_{gs}} \approx \frac{c'_{gc}}{\left[1 + \left(n'_s/n_{max}\right)^\gamma\right]^{1+1/\gamma}} \tag{19}$$

where V_{gs} is the intrinsic gate-source voltage, n_{max} is the maximum electron sheet density in the channel, n_s' is the ideal unified electron sheet density (see eq. (12)) and c_{gc}' is the corresponding unified channel capacitance of an infinitely deep potential well. When n_s' becomes comparable to or larger than n_{max}, c_{gs} will drop noticeably from its ideal value given by c_{gc}'.

At large gate voltage swings, the saturation in n_s is accompanied by an increase in the sheet density, n_t, of electrons in the AlGaAs layer owing to the real space transfer (see Fig. 7). This added charge contributes to the total differential gate capacitance per unit area, c_{gtot}, which can be represented as a parallel connection of c_{gc} and the capacitance c_{g1} associated with the AlGaAs layer. We can assume that the charge sheet density qn_t is located at a fixed distance d_1 from the gate, independent of the gate-source voltage (which is quite accurate for a delta-doped HFET). Then, this charge can be treated in full analogy with the charge density qn_s in the GaAs channel. Hence, by assuming that the onset of strong inversion in AlGaAs is characterized by a threshold voltage, V_{T1}, the capacitance c_{g1} can be written as

$$c_{g1} = \frac{c_{a1} c_{b1}}{c_{a1} + c_{b1}} \tag{20}$$

where

$$c_{a1} = \varepsilon_i / d_1 \tag{21}$$

is the contribution from the strong inversion regime,

$$c_{b1} = \frac{qn_{o1}}{\eta_1 V_{th}} \exp\left(\frac{V_{gt1}}{\eta_1 V_{th}}\right) \tag{22}$$

is the subthreshold contribution, and

$$q\,n_{o1} = \frac{\varepsilon_i\,\eta_1\,V_{th}}{2\,d_1} \tag{23}$$

is the sheet density of inversion charge in AlGaAs at the threshold for this layer, i.e., at $V_{gt1} \equiv V_{gs} - V_{T1} = 0$. In these expressions, ε_i is the dielectric permeability of the AlGaAs, η_1 is the subthreshold ideality factor for the AlGaAs channel, and V_{th} is the thermal voltage. The threshold voltage, V_{T1}, can be estimated from the maximum sheet charge density, qn_{max}, and the threshold voltage, V_T, of the interface channel:

$$V_{T1} \approx V_T + \frac{qn_{max}}{c_i} \tag{24}$$

This approach is similar to that of Lee et al. 1983 who considered the AlGaAs layer as a parasitic MESFET with a more positive threshold voltage than the HFET threshold.
Fig. 8 shows calculated values for $C_{gc} = LWc_{gc}$ based on eq. (19) and the total gate capacitance $C_{gtot} = LW(c_{gc} + c_{g1})$ based on eq. (20) for a typical HFET with nominal gate length $L = 1\,\mu m$ and gate width $W = 20\,\mu m$. We notice the drop in C_{gc} and the slight increase in C_{gtot} at large gate voltage swings.

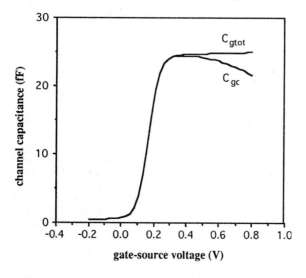

Fig. 8. Calculated gate-channel capacitance, C_{gc}, and total gate capacitance, C_{gtot}, for a typical HFET. (From Fjeldly and Shur 1991.)

The capacitance model described above can be used for calculating the channel contribution to the gate-source capacitance, C_{gs}, and the gate-drain capacitance, C_{gd}, by using an approximation similar to that used in the model by Meyer (1971):

$$C_{gs} = C_f + \frac{2}{3} C_{gc} \left[1 - \left(\frac{V_{sate} - V_{dse}}{2V_{sate} - V_{dse}} \right)^2 \right] \tag{25}$$

$$C_{gd} = C_f + \frac{2}{3} C_{gc} \left[1 - \left(\frac{V_{sate}}{2V_{sate} - V_{dse}} \right)^2 \right] \tag{26}$$

In eqs. (25) and (26), $V_{sate} = I_{sat}/g_{ch}$ is the effective extrinsic saturation voltage, I_{sat} is the saturation drain current, g_{ch} is the extrinsic channel conductance at low drain-source bias, and V_{dse} is an effective extrinsic drain-source voltage. V_{dse} is equal to V_{ds} for $V_{ds} < V_{sat}$ and is equal to $V_{sate} = I_{sat}/g_{ch}$ for $V_{ds} > V_{sate}$.
In order to obtain a smooth transition between the two regimes, we interpolate V_{dse} by the following equation:

$$V_{dse} = V_{ds} \left[1 + \left(\frac{V_{ds}}{V_{sate}} \right)^{m_c} \right]^{-1/m_c} \tag{27}$$

where m_c is a constant determining the width of the transition region between the linear and the saturation regime. The capacitance $C_f i$ is the side wall and fringing capacitance which can be estimated in terms of the capacitance of a metal line of length, W, as

$$C_f \approx \beta_c \, \varepsilon_s W \tag{28}$$

where β_c is on the order of 0.5 (see Gelmont et al. 1991).

CURRENT-VOLTAGE CHARACTERISTICS

In this Section, we will incorporate the description of subthreshold current into the HFET model discussed above. In the long channel limit, the subthreshold saturation current can be expressed as

$$I_{sat} = \frac{qW \, \mu_n V_{th} \, n_o}{L} \exp \left(\frac{V_{gt}}{\eta V_{th}} \right) \tag{29}$$

Here, μ_n is the low-field electron mobility, V_{th} is the thermal voltage, n_o is the channel sheet density of electrons at threshold, η is the subthreshold ideality factor, and $V_{gt} = V_{gs} - V_T$ is the gate voltage swing where V_{gs} is the extrinsic gate-source voltage and V_T is the threshold voltage. The sheet density of channel electrons at threshold is given by (Byun et al. 1989)

$$n_o = \frac{\varepsilon_i \, \eta V_{th}}{2q(d_i + \Delta d)} \tag{30}$$

where d_i is the thickness of the wide-gap semiconductor layer and Δd is a correction to this thickness related to the shift in the Fermi level in the inversion layer with respect to the bottom of the conduction band.

A unified expression for the saturation current, which has the correct asymptotic behavior both below and above threshold, is given in terms of eqs. (11), (12) and (13). Similarly, a unified expression for the electron sheet density in the channel, n_s, is given by eqs. (12) and (14).

Finally, the I–V characteristics can be expressed as:

$$I_d = \frac{g_{ch} V_{ds} \left(1 + \lambda V_{ds}\right)}{\left[1 + \left(V_{ds} / V_{sate}\right)^m\right]^{1/m}} \tag{31}$$

where

$$V_{sate} \approx I_{sat} / g_{ch} \tag{32}$$

is the effective, extrinsic saturation voltage, V_{ds} is the extrinsic drain-source voltage, m is a parameter that determines the shape of the characteristics in the knee region, $g_{ch} = g_{chi}/(1 + g_{chi}R_t)$ is the extrinsic linear channel conductance where g_{chi} is the intrinsic linear channel conductance, $R_t = R_s + R_d$ is the sum of the source and the drain series resistances, and λ is an empirical constant accounting for short channel effects.

An important effect which must be taken into account for an adequate description of the HFET behavior is the dependence of the threshold voltage on the drain-source voltage. This dependence (caused by the Drain Induced Barrier Lowering) is nearly linear, i.e.,

$$V_T = V_{To} - \sigma V_{ds} \tag{33}$$

where V_{To} is the threshold voltage at zero drain-source voltage and σ is a coefficient which depends on the gate voltage swing. Fjeldly and Shur 1991 proposed the following empirical expression for σ :

$$\sigma = \frac{\sigma_0}{1 + \exp\left(\dfrac{V_{gto} - V_{\sigma t}}{V_\sigma}\right)} \tag{34}$$

which gives $\sigma \to \sigma_0$ for $V_{gto} < V_{\sigma t}$ and $\sigma \to 0$ for $V_{gto} > V_{\sigma t}$. The voltage V_σ determines the width of the transition between the two regimes.

Fig. 9 shows the above-threshold HFET I–V characteristics and Fig. 10 shows the subthreshold HFET characteristics in a semilog plot.

As can be seen from the figures, this model reproduces quite accurately the experimental data in the entire range of bias voltages, over several decades of the current variation.

Another important factor determining HFET characteristics at high gate biases is the gate leakage current. The HFET model implemented in the AIM-Spice circuit simulator is based on the simplified equivalent ciruit shown in Fig. 11. A more detailed analysis of the gate leakage current in HFETs is given by Ponse et al. 1985, Ruden et al. 1988, Chen et al. 1988, Ruden et al. 1989a, Baek and Shur 1990, Ruden 1989, and Schuermeyer et al. 1991 and 1991a.

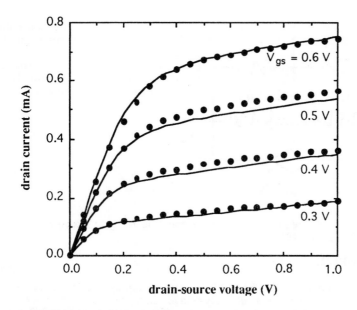

Fig. 9. Above-threshold experimental (symbols) and calculated (solid lines) $I–V$ characteristics for an HFET with nominal gate length $L = 1$ μm and width $W = 10$ μm. Other device parameters are: dielectric permittivity, $\varepsilon_s = 1.14 \times 10^{-10}$ F/m, $Al_{0.3}Ga_{0.7}As$ thickness, $d_i = 0.04$ μm, thickness correction, $\Delta d = 4.5 \times 10^{-9}$ m, mobility, $\mu_n = 0.4$ m²/Vs, saturation velocity, $v_s = 1.5 \times 10^5$ m/s, threshold voltage at zero drain bias, $V_{To} = 0.15$ V, subthreshold ideality factor $\eta = 1.28$, source and drain series resistances, $R_s = R_d = 60$ ohm, output conductance parameter, $\lambda = 0.15$ V⁻¹, $n_{max} = 2 \times 10^{16}$ m⁻², $m = \gamma = \delta = 3.0$, $\sigma_o = 0.057$, $V_\sigma = 0.1$ V, $V_{\sigma t} = 0.3$ V. (After Fjeldly and Shur 1991.)

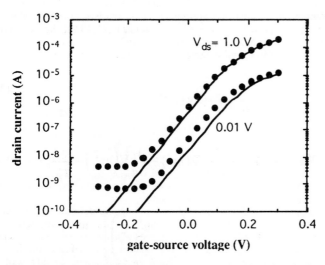

Fig. 10. Subthreshold experimental (symbols) and calculated (solid lines) $I–V$ characteristics for the same device as in Fig. 4.6.2. (After Fjeldly and Shur (1991).)

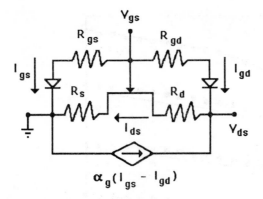

Fig. 11. Simplified FET equivalent circuit accounting for gate leakage current. (From Shur et al. 1992.)

2D HFET MODELING

The analytical models discussed so far in this Chapter are appropriate for circuit design, parameter extraction, design statistical analysis and quality control. Their succesful application is based on a meaningful parameter acquisition which allows us to obtain an excellent agreement with experimental data. However, a full two-dimensional simulation is indispensible for understanding device physics and many trade-offs involved in the device design. Such a simulation may be done at different levels. The simplest approach utilizing semi-empirical drift-diffusion equations describes many qualitative features of device characteristics but is not accurate enough for a quantitative description of short channel HFETs. Still, this approach may be quite useful for numerically challenging tasks, such as three-dimensional HFET modeling. Another example may be the analysis of the field distribution near the pinch-off for estimates of the breakdown voltage. The next level involves the use of the energy balance equations with parameters determined from the steady-state Monte Carlo simulations. This approach is a good compromise between a relatively inaccurate drift-diffusion simulation and a full blown and time consuming Monte Carlo simulation. It also makes it easier to implement more realistic boundary conditions than in the Monte Carlo approach.

The energy balance equation approach (see Shur 1976, Carnez et al. 1980, Feng and Hintz 1988, Sandborn et al. 1989, and Yamada and Tomita 1992) is based on the electron concentration, n, electron velocity, $\mathbf{v}$, and mean electron energy, w, given by

$$\frac{\partial n}{\partial t} + \nabla \cdot (n\mathbf{v}) = 0 \tag{35}$$

$$\frac{\partial \mathbf{v}}{\partial t} + \mathbf{v} \cdot \nabla \mathbf{v} = \frac{q}{m^*}\mathbf{E} - \frac{2}{3m^*n}\nabla(nw) + \frac{1}{3n}\nabla\left(nv^2\right) - \frac{\mathbf{v}}{\tau_p} \tag{36}$$

$$\frac{\partial w}{\partial t} + \mathbf{v} \cdot \nabla w = q\mathbf{v} \cdot \mathbf{E} - \frac{2}{3n}\nabla\left[nv\left(w - \frac{m^*}{2}v^2\right)\right] - \frac{w - w_o}{\tau_w} \tag{37}$$

where

$$n = n_\Gamma + n_L + n_X \tag{38}$$

$$\mathbf{v} = \left(n_\Gamma \mathbf{v}_\Gamma + n_L \mathbf{v}_L + n_X \mathbf{v}_X\right) / n \tag{39}$$

$$w = \left(n_\Gamma w_\Gamma + n_L w_L + n_X w_X\right) / n \tag{40}$$

Here the effective mass, m*, is given by

$$m^* = \left(n_\Gamma m_\Gamma + n_L m_L + n_X m_X\right) / n \tag{41}$$

and the effective momentum and energy relaxation times, τ_p and τ_w, are determined by fitting the results of a steady state Monte Carlo simulation (see Shur 1976). This model proved to be quite effective for simulation of submicron devices yielding excellent agreement with Monte Carlo simulations.

Recently, Silvaco offered a heterostructure simulator, called PRIZM, based primarily on the work done by a group at the University of Leuven. This simulator is to be replaced shortly by a new heterostructure simulator, called BLAZE, capable of simulating HEMTs in the frame of the 2D energy equation model.

MONTE CARLO SIMULATION OF HFETs

Within a semiclassical description, self-consistent simulation of semiconductor devices using Monte Carlo technique is clearly the most advanced approach (see, for example, Reggiani 1985, Fischetti and Laux 1988, Jacoboni and Lugli 1989, Laux et al. 1989 and 1990, Hess 1990, Jensen et al. 1991, and Brennan et al. 1991). The method provides a means to circumvent the inherent mathematical difficulties of solving the governing equation for charge transport, the Boltzmann Transport Equation (BTE). In addition, it can handle practically any geometry or material combination in a relatively straightforward manner. A great advantage is that it provides a unique insight into the device physics. This insight is a result of the nature of the method, relying on a detailed simulation of a large ensemble of charge carriers. Hence, at any time during a transient as well as in steady-state, the space and wavevector dependent distribution function and derived quantities can be found directly from the simulation.

However, the advantages of the Monte Carlo technique come at a heavy price in terms of the large amounts of computer time required for realistic simulations. Besides, such simulations require fairly accurate input of a large number of parameters related to band structure and scattering rates, of which many are not very well known. Other difficulties are related to the description of quantum effects, impact ionization, carrier-carrier scattering, ionized impurity scattering, and to high densities and degeneracy in the carrier gas.

In practice, one has to search for compromises between precision and the computer resources available. Hence, the role of self-consistent Monte Carlo simulations is primarily in helping to understand device physics and validate/reject new device concepts. We shall illustrate this point by two examples related to design aspects of HFETs.

The first HFET design aims at increasing the electron confinement in the GaAs channel of an AlGaAs/GaAs HFET as a means to reduce the short-channel effects (see Kanamori et al. 1992). For this purpose, a p-i-p+ doping profile is prescribed for the GaAs buffer, with the heavily doped p+ layer being removed by several hundred nanometers from the channel, allowing the application of a substrate voltage for threshold voltage adjustment. Although a one-dimensional analysis provides qualitative information on the properties of this device, it lacks the ability to predict how the electron transport depends on the details of the potential distribution and how this affects the short-channel behaviour. The Monte Carlo analysis of a device with a 0.3 μm gate length shows that the output conductance in saturation, a sensitive indicator of short channel effects, can be reduced by 80 % with the present design. This is mostly achieved by a reduction of the parallel conduction of electrons injected from the contacts into the buffer, as clearly illustrated in Fig. 12.

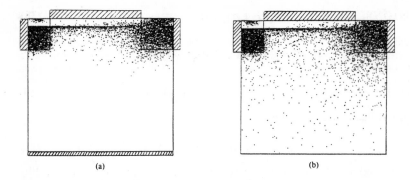

(a)   (b)

Fig. 12. Particle plots of the electron distribution in the *p-i-p*+ buffer HFET
(a) and in a conventional HFET (b). (From Kanamori et al. 1992, © 1992
IEEE.)

The second example concerns efforts to increase the speed performance of *p*-channel
HFETs for the purpose of developing complementary compound semiconductor logic. It
is well known that the average hole drift velocity in a GaAs channel is far inferior to that of
electrons, although the saturation velocity for holes and electrons are comparable. Hence,
the objective of Monte Carlo studies by Jensen et al. 1991, Lund et al. 1992 and Lund
1992 was to investigate *p*-HFET designs aimed at speeding up the hole transport by
redistributing the channel field such as to increse the accelerating field near the source.
One such design is the Variable Threshold HFET (VTHFET) where the threshold voltage
increases steadily from source to drain. The VTHFET concept can be realized by allowing
the lateral doping in the AlGaAs barrier layer to be position dependent. Using improved
hole scattering rates and valence band descriptions (Brudevoll et al. 1990, Brudevoll
1991), *p*-channel VTHFETs were compared to conventional HFETs by Monte Carlo
simulation. And, indeed, a speed advantage of the VTHFET was confirmed, althouh less
than predicted by simple analytical calculations. For a *p*-VTHFET with a 0.5 μm gate
length and a threshold voltage variation of 1.0 V, the cut-off frequency increased to 22.3
GHz from 17.5 GHz for a conventional *p*-HFET. Fig. 13 shows a comparison of the
average hole velocity versus position for the two devices, clearly indicating a velocity gain
in the VTHFET near the source.

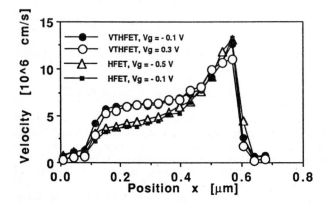

Fig. 13. Hole velocity versus position for a *p*-VTHFET and a conventional
p-HFET, both with a gate length of 0.5 μm. (From Lund et al. 1992.)

CONCLUSIONS

The analytical HFET model discussed in this Chapter accurately reproduces HFET *I-V* and *C-V* characteristics. The circuit simulator utilizing this model – called AIM-Spice and described by Lee et al. 1993 – can adequately simulate HFET integrated circuits. Aspects of 2D simulation of HFETs are discussed, and examples are given of Monte Carlo simulation used for investigating new HFETs design conceps.

ACKNOWLEDGEMENTS

This work has been supported by the Royal Norwegian Council for Scientific and Industrial Research, the NATO Scientific Affairs Division, and Office of Naval Research.

REFERENCES

Baek JH, Shur M (1990) Mechanism of Negative Transconductance in Heterostructure Field Effect Transistors. IEEE Trans. Electron Devices ED-37:1917-1921

Brennan K, Mansour N, Wang Y (1991) Simulation of Advanced Semiconductor Devices Using Supercomputers. Computer Physics Communications 67:73-92 (Proceedings of MSI Symposium on Supercomputer Simulation of Semiconductor Devices, Minneapolis, Nov. 1990)

Brudevoll T, Fjeldly TA, Baek J, Shur M (1992) Scattering Rates for Holes near the Valence Band Edge in Semiconductors. J. Appl. Phys. 67:7373-7382

Brudevoll T (1991) Monte Carlo Algorithms for Simulation of Hole Transport in Homogeneous Semiconductors. Dr. Ing. Thesis, Norwegian Institute of Technology, University of Trondheim, Norway

Byun Y, Lee K, Shur M (1990) Unified Charge Control Model and Subthreshold Current in Heterostructure Field Effect Transistors. IEEE Electron Device Letters EDL-11:50-53 (see erratum IEEE Electron Device Letters EDL-11:273)

Carnez B, Cappy A, Kaszinski A, Constant E, Salmer G (1980) Modeling of Submicron Gate Field-Effect Transistor Including Effects of Non-Stationary Electron Dynamics. J. Appl. Phys. 51:784-790

Chao PC, Shur M, Tiberio RC, Duh KHG, Smith PM, Ballingall JM, Ho P, Jabra AA (1989) DC and Microwave Characteristics of Sub-0.1 µm Gate-Length Planar-Doped Pseudomorphic HEMTS. IEEE Trans. Electron Devices ED-36:461-473

Chen CH, Baier S, Arch D, Shur M (1988) A New and Simple Model for GaAs Heterojunction FET Characteristics. IEEE Trans. Electron Devices ED-35: 570-577

Delagebeaudeuf D, Linh NT (1982) Metal-(n)AlGaAs-GaAs Two-Dimensional Electron GaAs FET. IEEE Trans. Electron Devices ED-29:955-960

Feng YK, Hintz A (1988) Simulation of submicron GaAs MESFET's using a full dynamic transport model. IEEE Trans. on Electron Devices ED-35:1419-1431

Feng M, Lau CL, Eu V, Ito C (1990) Does the Two-Dimensional Electron Gas Effect Contribute to High-Frequency and High Speed Performance of Field-Effect Transistors? Appl. Phys. Lett. 57:1233

Fischetti MV, Laux SE (1988) Monte Carlo Analysis of Electron Transport in Small Semiconductor Devices Including Band-Structure and Space-Charge Effects. Phys. Rev. B38:9721

Fjeldly TA, Shur M (1991) Unified CAD Models for HFETs and MESFETs. Proceedings of the 11th European Microwave Conference, Stuttgart, 1991, Workshop Volume:198-205

Gelmont, B, Shur M, Mattauch RJ (19919 Capacitance-Voltage Characteristics of Microwave Schottky Diodes. IEEE Trans. Microwave Theory and Technique 39:857-863

Hess K, Kizilyalli C (1986) Scaling and Transport Properties of High Electron Mobility Transistors. IEDM Technical Digest, Los Angeles:556-558

Hess K (1990) Supercomputer Images of of Electron Device Physics. Physics Today 43:34-42

Jacoboni C, Lugli P (1989) The Monte Carlo Method for Semiconductor Simulation. Springer, New York

Jensen GU, Lund B, Fjeldly TA, Shur M (1991) Monte Carlo Simulation of Short Channel Heterostructure Field Effect Transistors. Computer Physics Communications 67:1-61 (Proceedings of MSI Symposium on Supercomputer Simulation of Semiconductor Devices, Minneapolis, Nov. 1990)

Laux SE, Fischetti MV, Lee W (1989) Monte Carlo Simulation of Hot-Carrier Transport in Real Semiconductor Devices. Solid-State Electron. 32:1723

Laux SE, Fischetti MV, Frank DJ (1990) Monte Carlo Analysis of Semiconductor Devices: the DAMOCLES Program. IBM J. Res. Dev. 34:466

Lee K, Shur M, T. Drummond, Morkoç H (1983) Current-Voltage and Capacitance-Voltage Characteristics of Modulation-Doped Field Effect Transistors. IEEE Trans. Electron Devices ED-30:207-212

Lee K, Shur M, T. Drummond, Morkoç H (1984) Parasitic MESFET in (Al,Ga)As/GaAs Modulation Doped FET. IEEE Trans. Electron Devices ED-31:29-35

Lee K, Shur M, Fjeldly TA, Ytterdal T (1993) Semiconductor Device Modeling for VLSI. Prentice Hall, New Jersey

Lund B, Fjeldly TA, Shur M, Jensen G (1992) The Monte Carlo Technique as a Testing Ground for New Design Concepts. Proc. 1992 URSI Int. Symp. on Signals, Systems , and Electronics (ISSSE'92):605-609

Lund B (1992) Monte Carlo Simulation of Charge Transport in Semiconductors and Semiconductor Devices. Dr. Ing. Thesis, Norwegian Institute of Technology, University of Trondheim, Norway

Kanamori M, Jensen G, Shur M, Lee K (1992) Effect of p-i-$p+$ Buffer on Characteristics of n-Channel Heterostructure Field Effect Transistors. IEEE Trans. Electron Devices ED-32:226-233

Meyer JE (1971) MOS Models and Circuit Simulation. RCA Review 32:42-63

Morkoç H, Unlu H, Ji G (1991) Principles and Technology of MODFETs. John Wiley and Sons, New York

PRIZM (1991) User's Manual, Silvaco

Reggiani L, Ed., (1985), Hot-Electron Transport in Semiconductors. Springer, Berlin

Ponse F, Masselink WT, Morkoç H (1989) The Quasi-Fermi Level Bending in MODFETS and Its Effects on the FET Transfer Characteristics. IEEE Trans. Electron Devices ED-32:1017-1023

Ruden PP, Han CJ, Shur M (1988) Gate Current of Modulation Doped Field Effect Transistors. J. Appl. Phys. 64:1541-1546

Ruden PP (1989) Heterostructure FET Model Including Gate Leakage. IEEE Trans. Electron Devices ED-37:2267-2270

Ruden PP, Shur M, Akinwande AI, Jenkins P (1989a) Distributive Nature of Gate Current and Negative Transconductance in Heterostructure Field Effect Transistors. IEEE Trans. Electron Devices ED-36:453-456

Sandborn PA, Rao P, Blakey PA (1989) An Assessment of Approximate Nonstationary Charge Transport Models Used for GaAs Device Modeling. IEEE Trans. on Electron Devices ED-36:1244-1253

Schuermeyer FL, Shur M, Grider D (1991) Gate Current in Self-Aligned n-channel Pseudomorphic Heterostructure Field-Effect Transistors. IEEE Electron Device Lett. EDL-12:571-573

Schuermeyer FL, Martinez E, Shur M, Grider DE, Nohava J (1992) Subthreshold and Above Threshold Gate Current in Heterostructure Insulated Gate Field Effect Transistors, Electronics Letters 28:1024-1026

Shur M (1976) Influence of Non-Uniform Field Distribution on Frequency Limits of GaAs Field-Effect Transistors. Electron Letters 12:615-616

Shur M (1987) GaAs Devices and Circuits. Plenum, New York

Shur M (1990) Physics of Semiconductor Devices. Prentice Hall, New Jersey

Shur M, Fjeldly TA, Ytterdal, Lee K (1992) Unified GaAs MESFET Model for Circuit Simulations. Intern. J. of High Speed Electronics 3:201-233

Stern F (1972) Self-Consistent Results for n-type Si Inversion Layers. Phys. Rev. B-5:4891–4899

Subramanian S (1988) Frequency Dependence of Response of the DX Center in AlGaAs and its Influence on the Determination of the Band Discontinuity of GaAs/AlGaAs Heterojunctions. J. Appl. Phys. 64:1211-1214)

Yamada Y, Tomita T (1992) Accuracy of Relaxation Time Approximation for Device Simulation of Submicrometer GaAs MESFET's. Electronics Letters 28:393-395

HBT Modelling

Robert E. Miles
University of Leeds

INTRODUCTION

The Heterojunction Bipolar Transistor (HBT) is proving to be a remarkably versatile device with applications ranging from high frequency power generation to high speed logic. A particular attraction is that it can often be used directly in circuits that have been developed over many years for silicon technology representing a significant saving in design costs. Various different structures for the HBT have been suggested and fabricated based on both GaAs and InP substrates (more recently Ge/Si devices have also been developed). Although there are many variations in structure the essential feature which is common to them all is that the emitter has a wider energy gap than the base. Some of the variations will be mentioned later but from the modelling point of view there should not be a problem providing that the relevant parameters of the different regions are known.

As mentioned above the HBT is a bipolar device and therefore both electrons and holes must be included in any model. Compared to a simple MESFET model this tends to complicate matters because there is an extra variable involved together with the associated equations for current continuity. However, a compensating factor is that it is possible to obtain useful results from a one-dimensional HBT model although of course some important two dimensional effects such as current crowding and base spreading resistance will not be included.

This chapter will concentrate on models of the HBT based on the semiconductor device equations as derived from Maxwell's and Poisson's equations and current continuity. Monte-Carlo models (e.g. Juntao et al 1992) will not be discussed.

THE SEMICONDUCTOR EQUATIONS IN HETEROSTRUCTURES

In heterostructures the semiconductor properties, notably the energy gap E_G, dielectric constant ϵ, effective mass m^* and densities of states in the conduction and valence bands N_C and N_v, vary with position (Adachi 1985). As a result the simple semiconductor device equations (Sze 1981) have to be modified to take these variations into account. The changes may occur over a short distance (i.e. an abrupt junction) or be graded over a region of the device. The derivation of the modified device equations as applied to heterostructures is given in the next 3 sections.

N.B. In all that follows "q" is to be taken as 1.6×10^{-19}, i.e the **magnitude** of the electronic charge in coulombs. The appropriate attached sign (depending on whether electrons or holes are being considered) is included in the equations.

Poisson's Equation

Starting with Maxwell's equations with the symbols having their usual meanings

$$\nabla \times \mathscr{E} = -\frac{\partial B}{\partial t} \tag{1}$$

$$\nabla \times H = J + \frac{\partial D}{\partial t} \tag{2}$$

$$\nabla \cdot D = \rho \tag{3}$$

$$\nabla \cdot B = 0 \tag{4}$$

For any vector a, $\nabla \cdot (\nabla \times a) = 0$ and therefore from equation (4) we can write $B = \nabla \times A$. (A is known as the magnetic vector potential). Substituting into equation (1) for B we have

$$\nabla \times \left[\mathscr{E} + \frac{\partial A}{\partial t} \right] = 0 \tag{5}$$

and therefore from vector theory $\mathscr{E} + \partial A / \partial t$ can be expressed as the gradient of a scalar quantity which from energy considerations is $-\nabla \psi$, where ψ is the electrical potential.

In practice the term $\partial A / \partial t$ can be ignored except at very high frequencies and we then arrive at the familiar equation

$$\mathscr{E} = -\nabla \psi \tag{6}$$

Assuming that we can ignore piezo-electric effects and that the dielectric properties are independent of direction then $D = \epsilon_r \epsilon_o \mathscr{E}$ and equation (3) becomes

$$\nabla \cdot (\epsilon_r \epsilon_o \nabla \psi) = -\rho \tag{7}$$

where ϵ_r and ϵ_o are the relative permittivity and the permittivity of free space respectively.

In a uniform material the permittivity is constant resulting in the usual form of Poisson's equation but if the composition varies then equation (7) yields a modified Poisson equation which must be used at heterojunctions or in graded materials

$$\epsilon_r \nabla^2 \psi + \nabla \psi \nabla \epsilon_r = -\frac{\rho}{\epsilon_o} \tag{8}$$

the charge density ρ is given by:-

$$\rho = q(N_D^+ - N_A^- + p - n) \tag{9}$$

where N_D^+ and N_A^- are the ionised donor and acceptor densities, n and p are the electron and hole densities.

The Current Equations

By making some simplifying assumptions (Dekker 1962, Selberherr 1984) Boltzmann's transport equation can be reduced to the following two equations for electron and hole current densities respectively:-

$$J_n = n\mu_n \nabla E_{fn} \tag{10}$$

$$J_p = p\mu_p \nabla E_{fp} \tag{11}$$

J_n and J_p, μ_n and μ_p, and E_{fn} and E_{fp} are respectively the electron and hole current densities, mobilities and quasi Fermi levels (measured in Joules). Equations (10) and (11) apply to both homogeneous and heterogeneous structures at all levels of doping i.e. they include Fermi-Dirac Statistics. Notice also that the electron and hole current densities flow **up** their respective Fermi level gradients as depicted on a conventional energy band diagram.

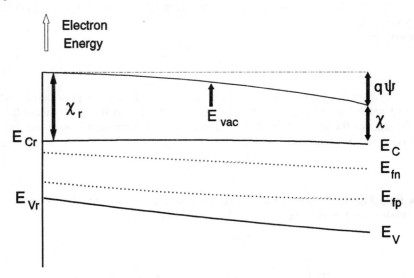

Figure 1. Energy band diagram for a graded heterostructure.

When a potential difference is applied across a semiconductor the vacuum level represents the distribution of potential (relative to a zero level at infinity) across the sample. If the sample is homogeneous the intrinsic Fermi level, the conduction band edge and the valence band edge are all parallel to the vacuum level and are therefore all equally good choices to represent the potential (remembering that when dealing with potential energy the zero energy level is quite arbitrary). Conventionally the intrinsic Fermi level is the one chosen. In a heterogeneous sample we must work with the vacuum level as none of the other levels are parallel to it. (In an abrupt heterojunction the intrinsic Fermi level and the band edges are likely to be discontinuous, a feature we do not expect of the potential as it implies an infinite or at least a very large electric field.) Fig. 1 shows the vacuum level E_{vac}, the conduction and valence band edges and the quasi

Fermi levels for electrons and holes in a graded heterostructure. The left hand side of the diagram is taken as a reference. It could for example be GaAs. The energy levels on this side have an "r" subscript to represent "reference". The electron affinity χ varies from left to right as does the energy gap E_G. The potential is measured relative to the vacuum level on the reference (left hand) side. As the zero of energy can be wherever we choose, it is convenient in Fig. 1 to let it be at E_{Cr}.

If for the time being we restrict ourselves to non-degenerate doping levels (i.e. where n/N_C and $p/N_V \ll 1$), then Boltzmann statistics apply and we can proceed as follows. At any point along the diagram in Fig. 1 the carrier densities are given by

$$n = N_C \exp\left[\frac{E_{fn} - E_C}{kT}\right] \tag{12}$$

$$p = N_V \exp\left[\frac{E_V - E_{fp}}{kT}\right] \tag{13}$$

or rewriting equations (12) and (13)

$$E_{fn} = E_C + kT \ln\left[\frac{n}{N_C}\right] \tag{14}$$

$$E_{fp} = E_V - kT \ln\left[\frac{p}{N_V}\right] \tag{15}$$

Now from Fig. 1

$$E_C = E_{Cr} + \chi_r - \chi - q\psi \tag{16}$$

and

$$E_V = E_C - E_G \tag{17}$$

giving

$$E_{fn} = E_{Cr} + \chi_r - \chi - q\psi + kT \ln\left[\frac{n}{N_C}\right] \tag{18}$$

$$E_{fp} = E_{Cr} + \chi_r - \chi - q\psi - E_G - kT \ln\left[\frac{p}{N_V}\right] \tag{19}$$

Substituting for E_{fn} and E_{fp} in equations (10) and (11) and differentiating (assuming a constant temperature)

$$J_n = n\mu_n \left[-q\nabla\psi - \nabla\chi - \frac{kT}{N_C}\nabla N_C + \frac{kT}{n}\nabla n \right] \tag{20}$$

$$J_p = p\mu_p \left[-q\nabla\psi - \nabla\chi - \nabla E_G - \frac{kT}{p}\nabla p + \frac{kT}{N_C}\nabla N_C \right] \tag{21}$$

(Note: Readers should satisfy themselves at this point that equations (20) and (21) reduce to the classical current density equations for uniform material.)

The equations for J_n and J_p show not only the usual electric field ($\mathscr{E} = -\nabla\psi$) dependence but also terms in N_C, N_V and χ associated with the spatial variations of material properties. The variations in N_C and N_V affect the diffusion currents while those in χ affect the drift currents.

By rearranging terms in equations (20) and (21) and substituting $\mathscr{E} = -\nabla\psi$

$$J_n = nq\mu_n \left[\mathscr{E} - \frac{1}{q}\nabla\chi - \frac{kT}{qN_C}\nabla N_C \right] + \frac{kT\mu_n}{q}\nabla n \tag{22}$$

$$J_p = pq\mu_p \left[\mathscr{E} - \frac{1}{q}\nabla\chi - \frac{1}{q}\nabla E_G + \frac{kT}{qN_V}\nabla N_V \right] - \frac{kT\mu_p}{qp}\nabla p \tag{23}$$

An alternative form for the current density equations can be obtained by defining two quantities θ_n and θ_p which are known as the band potentials:-

$$\theta_n = \frac{\chi - \chi_r}{q} + \frac{kT}{q} \ln\left[\frac{N_c}{N_{Cr}} \right] \tag{24}$$

$$\theta_p = \frac{\chi - \chi_r}{q} - \frac{kT}{q} \ln\left[\frac{N_V}{N_{Vr}} \right] + \frac{(E_G - E_{Gr})}{q} \tag{25}$$

Using the Einstein Relations in the drift/diffusion approximation i.e. $D = (kT/q)\mu$, we can write equations (22) and (23) as:-

$$J_n = -nq\mu_n\nabla(\psi + \theta_n) + qD_n\nabla n \tag{26}$$

$$J_p = -pq\mu_p\nabla(\psi + \theta_p) - qD_n\nabla p \tag{27}$$

Equations (26) and (27) have the same form as the current density equations for homogeneous material but with the addition of the band potentials which represent the effect on the current of the spatial variations in band gap and density of states.

Current Continuity

The equations for current continuity are unchanged from those in a uniform material and can be written as

$$\frac{\partial n}{\partial t} = \frac{1}{q} \nabla \cdot J_n + G \qquad (28)$$

$$\frac{\partial p}{\partial t} = -\frac{1}{q} \nabla \cdot J_p + G \qquad (29)$$

G, the generation/recombination rate per unit volume, is positive for generation and negative for recombination and can include a number of different processes. i.e.

$$G = G_{SRH} + G_{Auger} + G_{optical} + G_{surface} + G_{impact} \qquad (30)$$

G_{SRH} represents the Shockley-Read-Hall mechanism which is thermally activated mainly by recombination levels near the centre of the energy gap giving

$$G_{SRH} = -\frac{np - n_i^2}{\tau_n(p + n_i) + \tau_p(n + n_i)} \qquad (31)$$

where τ_n and τ_p, the electron and hole lifetimes, lie in the range 1 ns to 10 μs depending on the band gap of the material and its crystal quality.

Auger processes which require the interaction of a third particle are represented by:-

$$G_{Auger} = -(np - n_i^2)(nC_n + pC_p) \qquad (32)$$

C_n and C_p the Auger coefficients are of the order of $10^{-43} m^6 s^{-1}$.

Optical processes, which involve the absorbtion or emission of photons are given by:-

$$G_{optical} = -C_{opt}(np - n_i^2) \qquad (33)$$

where C_{opt} is the optical capture/emission rate.

Surface processes are similar to Shockley-Read-Hall but involve surface trap densities. Impact processes become important near breakdown.

SOLUTION OF THE DEVICE EQUATIONS

In this section the solution of the device equations will be discussed in the first instance for the case of low doping where Boltzmann statistics apply and then for arbitrary doping levels where Fermi-Dirac statistics must be used.

Low Doping - Boltzmann Statistics

In order to calculate the terminal characteristics of a heterostructure device the semiconductor equations (8), (26), (27), (28) and (29) must be solved for n, p, J_n, J_p and

ψ. If we also assume a low frequency approximation such that $\partial/\partial t = 0$ then equations (28) and (29) simplify to:-

$$\frac{1}{q}\nabla \cdot J_n = -G \tag{34}$$

$$\frac{1}{q}\nabla \cdot J_p = G \tag{35}$$

Apart from some simple cases this set of non-linear equations must be solved numerically and if required energy band diagrams can then be constructed using equations (14), (15), (16) and (17). Even though the above equations strictly speaking only apply to the steady state they can still be used for a quasi ac analysis up to about 15 GHz providing that displacement current is included.

Because of the wide numerical range of the parameters in the above equations accuracy is improved by scaling. The scaling scheme as suggested by De Mari (1962) has been found by experience to work well.

High Doping - Fermi-Dirac Statistics

In practical HBTs the doping levels, particularly in the base, can be very high ($> 10^{24}\text{m}^{-3}$ being typical) and the use of Boltzmann statistics can give rise to significant errors. Equations (12) and (13) are no longer adequate and the exponential terms need to be replaced with the Fermi Integral to become:-

$$n = \frac{2N_C}{\sqrt{\pi}} \cdot F_{1/2} \left[\frac{E_{fn} - E_C}{kT} \right] \tag{36}$$

$$p = \frac{2N_V}{\sqrt{\pi}} \cdot F_{1/2} \left[\frac{E_V - E_{fp}}{kT} \right] \tag{37}$$

With these more general equations we soon run into difficulties if we attempt to proceed as in equations (14) to (29). A better way is to substitute for n and p from equations (36) and (37) into equations (8), (10) and (11) and then solve the resulting equations together with the continuity equations (28) and (29) for J_n, J_p, E_{fn}, E_{fp}, and ψ. Carrier densities can then be found from equations (36) and (37). This technique has the advantage of being applicable to all carrier densities but there can be problems with the numerical solution as it appears to be less stable than for the non-degenerate case. However, when a Newton technique is used the solution can be obtained very quickly.

HETEROJUNCTION MODELLING

The usual introduction to HBT modelling is to start with an np heterojunction as may occur at the emitter base junction in the complete device. (The emitter is usually n-type because of the higher mobility of electrons). Typically the n-layer could be GaAlAs with a 20% Al content and the p-layer GaAs. Fig. 2 shows the energy band structure in the region of the junction.

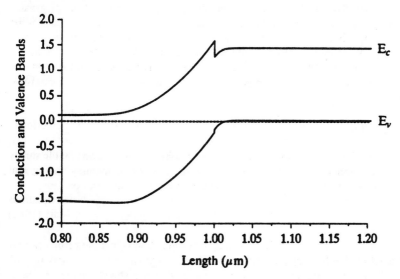

Figure 2. Band structure diagram for an np heterojunction.

Band Gap Discontinuity

The most apparent difference between this energy band diagram (Fig. 2) and that of a homojunction is the different value for the forbidden energy gap E_G on each side of the junction. The energy gap of $Ga_{1-x}Al_xAs$ depends on the molar fraction x and varies monotonically between 1.42 eV for GaAs and 2.16 eV for AlAs. In the figure the $Ga_{1-x}Al_xAs$ on the n-type side with x = 0.2 has an energy gap in the region of 1.65 eV and the GaAs on the p-type side has an energy gap of 1.42 eV. The bandgap change at the junction results in discontinuities in both the conduction and valence band edges as shown. The magnitudes of the discontinuities are responsible for the advantages that the HBT has over the BJT and it is therefore important to have accurate values in the model.

So far it has not been possible to calculate the band offsets theoretically and it is therefore necessary to rely on experimentally determined values. The accepted values change over a period of time as experimental techniques improve and it is therefore prudent to carry out a careful literature search before adopting a value for your model. It is however quite easy to change the offsets in a model as more accurate figures become available. At the time of writing the energy gap discontinuities in the $Ga_{1-x}A_xlAs/GaAs$ system (Ji et al 1987) have well accepted values given by

$$Q_v = \frac{\Delta E_v}{(\Delta E_C + \Delta E_v)} = 0.33 \tag{38}$$

where Q_v is independent of the molar fraction. A value of $Q_v = 0.3$ is also suggested by Ji et al for the $In_yGa_{1-y}As/GaAs$ system for y in the range 0.13 to 0.19 but there are complications due to the positions of the light and heavy hole valence bands.

Generation/Recombination

Assuming Schockley-Read-Hall generation/recombination a uniform lifetime of greater than 10 ns is typical of modern technology. One might expect a higher density of generation/recombination centres at the junction but in fact today's crystal growth technology is such that the densities are too low to have a noticeable effect. Calculations suggest that densities of $10^{17}m^{-2}$ would be required to have a visible affect on the electrical characteristics - a far higher density than would be expected in a carefully grown junction.

Boundary Conditions

The contacts to this simple one-dimensional device are ohmic which means that in effect the generation/recombination rate is very high at the surface, maintaining the carrier concentrations at their equilibrium levels (i.e. the excess carrier concentrations are zero).

Results

One advantage of modelling the one-dimensional heterojunction is that the results can be compared easily with either experiment or with an analytical model. Some typical comparisons of simulated and analytical I-V characteristics are shown in Fig. 3 where good agreement is seen at current levels above 0.1 Acm^{-2}. Unlike the analytical model, the simulation includes effects due to carrier recombination and generation which will affect the curves and there is also a possibility of numerical noise at low currents.

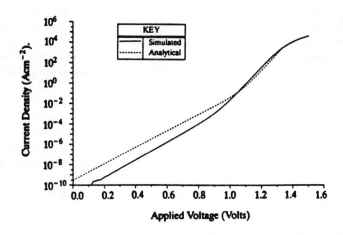

Figure 3. Comparison of numerical and analytical model for a GaAs pn junction diode.

Factors Affecting HBT Performance

A one-dimensional heterojunction model can also be used to investigate some important features of the HBT such as graded junctions and the effect of a misalignment of the alloy and doping boundaries at the emitter/base junction.

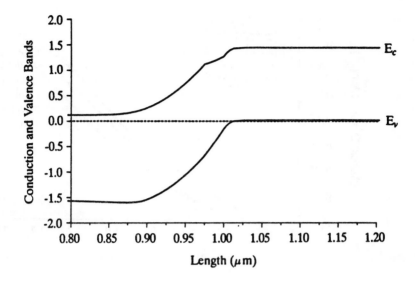

Figure 4. Band diagram for a heterojunction with a junction grading over 25 nm. (cf Fig. 2)

The band diagram in Fig. 2 shows the well known "spike and notch" structure in the conduction band at the junction. The spike is thought to impede the flow of electrons across the junction and the notch can trap electrons and thereby increase the local recombination rate. Both of these effects would act to degrade the properties of the junction but the spike and notch can be removed by grading the alloy composition on the n-type (i.e. emitter) side. In practice grading has been shown to improve the high frequency operation and current gain of the HBT. The effect of grading the alloy composition over different lengths is shown in the band structure diagrams in Fig. 4. For a grading length of 25 nm or more the spike and notch are completely removed but simulations show that they do show up again under forward bias conditions. The effect of the grading on the forward I-V characteristic is shown in Fig. 5.

During the growth of HBT layers it is possible for some of the impurities in the highly doped base to diffuse into the emitter. This causes a misalignment of the doping and alloy junctions which can reduce the emitter injection efficiency. The results of a simulation of this effect (Holder et al 1990) are shown in Fig. 6 where it can be seen that a significant reduction in emitter injection efficiency can occur. It can however be shown that grading the junction reduces the influence of the misalignment - another reason for grading the junction.

HBT MODELS

In modelling the complete HBT one can choose a one or two-dimensional simulation. Obviously the one-dimensional case will not include two-dimensional effects such as current crowding and base resistance. However, a study of the current flow in a two-dimensional simulation shown in Fig. 7 shows that much of the device has one-dimensional symmetry. It should be possible to inject current into the nodes of the

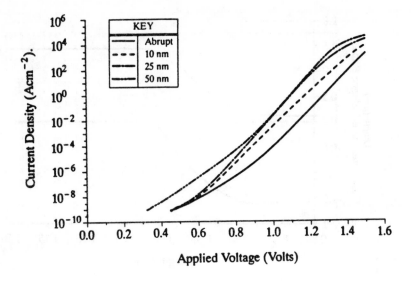

Figure 5. The effect of alloy grading on the I-V curves of a heterojunction diode.

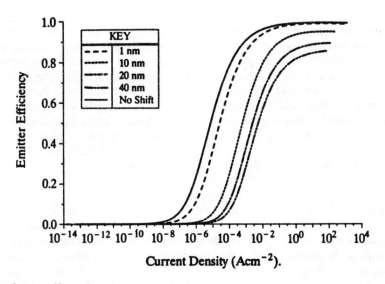

Figure 6. The effect of doping/alloy misalignment on the emitter injection efficiency.

base region of a one-dimensional model in such a way as to include something of the two-dimensional flow and hence produce a quasi two-dimensional model as has been developed for the MESFET and HEMT (see Chapter 12).

The first thing to do in a full two-dimensional simulation is to decide on the choice of device geometry. Two typical geometries are shown in Fig. 8, where 8(a) is a mesa etched device and 8(b) is a planar device produced by ion implantation and proton

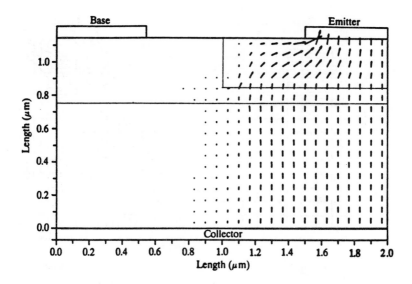

Figure 7. Current flow in a 2-D HBT simulation.

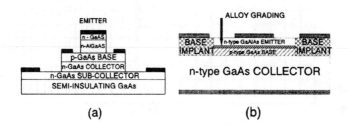

Figure 8. HBT structures. (a) mesa etched, (b) ion implanted/proton isolation.

isolation. Whichever is chosen, the electric fields are highest at the junctions and it is therefore an advantage to solve the equations on a finer grid in these regions. For the case of the planar device Fig. 9 shows a self refining grid which adds extra nodes where required in the high field regions. This can make a significant difference to the memory requirements during simulation. Also from the geometrical symmetry of the HBT it can be seen that we only need to simulate one half of the device.

The important feature in modelling the HBT is to identify the regions of high recombination as these can have a significant influence on the device characteristics. For example the proton isolation region in device (b) can be shown to reduce the current gain from the expected values of several thousand to below 100, a figure which is not uncommon in integrated circuits. Similarly, surface recombination at the sidewalls of the base reduces the current gain of the mesa device. All of the contacts to the HBT are

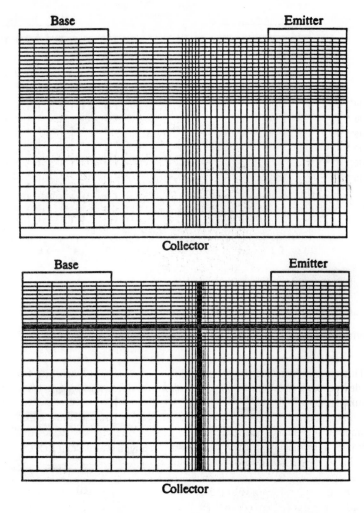

Figure 9. Self-refining solution grid for an HBT (a) initial choice (b) after 3 iterations.

ohmic with the boundary condition as discussed.

One of the advantages of computer simulation is that conditions existing in different parts of the device can be plotted out and visualised as shown in Fig. 10. This can help considerably in understanding the processes that determine the device performance.

CONCLUSION

To date, HBT models have been largely drift/diffusion based, using Boltzmann statistics, confined to the steady state and restricted to uniform temperatures. There are good reasons for these limitations mainly to do with computing resources. When modelling the HBT it is necessary to simulate a large area of material (cf a MESFET) and this together with it being a bipolar device means that more equations must be solved at a larger number of grid points. As computing power increases we will no doubt see more of these effects included in the models.

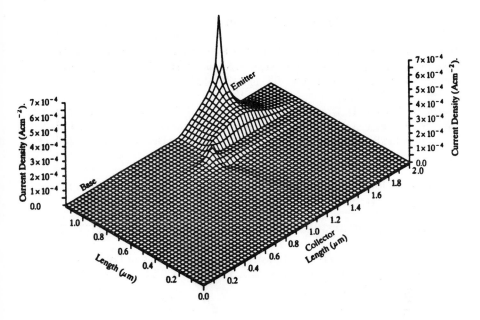

Figure 10. Current contour map for an HBT showing crowding effects at the emitter.

This chapter has described the fundamentals of HBT modelling based on the steady state device equations and valid up to about 15 GHz. At higher frequencies the full continuity equation and displacement current must be included and consequently few rf simulations have appeared because the extra terms in the equations make large demands on computer resources. For the same reasons few energy models of the HBT exist.

Power HBTs are also known to operate at high temperature, with values of over 300 C being suggested (Webb, P.W. and Russel, I.A.D. 1991) This is hot enough to have a significant effect on the operating characteristics. Equations (20) and (21) implicitly assume that temperature is uniform but a non-uniform temperature could be included at this point. This however is again likely to cause computing problems.

It is clear therefore that the HBT is an important device in modern electronics and as such there is a consequent need to understand its operation in more detail. Computer simulation will no doubt play an important part in this process.

REFERENCES

Adachi S (1985) GaAs, AlAs and Al$_x$Ga$_{1-x}$As: Material parameters for use in research and device applications. J. Appl. Phys, Vol. 58, pp R1-R28.

Dekker, A.J., (1962) Solid State Physics, Macmillan, pp 275-283.

Holder D J, Miles R E, Snowden C M (1991) Two-dimensional simulation of GaAs/AlGaAs heterojunction bipolar transistors. Gallium Arsenide and Related Compounds 1990, IOP Conference Series, Ed K. F. Singer, pp 377-382.

De Mari A (1968) An accurate numerical steady-state one dimensional solution of the p-n junction. Solid-State Electronics, Vol. 11, pp. 35-58.

Ji G, Huang D, Reddy U K, Unlu H, Henderson T S, Morkoc H (1987) Determination of band offsets in AlGaAs/GaAs and InGaAs/GaAs multiple quantum wells. J. Vac. Sci. Technol, Vol. B5, pp. 1346-1352.

Juntao H, Pavlidis D, Tomizawa K (1992) Monte Carlo Studies of the Effect of Emitter Junction Grading on the Electron Transport in InAlAs/InGaAS Heterojunction Bipolar Transistors. IEEE Trans. Electron Devices Vol. 39, pp 1273-1281.

Selberherr S (1984) Analysis and Simulation of Semiconductor Devices. Springer-Verlag, pp 20-21.

Sze S (1981) Physics of Semiconductor Devices (2nd Edition), (Wiley) pp 50-51.

Webb P W, Russel I A D (1991) Transient Thermal Modelling of Gallium Arsenide Bipolar Heterojunction Power Transistors. Proc. IMACS'91, Dublin, Vol. 4, pp 1678-1679.

Gunn Diode and IMPATT Diode Modelling

Michael Shur
University of Virginia

Transferred-electron devices (utilizing the bulk negative resistance of gallium arsenide, indium phosphide or related compounds) and avalanche devices that use impact ionization in high electric field -- are the most powerful solid-state sources of microwave energy. In this Chapter, we discuss different approaches to modeling these devices. First, we consider the mechanism responsible for the negative differential mobility in GaAs. Then we discuss and compare a full blown self-consistent Monte Carlo simulation of transferred electron devices with a much simpler (but a far less accurate) approach based on the drift-diffusion equation. Then we consider IMPATT devices which use phase delays related to avalanche breakdown and carrier drift in a semiconductor diode and discuss additional problems involved in modeling IMPATT devices.

NEGATIVE DIFFERENTIAL MOBILITY

In high electric fields, the electron velocity, v, in GaAs, InP, and some other compound semiconductors decreases with an increase in the electric field, F, so that the differential mobility $\mu_d = \dfrac{dv}{dF}$ becomes negative (see Fig. 1). Ridley and Watkins (1961) and Hilsum (1962) were first to suggest that such a negative differential mobility in high electric fields is related to an electron transfer between different minima (valleys) of the conduction band in GaAs. When the electric field is low, electrons are primarily located in the central valley of the conduction band. As the electric field increases, many electrons gain enough energy for the intervalley transition into higher satellite valleys. The electron effective mass in the satellite valleys is much greater than in the central valley. Also, the intervalley transition is accompanied by an increased intervalley scattering. These factors result in the decrease of the electron velocity in high electric fields (see Fig. 1).

In more details, the mechanism of negative differential conductance proposed by Ridley, Watkins, and Hilsum may be explained using a simple "two-valley" model. In this model, satellite valleys of the conduction band are represented by one equivalent minimum (by one "upper valley"). The low "valley" is the lowest central minimum of the conduction band. When the electric field is low, practically all electrons are in the lowest minimum of the conduction band and the electron drift velocity v is given by

$$v_1 = \mu F \tag{1}$$

where μ is the low-field mobility and F is the electric field.

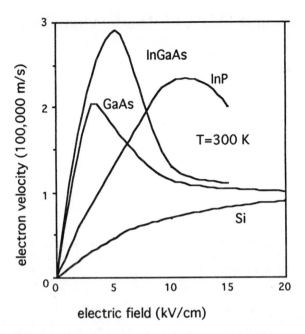

Fig. 1. Electron drift velocities in silicon and in several compound semiconductors at room temperature. The curve for $In_xGa_{1-x}As$ is for $x = 0.53$ (from Shur (1990)).

In a higher electric field electrons are "heated" by the field and some carriers may gain enough energy to transfer into upper valleys where the electron drift velocity is approximately equal to the electron saturation velocity, v_s :

$$v_2 \approx v_s \qquad (2)$$

(Strictly speaking, the velocity of electrons in upper valleys is not constant and may depend on the electric field. However, the assumption $v_2 \approx v_s$ allows us to reproduce correctly the velocity saturation in high electric fields.)

The current density is given by

$$j = qv_1n_1 + qv_2n_2 \qquad (3)$$

where n_1 is the electron concentration in the lowest valley and n_2 is the electron concentration in the upper valley:

$$n_1 + n_2 = n_o \qquad (4)$$

where $n_o \simeq N_d$ where N_d is the ionized donor density. The average electron velocity can be defined as

$$v = \frac{n_1}{n_o}v_1 + \frac{n_2}{n_o}v_2 \qquad (5)$$

Using the results of a Monte Carlo calculation one can check that this simplified model is quite adequate for GaAs at 300 K if we assume that the fraction of electrons in the upper valleys, $\xi = n_2/n_o$, is given by

$$\xi = \frac{A\left(\dfrac{F}{F_s}\right)^t}{1 + A\left(\dfrac{F}{F_s}\right)^t} \tag{6}$$

where $F_s = v_s/\mu$. Eq. (5) can be now rewritten as

$$v(F) = v_s\left[1 + \frac{\dfrac{F}{F_s} - 1}{1 + A\left(\dfrac{F}{F_s}\right)^t}\right] \tag{7}$$

The comparison with the results of the Monte Carlo calculation (see Xu and Shur (1987)) shows that such an interpolation is accurate within 10%-15%. From eqs. (6) and (7) we find that once $d\xi/dF$ exceeds a critical value

$$\frac{d\xi}{dF} > \frac{1-\xi}{F - F_s} \tag{8}$$

the differential mobility becomes negative. Eq. (8) shows that a negative differential mobility is the consequence of a rapid increase in the intervalley transfer with an in increase in the electric field.

GUNN DIODE PHYSICS

A negative differential resistance may lead to a growth of small space charge fluctuations. A simplified equivalent circuit of a uniformly doped semiconductor may be presented as a parallel combination of the differential resistance

$$R_d = \frac{L}{q\mu_d n S} \tag{9}$$

and the differential capacitance:

$$C = \frac{\varepsilon S}{L} \tag{10}$$

Here S is the cross-section of the sample and L is the sample length. Hence, the equivalent RC time constant determining the evolution of the space charge is given by

$$\tau_{md} = R_d C_d = \frac{\varepsilon}{q\mu_d n} \tag{11}$$

(τ_{md} is called is the differential dielectric relaxation time.) In a material with a positive differential conductivity, a space charge fluctuation, ΔQ, decays exponentially with this time constant. However, if the differential conductivity is negative the space charge fluctuation may actually grow with time. This process can be analyzed using the Poisson equation

$$\frac{\partial F}{\partial x} = \frac{q(N_d - n)}{\varepsilon} \tag{12}$$

and the semi-empirical drift-diffusion equation for the total current

$$I = qnv(F) + qD_n(F)\frac{\partial n}{\partial x} + \varepsilon\frac{\partial F}{\partial t} \tag{13}$$

Here F is the electric field, x is the space coordinate, n is the electron concentration, N_d is the concentration of shallow ionized donors, ε is the dielectric permittivity, q is the electronic charge, I is the total current density, and μ is the low field mobility. A small-signal analysis is based on seeking the solutions of eqs. (12) and (13) in the following form

$$n = n^{(0)} + n^{(1)}\exp[i(\omega t - kx)] \tag{14}$$

$$F = F^{(0)} + F^{(1)}\exp[i(\omega t - kx)] \tag{15}$$

where $n^{(1)} \ll n^{(o)}$ and $F^{(1)} \ll F^{(o)}$ and the wave vector $k = 2\pi/\lambda$ where λ is the wavelength. The substitution of eqs. (14) and (15) into eqs. (12) and (13) yields two uniform algebraic equations for $n^{(1)}$ and $F^{(1)}$ which have a nonzero solution if and only if the system determinant is zero. This leads to the following dispersion equation:

$$\omega = -kv\left(F^{(0)}\right) - i\left[\frac{1}{\tau_{md}\left(F^{(0)}\right)} + D_nk^2\right] \tag{16}$$

Eq. (16) may be used to describe the evolution of a sinusoidal fluctuation of electron concentration and related variation of the space charge and electric field (the space-charge waves). The real part $\mathrm{Re}(\omega) = -kv(F_o)$ determines the wave frequency. The imaginary part $\mathrm{Im}(\omega) = -1/\tau_{md} - D_nk^2$ determines the attenuation or growth time constant of the wave for $\mathrm{Im}(\omega) < 0$ and $\mathrm{Im}(\omega) > 0$, respectively. A negative differential Maxwell relation time $(\tau_{md} < 0)$ is a necessary condition of instability. Neglecting diffusion and assuming a real value of frequency ω, we find from eq. (16)

$$\mathrm{Re}(k) = -\frac{\omega}{v} \qquad \mathrm{Im}(k) = -\frac{1}{\tau_{md}v} \tag{17}$$

As can be seen from Eq (17), when $\tau_{md} < 0$, the wave grows in the direction of propagation, i.e., from the cathode to anode. The magnitude of this increase is proportional to $\exp[\mathrm{Im}(k)L]$. If $\mathrm{Im}(k)L \ll 1$ and, fluctuation growth over the wavelength $\lambda = 2\pi/|k|$, is small. Under such conditions the diode may operate as an amplifier. In a practical device, such an amplification occurs if an integer number of wavelengths, $\lambda =$

$2\pi/k$, of the space-charge wave can fit inside the sample. Eq. (17) shows that this condition is fulfilled when frequency ω is equal to transit time frequency, $\omega_T = 2\pi f_T = 2\pi v/L$, or to its harmonics. The amplification at higher harmonics of the transit frequency (corresponding to several wavelengths fitting within a sample) may be suppressed by diffusion since higher frequencies correspond to higher values of k. Hence, the diffusion term which is proportional to $D_n k^2$ becomes more important for higher harmonics (see eq. (16)).

From the condition $\text{Im}(k)L \ll 1$ and eq. (17) we find that the amplification regime takes place when

$$n_o L < \frac{\varepsilon v \left(F^{(0)} \right)}{q |\mu_d|} \tag{18}$$

For $v \simeq 10^5$ m/s, $|\mu_d| \simeq 0.15$ m^2/vs, $\varepsilon = 1.14 \times 10^{-10}$ F/m, we obtain $n_o L < 1.5 \times 10^{11}$ cm^{-2}. More accurate calculations yield $n_o L \leq 10^{12}$ cm^{-2}. This condition (first introduced by Kroemer (1965)) is called the Kroemer criterion.

When the fluctuation growth, caused by the negative differential conductivity, is large (i.e. $\text{Im}(k)L \gg 1$), the fluctuation develops very rapidly over a small fraction of the total sample length. Using elegant measurements of time-dependent potential distributions in GaAs samples, John Gunn (1963) proved that, under such conditions, propagating regions of high electric field -- called high field domains -- periodically form at the cathode contact or nearby, propagate across the sample, and annihilate at the anode (see Fig. 2).

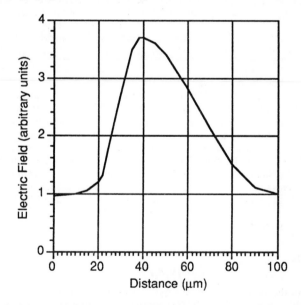

Fig. 2. Measured field distribution in propagating high field domains (from J. B. Gunn (1966)).

The mechanism of the domain formation may be described as follows. Let us assume that the average electric field in the sample is higher than the peak field, F_p, corresponding to the maximum value of the electron drift velocity, and that a small fluctuation of the electric field is present near the cathode. Such a fluctuation may exist, for example, as a consequence of a smaller doping fluctuation. The higher field in this region leads to a smaller electron velocity because of the negative slope of the v vs. F curve.

Hence, the electrons in front of and behind the high field region move faster than inside the region. This leads to the electron depletion in the leading edge and to the electron accumulation in the trailing edge of the fluctuation. The resulting dipole layer increases the fluctuation electric field, intensifying the fluctuation growth while it propagates toward the anode. If the bias voltage is kept constant the growth of the fluctuation can only come at the expense of the electric field, F_r, outside the fluctuation. This process continues until the electron velocities inside and outside the high field domain become equal and close, in most cases, to the electron saturation velocity. Thus, the fluctuation develops into a stable propagating high field domain. The decrease of F_r leads to a current drop during high field domain formation. The annihilation of the high field domain when it reaches the anode causes a temporary increase in the current, up to the peak value corresponding to the peak electron velocity (see Fig. 1). After the domain annihilation at the anode, the next domain nucleates near the cathode and the process repeats itself. Most often just one high field domain forms in the sample because the voltage drop across the domain results in a field smaller than F_p everywhere except within the domain. As mentioned above, in most cases, domains form near the cathode and propagate through the sample with a velocity close to the electron saturation velocity, v_s. The electrons outside the domain also move with the saturation velocity. Hence, the current density in a sample with a domain is equal to $j_s = qn_o v_s$. When the domain is extinct, the current density is equal to the peak current density, $j_p = qn_o v_p$ where v_p is the peak velocity. During the domain annihilation and the formation of a new domain the current density rises from j_s to j_p. The domain transit time is equal to L/v_s. Once formed, the domain does not disappear if the applied voltage is decreased below the threshold voltage during the domain propagation. The reason is that the field within the domain remains higher than F_p even though the average field may be below the peak field unless the voltage drop is too big. For large values of $n_o L \geq 10^{12}$ cm^{-2}, the domain sustaining voltage is close to $F_s L$ and the electric field outside the stable domain is close to F_s. The domain field depends on $n_o L$ and the applied voltage and may vary from approximately 30 to 200 kV/cm. It is limited only by the critical field for avalanche breakdown. The time of stable domain formation may be considerably larger than the initial growth time, $3|\tau_{md}|$. It decreases with an increase in n_o, and may vary from a few picoseconds to a few hundred picoseconds. The domain shape is also dependent on the doping density. When the doping density is large (much larger than ~10^{15} cm^{-3} for GaAs) a small relative change in the equilibrium electron concentration is sufficient to produce a large space charge, supporting a high domain field. In this case the domain shape is nearly symmetrical. In the opposite limiting case ($n_o \ll 10^{15}$ cm^{-3}) the leading edge is totally depleted of carriers, creating a positive space charge density $qn_o = qN_d$. The density of electrons in the accumulation layer (i.e., in the trailing domain edge) is limited only by diffusion processes and can be much larger than n_o. The resulting field distribution is very similar to the field distribution in a p^+-n junction moving with a domain velocity which is close to the electron saturation velocity, v_s. In the nearly depleted leading edge the drift and diffusion currents are small and the current density is determined by the displacement current $j = \varepsilon \dfrac{\partial F}{\partial t}$ so that the total current is given by

$$I = qv_s n_o(x) S(x) \tag{19}$$

where $x = v_s t$ is the coordinate of the moving domain. This means that the current waveform should reproduce the $n_o(x)S(x)$ profile. This domain property may be utilized for implementing numerous analog and logic functions (see, for example, Shoij (1967) and Engelbrecht (1967)).

Stable propagating domains exist in relatively long samples which operate as microwave oscillators at relatively low frequencies (10 to 40 GHz or less). In short samples, operating as microwave oscillators at higher frequencies (50 GHz and above), a typical form of the instability is the propagation of so-called accumulation layers (see Fig. 3). The theory describing this mode of operation was developed by D'yakonov et al

(1981). It shows that the propagation velocity of accumulation layers may be even larger than the peak velocity in GaAs.

A more detailed discussion of different transferred electron devices (including microwave generators, amplifiers, and functional devices) and their characteristics was given by Shur (1987), Ch. 5 and 6.

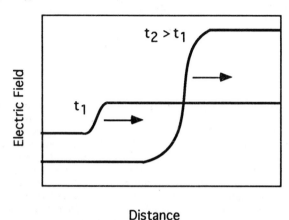

Distance

Fig. 3. Propagation of an accumulation layer in a Gunn diode.

APPROACHES TO GUNN DIODE MODELING

As follows from the discussions in the previous Section, semi-empirical drift-diffusion eq. (13) for the device current allows us to obtain an adequate qualitative and, in certain cases, a quantitative description of the Gunn effect. We can correctly identify the conditions for a small signal and large signal operation, describe stable Gunn domains, determine the characteristic times of the domain growth and dissolution, identify and describe other forms of instabilities in the transferred electron diodes, such as accumulation layers, predict the current waveform in samples with a non-uniform cross-section, find a small signal impedance, describe (to a limited extent) the role of contacts, and model the device-circuit interaction (see, for example, Shur (1987).) However, one should clearly understand important and severe limitations of such an approach.

First, this approach postulates that even in a sample with a very non-uniform field distribution both the electron velocity, v, (or electron mobility $\mu = v/F$) and the electron diffusion coefficient, D_n, only depend on the electric field, F. This assumption may be justified in long samples where the electron mean free path, $\lambda = v_T\tau_m$ (here τ_m is the momentum relaxation time) is very small compared with the high field domain size. One can even argue that the validity of eq. (13) requires the domain dimensions to be much larger than $v_T\tau_e$ where τ_e is the energy relaxation time. This latter length can be easily as large as 1000 Å at room temperature and is comparable to the characteristic domain accumulation layer size in short samples (on the order of a micron or so).

Second, the only characteristic time constants involved in eqs. (12) and (13) are related to the differential Maxwell dielectric relaxation time and electron transit time. In fact, the characteristic momentum and energy relaxation times also effect the transient response and device behavior at frequencies as low as 30 GHz (see Rees (1969)).

Third, as we discussed above, the physics of the Ridley-Watkins-Hilsum-Gunn effect is linked to the intervalley transitions. Even in a long sample, a more rigorous approach should distinguish between electrons in different valleys and describe electrons in each valley by a separate space and time dependent distribution function.

Fourth, in rapidly and abruptly varying electric fields inside of a transferred electron device, the electron energy of chaotic motion is a function of distance. This leads to an electron thermal diffusion current. Eq. (13) does not account for such a current. An attempt to account for the thermal diffusion current by replacing the $D_n(F)\partial n/\partial x$ term with the $\partial D_n(F)\partial n/\partial x$ term in eq. (13) may be too crude and may lead to unphysical results.

In his pioneering paper, Bløtekjær (1970) tried to address all these problems by deriving a set of simplified transport equations by evaluating the moments of the Boltzmann transport equation and assuming in the process that the electron distribution function in each valley can be approximated by a displaced Maxwell distribution function:

$$ f = A\exp\left(-\frac{m\left(v^2 - v_d^2\right)}{2k_B T_e} \right) \tag{20} $$

where v_d is the electron drift velocity, m is the electron effective mass, v is the electron velocity, k_B is the Boltzmann constant, and T_e is the effective electron temperature for a given valley. His work still serves as the basis of the so-called energy balance equations used in modeling of all kinds of semiconductor devices, including the transferred electron diodes. To a certain extent, this approach has been incorporated into some commercial simulators (see Silvaco PRIZM manual (1991)). However, the assumption of the displaced Maxwell distribution function is not very realistic. The electron transfer from upper valleys back to the central valley of the conduction band strongly affects the tail of the distribution function at high energies and makes the distribution function to be very different from a Maxwellian distribution. For this reason, many modern simulations of transferred electron devices rely (either directly or indirectly) on the Monte Carlo simulations. In this method, the electron transport is simulated as a series of free electron flights in an accelerating electric field. These flights are randomly interrupted by scattering events which reproduce all relevant scattering mechanisms, including, of course, intervalley scattering. Jacoboni and Lugli (1989) gave a good description of the Monte Carlo technique applied to simulation of semiconductor material properties and semiconductor devices. Jensen et al (1991) reviewed more recent work in this area. Shur (1976) proposed to use the results of the steady state Monte Carlo simulations to extract the parameters of a simplified version of the energy balance model and demonstrated that the results were in good agreement with the transient Monte Carlo calculations.

Tait and Krowne (1988) extracted the dependencies $\mu(F)$ and $D(F)$ from the Monte Carlo calculations for GaAs and InP (see Fig. 4) and used these dependencies to simulate transferred electron devices.

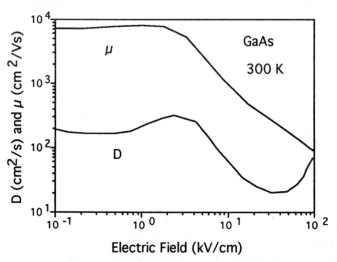

Fig. 4. Electron mobility and diffusion coefficient for GaAs extracted from steady state Monte Carlo simulations (after Tait and Krowne (1988)). © Copyright IEEE 1988.

Fig. 5 compares the waveforms of a 3 μm GaAs and InP transferred electron diodes obtained using the drift diffusion and Monte Carlo approaches by Tait and Krowne (1988). As can be seen from the figure, the qualitative agreement is fairly reasonable though quantitative differences are quite important. All in all, it seems that the simplified approach may be useful for devices with sizes greater than 3 micron operating at frequencies below 30 GHz or so. For shorter devices or higher frequencies such an analysis may still retain a limited ability to describe at least some devices features but only in a qualitative way. Under such conditions, Monte Carlo simulations (or energy balance equations with parameters extracted from the Monte Carlo simulations) should be used. The problems with the Monte Carlo simulations are (1) large amounts of computer time required and (2) the need to know numerous parameters characterizing different scattering mechanisms. The first problem is rapidly becoming less important with an advent of high speed workstations. However, the second problem is here to stay, especially for new emerging semiconductor materials even though there is a slow but sure movement towards adopting some "standard" Monte Carlo parameters for different compounds.

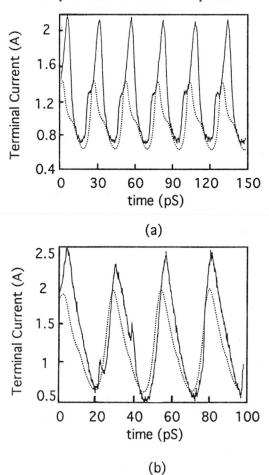

(a)

(b)

Fig. 5. Current waveforms of a 3 μm long GaAs (a) and InP (b) transferred electron diodes obtained using the drift diffusion and Monte Carlo approaches (after Tait and Krowne (1988)). © Copyright IEEE 1988.

The use of the energy balance equations poses another problem. These equations utilize average electron energy (electron temperature) and make use of energy dependent momentum and energy relaxation times and energy dependent effective masses. The problem is how to extract these parameters from steady state Monte Carlo simulations. Curow (1992) compared different recipes for such an extraction.

All in all, the modeling techniques for transferred electron devices (or any other semiconductor devices, for that matter) are not completely and accurately derived from the first principles. Hence, their accuracy is somewhat limited and can always be questioned. Nevertheless, device modeling still retains an awesome predictive power and is indispensable for an intelligent device design and analysis of device operation.

Figs. 6 illustrate the application of the drift-diffusion technique to modeling a new proposed transferred electron device - Modulated Impurity Concentration Transferred Electron Device (MICTED) - which is expected to operate more efficiently at the second harmonic (see Figs. 7 and 8). In this device, the increased doping concentration in the central section of the device leads to a much higher second harmonic content in the current waveform.

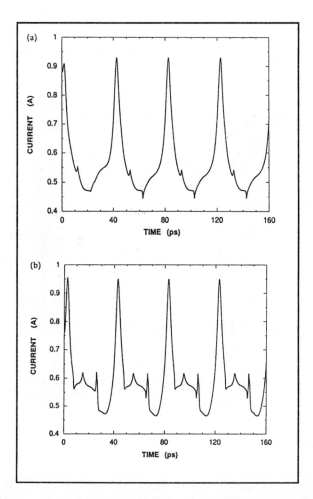

Fig. 6. Current waveforms of a conventional 3 μm InP TED (a) and of a comparable 3 μm InP MICTED (b) obtained using the drift diffusion approach (after Jones et al. (1992)).
© Copyright IEEE 1992

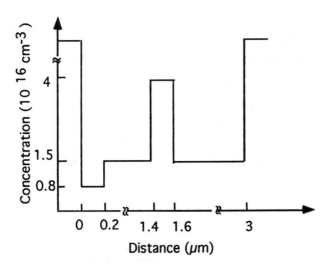

Fig. 7. Doping profile of a 3 μm InP MICTED (after Jones et al. (1992)).
© Copyright IEEE 1992.

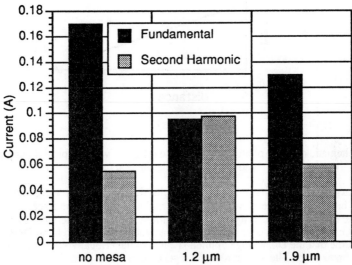

Fig. 8. Computed efficiencies for the fundamental and second harmonic for MICTED and comparable conventional transferred electron device with no mesa (after Jones et al. (1992)). © Copyright IEEE 1992

IMPATT and TRAPATT DIODES.

The Transferred Electron Devices (TEDs) discussed above utilize bulk negative differential resistance related to the intervalley transition in GaAs and other compound semiconductors. Another family of power microwave devices which includes IMPATT and TRAPATT diodes, relies on the phase shifts between the device current and applied voltage creating a "dynamic" negative differential resistance.

In an IMPact ionization Avalanche Transit Time (IMPATT) diode two mechanisms create additive phase delays -- avalanche breakdown (generating electron-hole pairs) and carrier drift across a special "drift" section of the device. Fig. 9 shows schematic diagram of an IMPATT diode along with at typical doping profile and electric field distribution.

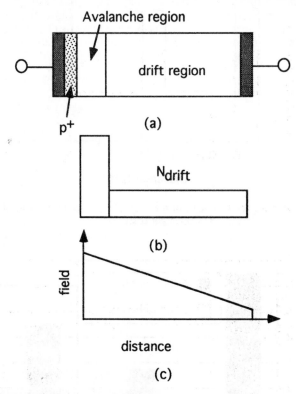

Fig. 9. Schematic diagram of an IMPATT diode (a) along with a typical doping profile (b) and electric field distribution (c).

When a large enough voltage is applied to the device, carriers are generated by the avalanche breakdown process in the high field region at the p^+-n interface. These carriers move in the electric field creating the electric current at the p^+-n interface. Let us consider the situation when the sum of the DC bias and ac voltage is applied to an IMPATT diode. Let us assume that the DC voltage applied is equal to the critical voltage needed to initiate avalanche breakdown. The impact ionization generation rate is given by

$$G = j_n \exp\left[\left(-\frac{F_o}{F}\right)^m\right] \tag{21}$$

where F_o is the characteristic field of the impact ionization, m is a constant (typically between 1 and 2), and j_n is the electron current density. Hence, when the ac voltage increases the total voltage applied to the device, the generation rate rises sharply. When the ac voltage decreases from the peak value, the generation rate decreases. The rate of current increase is proportional to the generation rate. Hence, the current continues to rise reaching the peak when the ac voltage drops to zero. This means that the phase of the generated current is delayed by 90° with respect to the ac voltage, i. e. the avalanche breakdown creates a phase delay of 90° between the ac voltage and current. An additional

90° (or so) phase delay is created by utilizing carrier drift across the drift region. (The phase shift introduced by the drift is equal to ωT_{tr} where T_{tr} is the transit time of carriers across the drift region and ω is the oscillation frequency.) As a consequence the total phase delay of the electron current with respect to the applied voltage is close to 180°. When the phase shift is 180°, the current decreases when the voltage increases. This corresponds to a dynamic negative differential resistance (the first Fourier component of the current, i_ω, has the sign opposite to the first Fourier component of the voltage, V_ω).

Read (1958) and Tager et al. (1959) were first to propose the idea of an IMPATT device and, therefore, this device is frequently called the Read diode. Johnston et al. (1965) were first to report IMPATT oscillations in an avalanching silicon diode in a microwave cavity. Different modifications of the Read diode have been proposed in order to optimize the generated current pulse and increase the breakdown voltage and, hence, the generated microwave power. In practice, almost any reversed biased p-n junction may operate as an IMPATT diode. Misawa (1966) showed that even a p-i-n structure, where the electric field in the undoped region is nearly constant, may still operate as an IMPATT device.

Prager et al. (1967) discovered a completely new and more efficient operation regime of IMPATT diodes operating at relatively low frequencies. Computer simulations of Scharfetter et al. (1968) showed that in this new regime, during part of the cycle, the drift region of the diode is filled by a high-density electron-hole plasma that is "trapped" within the device. In this state the device impedance is very small and, hence, the voltage across the device is low. The electrons and holes are removed from the device via electron and hole transit to the contacts. Hence, this new regime is called the TRApped Plasma Avalanche Triggered Transit (TRAPATT) regime and avalanche diodes operating in this regime are called TRAPATT diodes.

The physics of this new regime may be understood by considering the electric current in the undoped (drift) region. The total current through the device should be continuous. The avalanche current in the avalanche region is carried by electrons and holes. However, there are no carriers in the drift region, and, hence, the current in this region must be the displacement current with the current density $\varepsilon \partial F / \partial t = \varepsilon \omega F$, where F is the electric field. The maximum value of the displacement current is $\varepsilon \omega F_{max}$ where F_{max} is maximum field in the structure. But what happens if the avalanche current is greater than $\varepsilon \omega F_{max}$? Then the current continuity is maintained by the electric field redistribution in such a way that the avalanche region moves towards the anode, leaving high density electron-hole plasma behind ("trapped" plasma).

Modeling of IMPATT and TRAPATT diodes involves the solution of the same basic semiconductor equations as for transferred electron devices, with an additional important complication -- the need to simulate impact ionization processes. For these devices, an empirical approach utilizing empirical generation rates (see eq. (21)) may actually have an advantage. Indeed, the Monte Carlo simulations for impact ionization processes (see, for example, Brennan et al. (1991)) are quite difficult for several reasons. First, the impact ionization is caused by very energetic ("lucky") electrons belonging to the tail of the distribution function. An accurate calculation of the distribution function tail may strain even supercomputer resources (see Gelmont et al. (1992) for an example of the impact ionization rate calculation relying on the distribution function computed using the Monte Carlo technique). Second, these highly energetic electrons occupy the energy states for which little information is available. Third, the impact ionization processes create both electrons and holes; hence, the number of particles varies, and the two distributions have to be described self-consistently. All these challenges have to be properly addressed for the Monte Carlo simulation of IMPATT diodes.

The empirical approach has its own pitfalls. In addition to difficulties mentioned above in connection with our discussion of transferred electron devices, we now need to know empirical ionization rates by both electrons and holes. There are different opinions on how accurately these ionization rates can be determined (primarily from experiments involving the breakdown in p-n junctions). Consequently, the IMPATT and TRAPPATT diode simulation can primarily elucidate the device physics and basic principles of device operation. It can be also applied for comparative estimates of device efficiency for different doping profiles and device geometry. The basic approach to such a simulation has remain unchanged from that by described Gummel (1964) in his pioneering paper on semiconductor device modeling.

REFERENCES

K. Bløtekjær, Transport Equations for Two-Valley Semiconductors, IEEE Trans. Electron Devices, ED-17, No. 1, pp. 38-47, Jan. (1970)

K. H. Brennan, N. Mansour, and Y. Wang, Simulation of Advanced Semiconductor Devices Using Supercomputers, Computer Physics Communications, vol. 67, No. 1, pp. 173-92 (1991)

M. Curow, Applied Physics Letters, (1992)

M. I. D'yakonov, M. E. Levinshtein, and G. S. Simin, Soviet Phys. Semicond., 13, No. 7, pp. 1365-1368 (1981)

R. S. Engelbrecht, Bulk Effect Devices for Future Transmission Systems, Bell Lab. Rec., 45, No. 6, pp. 192-201 (1967)

B. Gelmont, K.-S. Kim, and M. S. Shur, Theory of Impact Ionization and Auger Recombination in HgCdTe, Phys. Rev. Lett., 69, No. 8, pp. 1280-1283, August (1992)

H. Gummel, A Self-consistent Iterative Scheme for One-dimensional Steady State Transistor Calculations, IEEE Trans. Electron Devices, ED-11, pp. 455-465 (1964)

J. B. Gunn, Microwave Oscillations of current in III-V semiconductors, Solid State Communications, 1 (4), pp. 88-91 (1963)

J. B. Gunn, Electron Transport Properties Relevant to Instabilities in GaAs, J. Phys. Soc. Jap., Supplement, pp. 509-513 (1966)

C. Hilsum, Transferred Electron Amplifiers and Oscillators, Proc. IRE, 50,(2), pp. 185-189 (1962).

C. Jacoboni and P. Lugli, The Monte Carlo Method for Semiconductor Simulation, Springer Series on Computational Microelectronics, ed. S. Selberherr, Springer-Verlag, Wien, New York (1989)

G.U. Jensen, B. Lund, M. Shur, and T.A. Fjeldly, Monte Carlo Simulation of Short Channel Heterostructure Field Effect Transistors, Computer Physics Communications, vol. 67, No. 1, pp. 1-61 (1991)

R. L. Johnston, B. C. DeLoach, Jr., and B. G. Cohen, Bell Syst. Tech. J., 44, p. 369 (1965)

S. H. Jones, G. B. Tait, and M. Shur, Modulated-Impurity-Concentration Transferred-Electron Devices Exhibiting Large Harmonic Frequency Content, Microwave and Optical Technology Letters, vol. 5, No. 8, pp. 354-359, July (1992)

H. Kroemer, Proc. IEEE, 53, No. 9, p. 1246 (1965)

T. Misawa, IEEE Trans. Electron Devices, ED-13, No. 9, p. 137 (1966)

H. J. Prager, K. K. N. Chang, and S. Weisbrod, High Power, High Efficiency Silicon Avalanche Diodes at Ultrahigh Frequencies, Proc. IEEE, 55, p. 586 (1967)

PRIZM manual, Silvaco (1992)

W. T. Read, Bell Syst. Tech. J., 37, p. 401 (1958)

H. D. Rees, Hot Electron Effects at Microwave Frequencies in GaAs, Solid State Comm., 7, No. 2, pp. 267-269 (1969)

B. K. Ridley and T. B. Watkins, The Possibility of Negative Resistance, Proc. Phys. Soc., 78 (8), pp. 293-304 (1961)

D. L. Scharfetter, D. J. Bartelink, H. K. Gummel, and R. L. Johnson, IEEE Trans. Electron Devices, ED-15, p. 691 (1968)

M. Shoji, Functional Bulk Semiconductor Oscillators, IEEE Trans. Electron Devices, ED-14, No. 9, pp. 533-546 (1967)

M. Shur, Influence of Non-Uniform Field Distribution in the Channel on the Frequency Performance of GaAs FETs, Electronics Letters, 12, No. 23, pp. 615-616 (1976)

M. Shur, GaAs Devices and Circuits, Plenum Publishing, New York (1987)

M. Shur, Physics of Semiconductor Devices, Prentice Hall, New Jersey (1990)

G. B. Tait and C. M. Krowne, Efficient Transferred Electron Device Simulation Method for Microwave and Millimeter Wave CAD Applications, Solid-State Electronics, vol. 30, No. 10, pp. 1025-1036 (1987)

G. B. Tait and C. M. Krowne, Large Signal Characterizations of Unipolar III-V Semiconductor Diodes at Microwave and Millimeter-Wave Frequencies, IEEE Trans. Electron Devices, ED-35, No. 2, pp. 223-229 (1988)

A. S. Tager, A. I. Mel'nikov, G. P. Kobel'kov, and A. M. Tsebiev, Discovery Diploma #24, priority date No. 27, 1959 (in Russian), see also A. S. Tager and V. M. Vald-Perlov, Lavinnoproletnye Diody (in Russian), Soviet Radio (1968)

J. Xu and M. Shur, IEEE Trans. Electron Devices, ED-34, No. 8, pp.1831-1832 (1987)

Introduction to Quantum Modelling

Robert E. Miles
University of Leeds

INTRODUCTION

Although we all live in a quantum mechanical world we have to look very hard to see quantum effects. This is because quantum phenomena such as the discretisation of energy levels and the wave nature of matter only become apparent at the sub-microscopic level. If we solve Schrödinger's equation for electrons confined in an infinitely deep one-dimensional potential well then we find that at room temperature the separation between energy levels will only become comparable to the thermal energy for a well with a width of less than 4 nm. In modern semiconductor device processing these dimensions can be readily achieved in the vertical direction where crystal growth can be controlled to atomic dimensions and are now being approached by the lithographic process in the horizontal direction. We would therefore expect to observe, and perhaps exploit, quantum phenomena in present day structures. In practice the familiar semiconductor devices such as MESFETs still appear to operate in a an essentially classical way even when their dimensions are very small (e.g. gate lengths < 30 nm). One of the reasons why quantum effects are not always apparent in semiconductors is that changes in potential tend to occur over a Debye Length which smooths things out over relatively large distances. Although a number of structures have been proposed and fabricated to study the physics of quantum effects most of these only operate at very low temperatures (4 K) where they are of little use as practical devices. However, there is one device, the Double Barrier Resonant Tunnelling Diode (DBRTD), that operates on purely quantum mechanical principles at room temperature and it can also be shown that the inclusion of quantum effects in HEMT models does make a noticeable difference to the device characteristics. It is still early days for quantum devices and quantum modelling but as device dimensions continue to shrink and quantum phenomena are better understood it is expected that these new device concepts will become important.

This chapter will outline how quantum mechanical ideas can be applied to device modelling based on the effective mass form of the Schrödinger equation. The transmission of electrons across potential barriers will be discussed leading to a description of the DBRTD. We will then go on to discuss how a self consistent solution of the semiconductor device equations and Schrödinger's equation may be achieved in one and two-dimensional HEMT simulations.

THE EFFECTIVE MASS SCHRÖDINGER EQUATION

As this chapter is about applications of quantum modelling the background to the effective mass form of the Schrödinger equation will not be discussed. For the interested reader the book on Quantum Phenomena by Datta (1989) is highly recommended. In the effective mass form, Schrödinger's equation can be written as:-

$$ih\frac{\partial}{\partial t}\Psi = -\frac{h^2}{2m^*}\nabla^2\Psi + E_C\Psi \tag{1}$$

where h, the reduced Planck's constant, is $h/2\pi = 1.06 \times 10^{-34}$Js and m^* is the effective mass of an electron which in this chapter will be taken as a constant independent of the material. E_C is the potential energy which will be taken as the edge of the conduction band. $\Psi(E,r,t)$ is a complex function of energy, position and time known as the wave function. In practice, because of the imposed boundary conditions, Ψ will only exist at certain values of E as for example in the energy levels of an atom but the levels may be so closely spaced that they form a continuum as occurs in the energy bands of a crystal. The product $\Psi^*\Psi$ (where Ψ^* is the complex conjugate of Ψ) is interpreted as the probability that an electron exists at a point in space at time t. As the probability that the electron exists somewhere in space is unity then

$$\int \Psi^*\Psi dr = 1 \tag{2}$$

where the integral is taken over all space.

The electron density at point r is given by:-

$$n = \langle \Psi^*\Psi \rangle \tag{3}$$

Where $\langle \rangle$ means a summation over all occupied energy levels i.e. all electrons.

The solution to equation (1) can be written as

$$\Psi(r,t) = \psi(r)\exp\left[-\frac{iEt}{h}\right] \tag{4}$$

where ψ is a function of position only and E is the total energy of an electron i.e. potential and kinetic energy. Differentiating equation (4) and substituting into equation (1) gives

$$\frac{h^2}{2m^*}\nabla^2\psi + (E - E_c)\psi = 0 \tag{5}$$

Note that equation (5) contains only the spatial part of the wave function.

For the case where $E_c (= E_{c0})$ is constant in all directions (i.e. as far as a particular electron is concerned, the background charge of the other electrons and the atomic nuclei is smeared out uniformly) a solution to equation (5) is

$$\psi = C\exp(\pm ik_x x)\exp(\pm ik_y y)\exp(\pm ik_z z) \tag{6}$$

where k_x, k_y and k_z are related to the momenta p_x, p_y and p_z in the x, y and z directions by the de Broglie equation $hk = p$ and C is a constant chosen to satisfy equation (2).

Substituting into equation (5) with $E_c = E_{c0}$ gives

$$E = E_{c0} + \frac{h^2}{2m^*}(k_x^2 + k_y^2 + k_z^2) \tag{7}$$

By analogy with classical mechanics p/m^* is the velocity v of an electron, so equation (7) is just saying that the total energy E of an electron is equal to the potential energy E_{C_0} plus the kinetic energy $\frac{1}{2}m^*v^2$.

If we substitute equation (6) back into equation (4) it can be seen that the wave function represents waves which are travelling backwards and forwards in the x, y and z directions. As stated above the allowed values of the k's will depend on the boundary conditions i.e. the shape of the potential well which confines the electrons. If in equation (7) $E > E_{C_0}$, k_x, k_y and k_z are real and we have travelling waves but if $E < E_{C_0}$ one or more of the k values will be imaginary and the waves will decay exponentially in the corresponding direction.

Current density can be calculated from the wave functions using

$$J = -\frac{iqh}{2m^*} \langle (\nabla\psi)^*\psi - \psi^*(\nabla\psi) \rangle \tag{8}$$

and if we use the wave functions from equation (6) in (8) we find that

$$J = -\frac{qhk}{m^*} \langle |C^2| \rangle \tag{9}$$

From equation (3), $n = \langle |C^2| \rangle$ and therefore equation (9) can be written

$$J_n = -\frac{nqhk}{m^*} \tag{10}$$

which with $\hbar k/m^* = v$ gives the familiar classical result

$$J_n = -nqv \tag{11}$$

Solution of Schrödingers Equation in the Region of a Potential Step

The solution of the Schrödinger equation outlined in the previous section is for a uniform sample of semiconductor. If E_C varies with position equation (5) can be very difficult to solve, but if the variation is in only one dimension then some useful results can be obtained.

Let us look at the situation illustrated in Fig. 1 which shows a potential step in the z direction only. This could for example be a GaAs/AlGaAs boundary where the step would be about 0.3 eV for an Al mole fraction of 30% and there is no band bending. On each side of the step the potential is uniform. If the step is at $z = 0$ then for $z < 0$, $E_C = E_{C1}$ and for $z > 0$, $E_C = E_{C2}$. In this example the material is uniform in the x and y directions and the wave function will therefore have the same exponential form as in equation (6) in these directions.

If we let the wave function have the form $\phi(z)$ in the z direction then we can write

$$\psi = Cexp(ik_x x)exp(ik_y y)\phi(z) \tag{12}$$

Substituting this into equation (5) gives

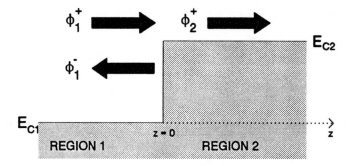

Figure 1. One-dimensional potential step.

$$\frac{\partial^2 \phi}{\partial z^2} + \frac{2m^*}{\hbar^2}\epsilon\phi = 0 \tag{13}$$

where

$$\epsilon = \epsilon_1 = E - E_{C1} - \frac{\hbar^2}{2m^*}(k_x^2 + k_y^2) \; ; \quad z < 0$$
$$\epsilon = \epsilon_2 = E - E_{C2} - \frac{\hbar^2}{2m^*}(k_x^2 + k_y^2) \; ; \quad z > 0 \tag{14}$$

Solving equation (13), $\phi(z) = \exp(\pm ik_1 z)$ for $z < 0$ and $\phi(z) = \exp(\pm ik_2 z)$ for $z > 0$. where

$$k_1 = \left[\frac{2m^*\epsilon_1}{\hbar}\right]^{\frac{1}{2}} \qquad k_2 = \left[\frac{2m^*\epsilon_2}{\hbar}\right]^{\frac{1}{2}} \tag{15}$$

The positive sign in the exponentials is identified with a wave travelling in the positive z direction and the negative sign with one travelling in the negative direction.

If we assume a wave $\phi_1^+ = \exp(ik_1 z)$ travelling to the right and incident on the left of the step in Fig. 1 then a fraction r_{11} will be reflected back by the step giving a wave $\phi_1^- = r_{11}\exp(-ik_1 z)$ travelling back towards the left. (The superscripts + and - refer to waves travelling in the positive and negative z directions respectively and the subscripts refer to the region in which the wave is travelling - see Fig. 2.) A fraction t_{12} will be transmitted across the step and hence a wave $\phi_2^+ = t_{12}\exp(ik_2 z)$ will travel on from the step to the right. (The subscripts on r and t refer to reflection back into side 1 and transmission from side 1 to side 2.) The wave function must be continuous at the step and therefore ϕ and $d\phi/dz$ must be continuous. This leads to

$$1 + r_{11} = t_{12} \qquad k_1(1 - r_{11}) = t_{12}k_2 \tag{16}$$

or

$$r_{11} = \frac{(k_1 - k_2)}{(k_1 + k_2)} \qquad t_{12} = \frac{2k_1}{(k_1 + k_2)} \tag{17}$$

Using equation (8) for $z < 0$ where $\phi = \exp(ik_1z) + r_{11}\exp(-ik_1z)$, the current in region 1 is

$$J = -\frac{|C|^2q\hbar k_1}{m^*}(1 - |r_{12}|^2) = J_i - J_r \tag{18}$$

where J_i and J_r are the incident and reflected currents respectively.
For $z > 0$ where $\phi = t_{12}\exp(ik_2z)$ the transmitted current J_t in region 2 is

$$J_t = -\frac{|C|^2q\hbar k_2}{m^*}|t_{12}|^2 \tag{19}$$

R the fraction of current reflected is $J_r/J_i = |r_{11}|^2$

and T the fraction transmitted is $J_t/J_i = |t_{12}|^2 k_2/k_1$.

Similar equations for a wave incident on the right of the step can be derived with a reflection factor r_{22} and transmission factor t_{21} as above but with k_1 and k_2 interchanged.

Scattering Matrices

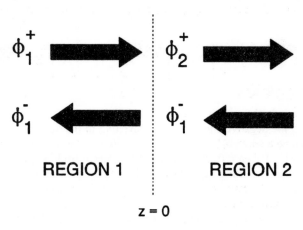

Figure 2. Electron waves incident on both sides of a scattering feature.

If we generalise and consider a one-dimensional scattering centre with waves incident on both sides as shown in Fig. 2 we can write

$$\phi_2^+ = t_{12}\phi_1^+ + r_{22}\phi_2^-$$
$$\phi_1^- = r_{11}\phi_1^+ + t_{21}\phi_2^-$$
(20)

Equations (20) can be combined in a matrix form as

$$\begin{bmatrix} \phi_2^+ \\ \phi_1^- \end{bmatrix} = \begin{bmatrix} t_{12} & r_{22} \\ r_{11} & t_{21} \end{bmatrix} \begin{bmatrix} \phi_1^+ \\ \phi_2^- \end{bmatrix}$$
(21)

The matrix elements in equation (21) for scattering at a step are given in the previous section. It is also possible to treat the propagation of a wave along a region of constant potential as a scattering event in which case all that happens is that the waves undergo a phase shift with no change in amplitude. In this case $t_{12} = t_{21} = \exp(ikd)$, $r_{11} = r_{22} = 0$ and for a free propagation equation (21) becomes:-

$$\begin{bmatrix} \phi_2^+ \\ \phi_1^- \end{bmatrix} = \begin{bmatrix} e^{ikd} & 0 \\ 0 & e^{ikd} \end{bmatrix} \begin{bmatrix} \phi_1^+ \\ \phi_2^- \end{bmatrix}$$
(22)

where k takes the appropriate value depending on the potential energy E_c and d is the length of the region.

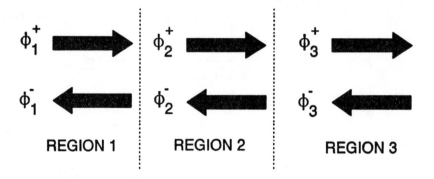

Figure 3. Sequential scattering.

If the scattering event represented in equation (21) is immediately followed by a second scattering from region 2 to region 3 (see Fig. 3) then a similar equation can be written with subscripts 2 and 3, i.e. t_{12} becomes t_{23} and r_{22} becomes r_{11} etc. The individual scattering matrices for the two events can be cascaded to give the total scattering matrix from region 1 to region 3:-

$$
\begin{bmatrix} \phi_3^+ \\ \phi_1^- \end{bmatrix} = \begin{bmatrix} t_{13} & r_{33} \\ r_{11} & t_{31} \end{bmatrix} \begin{bmatrix} \phi_1^+ \\ \phi_3^- \end{bmatrix} \tag{23}
$$

where

$$
\begin{aligned}
t_{13} &= t_{23}(1 - r_{21}r_{23})^{-1}t_{12} \\
r_{13} &= r_{12} + t_{21}r_{23}(1 - r_{21}r_{23})^{-1}t_{12} \\
r_{31} &= r_{32} + t_{23}(1 - r_{21}r_{23})^{-1}r_{21}t_{32} \\
t_{31} &= t_{21}(1 - r_{21}r_{23})^{-1}t_{32}
\end{aligned} \tag{24}
$$

See Datta pp 31-32 for a derivation of equations (24).

THE DOUBLE BARRIER RESONANT TUNNELLING DIODE

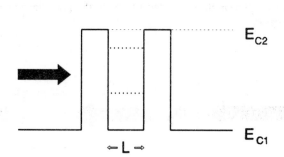

Figure 4. The double barrier resonant tunnelling structure with 2 quasi-bound states in the potential well.

Using the scattering matrices for potential steps and free propagations as calculated earlier and the rules for cascading as discussed above we can now calculate transmission probabilities for any sequence of steps linked by free propagations. In the particular example of a double barrier as illustrated in Fig. 4 where if we assume that there is no defect scattering a wave undergoes four transmissions at steps and three free propagations as it travels through the structure. Some careful algebra will lead to the following expression for the transmission probability

$$
T_{DB} = \left[1 + \frac{4r_{SB}^2}{t_{SB}^4}\sin^2 k_1 L \right]^{-1} \tag{25}
$$

In equation (25) L is the width of the potential well and r_{SB} and t_{SB} are the wave function reflection and transmission coefficients for a single barrier.

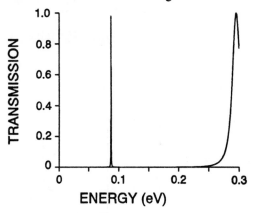

Figure 5. The transmission probability of a double barrier as a function of incident energy.

The expression in equation (25) is illustrated in Fig. 5 where it can clearly be seen that at certain electron energies (i.e. wavelengths) the double barrier structure has a transmission probability of 1, i.e. it is transparent. These resonances occur when the energy of the incident electrons is equal to that of a quasi-bound state in the potential well between the barriers (see Fig. 4). This is a purely quantum mechanical result. The fall in transmission probability after the resonant peak gives rise to the negative differential resistivity that is observed in the current-voltage characteristics of these structures (see Fig. 6).

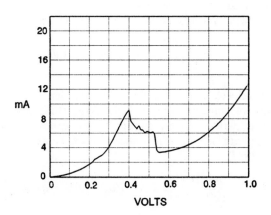

Figure 6. A typical experimental current/voltage characteristic of a DBRTD.

The resonant peaks in Fig. 5 are very sharp because they correspond to a mono-energetic beam of electrons incident on the double barrier. In a real device the incident electrons will have an energy distribution (Fermi-Dirac or Boltzmann) and the total transmission will correspond to an integration over all of the incident electrons. This will have the effect of broadening the peaks as is more typical of observation (Fig. 6). There may also be some thermal broadening of the energy levels in the potential well due to the effect of lattice vibrations on the well width and barrier thickness.

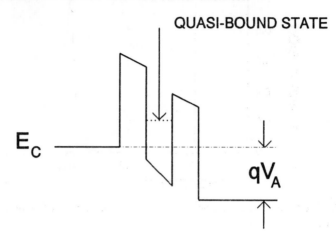

Figure 7. Resonant tunnelling structure with applied bias V_A.

It should also be realised that equation (25) applies only to the case where there is no voltage drop across the double barrier and that electrons are incident on one side of the structure only. In reality, under voltage bias the double barrier will look more like Fig. 7 where the potential falls across all three regions. It can be seen intuitively that electrons incident on the right must have a higher energy than those on the left to transit through the barrier to the quasi-bound state. A smaller fraction of the electrons on the right will therefore cross the double barrier structure giving a net electron current flow towards the right. In order to find the current-voltage relationship we must take account of electrons incident on each side of the device.

The current J^+ flowing towards the right can be expressed as

$$J^+ = -\frac{q}{4\pi^3} \int_{k_x k_y k_z} \frac{\hbar k_z}{m^*} T^+(k_z) f(k) dk_x dk_y dk_z \tag{26}$$

where $\hbar k_z / m^*$ is the velocity normal to the double barrier, T^+ is the transmission probability in the positive z direction and $f(k)$ is the Fermi-Dirac probability function. A similar equation can be written for the current J^- flowing towards the left. The net current flow J will therefore be $J^+ - J^-$ which when converted into an integral over energy becomes (Chou et al 1987)

$$J^+ - J^- = -\frac{qm^*}{2\pi^2\hbar^3} \int_{E_z = 0}^{\infty} T(E_z) \left[\int_{E_L} [f(E) - f(E + qV)] dE_L \right] dE_z \tag{27}$$

where E_L is the energy in the longitudinal direction i.e. parallel to the double barrier. Equation (27) holds because at a given energy, $T^+(E) = T^-(E)$. Integrating over E_L yields

$$J = -\frac{qm^{*}k_{B}T}{2\pi^{2}\hbar^{2}} \int_{k_{z}=0}^{\infty} T(E_{z})\ln \left[\frac{1 + \exp\left[\dfrac{E_{F} - E_{L}}{k_{B}T}\right]}{1 + \exp\left[\dfrac{E_{F} - E_{L} - qV}{k_{B}T}\right]} \right] dE_{z} \qquad (28)$$

where k_{B} is Boltzmann's constant, T the absolute temperature and the product $k_{B}T$ is the thermal energy. The logarithmic term in the integral is known as the supply function. With a knowledge of the transmission probability $T(E)$ the current-voltage characteristic can be determined.

Close to resonance the electron density builds up in the potential well (i.e. the wave function amplitude increases near resonance) and this will have a significant effect on the potential distribution across the double barrier. A better transmission probability function than that in equation (25) would therefore be obtained with a self consistent solution of the Schrödinger equation with Poisson's equation.

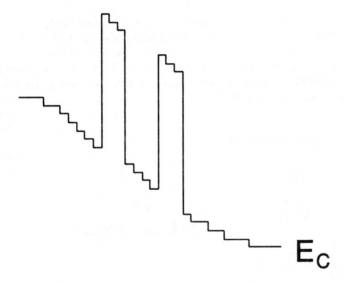

Figure 8. A potential distribution discretised into finite steps.

One approach (Ando Y and Itoh T 1986) to this self consistent solution is to discretise the potential into a series of steps as shown schematically in Fig. 8. We can then cascade the scattering matrices for each of the steps and horizontal regions as discussed above and obtain the wave functions for any arbitrary potential distribution providing a sufficient number of steps are used. The wave functions and hence the electron densities can be calculated giving an updated potential distribution. This process can be repeated until the solution converges.

All that has been said above refers to the double barrier structure in isolation. In a real device there will be other layers on each side of the double barrier. (The precise details of these layers have a significant impact on the overall electrical characteristics.) A conduction band edge profile with an accumulation region on one side of the double barrier and a depletion region on the other as illustrated in Fig. 9 will therefore be more

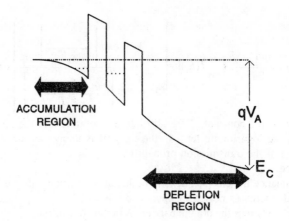

Figure 9. Conduction band edge profile in a DBRTD.

representative of a real device. As a first approximation the current through these regions can be treated classically and combined self consistently with equation (28) but there is evidence to suggest that quantisation also occurs in the potential well resulting from the band bending in the accumulation region (see Fig. 9). To the authors knowledge a complete self consistent model for the double barrier resonant tunnelling diode including the contact layers has not been published.

QUANTUM MODELLING IN HEMTS

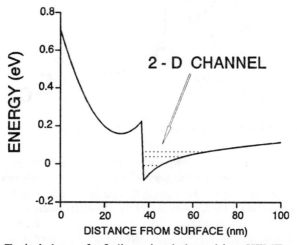

Figure 10. Typical shape of a 2-dimensional channel in a HEMT.

In the HEMT the conducting electrons are confined to move in a two-dimensional channel which is formed at the heterojunction between two semiconductors of differing band gap. The original devices were GaAs/AlGaAs lattice matched systems but more recently improved performance has been obtained from pseudomorphic heterojunctions such as AlGaAs/InGaAs/GaAs.

We would expect to observe quantum effects due to the confinement of the electrons in the two-dimensional channel. However calculations show that not all of the electrons are constrained to flow in the channel. At high gate bias the channel becomes very shallow and a significant fraction of the electrons have enough energy to move into bulk energy levels and form a parallel conducting path. The problem is therefore to combine quantum and classical effects in the one device.

A typical shape for the two-dimensional channel is shown in Fig. 10. As can be seen, the channel potential well has an approximately triangular shape getting wider towards the top. As the well widens the spacing of the energy levels decreases, a feature that can be made use of to combine quantum and classical behaviours. If the spacing of the energy levels is greater than the thermal energy then we treat them as distinct two-dimensional levels. On the other hand, once the level spacing falls to less than $k_B T$ their individual energies will be blurred out and we assume classical three-dimensional behaviour. This is illustrated in Fig. 10 which shows three localised energy levels in the channel. Above the two-dimensional levels we assume the usual three-dimensional density of states function for bulk material. The discrete energy levels come from a self consistent solution of Schrödinger's and Poisson's equation. When performing this calculation there are two important factors to take into account. The first is the boundary condition on the wave equation and the second is the occupation of the energy levels.

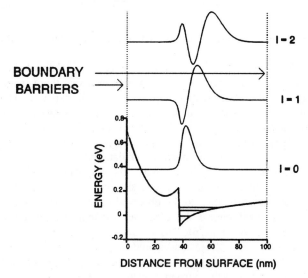

Figure 11. Wave functions for the energy levels in a HEMT channel.

When solving Schrödinger's equation in the HEMT we will again use equations (12) and (13) to find a function $\phi(z)$ for the energy levels in the two-dimensional channel. This can only be done analytically if the channel is an infinitely deep potential well or if it has a triangular shape (Yoshida 1986) but this is not usually the case. Equation (13) must therefore be solved numerically with the imposition of some boundary conditions. This is a difficult problem because unless there are some very high (infinite?) potential barriers to confine the electrons the functions $\phi(z)$ will extend to infinity. This situation can be approximated (as illustrated in Fig. 11) by assuming infinite barriers at the gate/semiconductor contact and at the edge of the semi-insulating substrate. As we are only interested in calculating the energy levels of electrons confined to the channel then the non-physical infinite boundaries will have little effect. The wave functions corresponding to the three energy levels are also shown in Fig. 11 where it can be seen that they have decayed to zero long before the they reach the artificial boundaries.

In order to calculate the potential distribution in the device it is necessary to know the occupation of both the two and three-dimensional energy levels. Obviously we cannot calculate the position of the 2-dimensional levels in the channel until the potential distribution has been calculated. We therefore have to arrive at a self consistent solution. Unlike the three-dimensional energy levels in the conduction band, the spacing of the two-dimensional levels depends on the potential profile but a simplifying factor is that **each** level has an energy independent density of states (Datta, 1989, pp 99-102) given by

$$N_{2D} = \frac{m^*}{\pi \hbar^2} \tag{29}$$

The electron density at a point is then given by $n = n_{2D} + n_{3D}$

$$= \sum_l \frac{m^* k_B T}{\pi \hbar^2} |\phi_l(z)|^2 \ln(1 + e^{(\epsilon_l - E_F)/k_B T}) + N_C \frac{2}{\sqrt{\pi}} F_{\frac{1}{2}}[(E_F - E_C)/k_B T] \tag{30}$$

where N_C is the usual density of states at the conduction band edge and $F_{\frac{1}{2}}$ is the Fermi function.

For a one-dimensional solution in a direction perpendicular to the channel (i.e. a slice across the device) it can, to a first approximation, be assumed that there is no current flow in this direction (i.e. the electron quasi Fermi level is horizontal). The set of equations then contains only one variable which is most commonly taken as the potential. The density of mobile electrons can be found from equation (30). Fig. 10 is a one-dimensional solution calculated in this way. Fig. 12 compares the electron distribution in a HEMT for classical and quantum mechanical calculations. It can be seen that the overall amount of charge is similar in each case but is more spread out when quantum mechanics is used. This behaviour goes some way to showing why a classical model of the HEMT is a reasonable approximation to the observed electrical behaviour.

The one-dimensional slices discussed above can be used in a "charge control" two-dimensional HEMT model (Yoshida 1986). The mobile charge in the channel can be calculated for a range of gate bias voltages and stored in a look-up table. These values can then be used in a self-consistent way to find the conducting charge under the gate as a function of position and hence the current flowing in the device. In this kind of solution the electron Fermi level will be horizontal in a direction perpendicular to the gate and will therefore not model any lateral current flow. (Strictly speaking no current will even flow in and out of the device but this problem can be overcome from current continuity considerations.) This model has the distinct advantage that Schrödinger's equation does not need to be solved in two dimensions.

For a more complete two-dimensional simulation it is necessary to allow the electron Fermi level to vary in two dimensions to account for electrons that have enough energy to escape from the channel and form a parallel conducting path. In this solution scheme the electron quasi Fermi level and the potential are chosen as the independent variables to facilitate the calculation of the two-dimensional energy levels. The shape of the

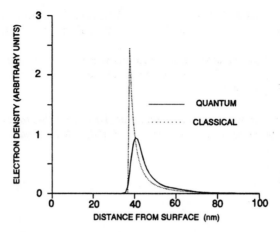

Figure 12. Electron densities in a HEMT as calculated from classical and quantum mechanics.

channel only varies slowly in the direction of current flow so we ignore quantisation in this direction. It is expected that a full two-dimensional self consistent Schrödinger/device equation solution will soon appear in the literature.

CONCLUSION

This chapter has discussed the application of quantum mechanics to two rather different semiconductor devices. For the first, the double barrier resonant tunnelling diode, the operation is purely quantum mechanical but in the HEMT quantum mechanics is almost incidental. The HEMT has proved itself to be a practical device and is currently being produced in large quantities for various applications. The DBRTD is still not much more than a laboratory curiosity but it has been shown to have the potential of operating at very high frequencies (Brown et al 1991). If, as seems inevitable, we are to exploit the higher frequency bands the question arises as to whether we should be looking towards new device principles based on quantum effects or trying to develope existing technologies?

REFERENCES

Ando Y and Itoh T (1987) Calculation of transmission current across arbitrary potential barriers. J. Appl. Phys. Vol. 61, pp 1497-1502

Brown E R, Söderström J R, Parker C D, Mahoney L J, Molvar K M, McGill, T C 1991 Oscillations up to 712 GHz in InAs/AlSb resonant-tunneling diodes. Appl. Phys. Lett, Vol. 58, pp 2291-2293

Datta S, (1989) Quantum Phenomena, Addison Wesley Modular Series on Solid State Devices.

Chou S Y, Wolak E and Harris J S (1987) Resonant tunnelling of electrons of one or two degrees of freedom. Appl. Phys. Lett, Vol. 52, pp 657-659

Yoshida J (1986) Classical Versus Quantum Mechanical Calculation of the Electron Distribution at the n-AlGaAs/GaAs Heterointerface. IEEE Trans. Electron Devices, Vol. ED-33, pp 154-156.

Acknowledgement: I would like to thank Rob Drury for the robust and stimulating discussions during the writing of this chapter. It did us both good!

Modelling of Distributed Feedback Lasers

Geert J.I. Morthier and Roel G. Baets
University of Gent

This chapter describes how laser diodes for optical communication, with their complex structure and physical interactions can be simulated. These lasers are usually of such a nature that variations in the lateral and transverse direction, as well as the electronic transport problem can be treated in a most simplified way, while longitudinal and spectral variations need to be taken into account in a detailed way .

In a first part we will show here how a mathematical description that includes longitudinal and spectral variations can be derived from first principles. The second part concentrates on the numerical implementation of this mathematical model and includes the description of the computer model CLADISS and some examples.

I DERIVATION OF A LONGITUDINAL DESCRIPTION

I.1 From Practical Device To Model

I.1.1 Geometry of a laser diode

Fig. I.1 shows the schematic cross section of an etched-mesa buried heterostructure laser, with its typical dimensions. Many other geometrical configurations can be found in literature (see e.g. [1], [2], [3] and [4]), but the lateral/transverse geometry always seems to have a similar degree of complexity. The active layer typically has a width w of 2µm, a thickness d of 0.1-0.2µm and a length L of 300-1000µm.

The double heterojunction formed by the p-InP layer, the InGaAsP layer and by the n-InP layer helps to confine the carriers in the active InGaAsP layer, where they can recombine (spontaneous and stimulated emission) to produce light. Obviously, an efficient carrier confinement can only be obtained when the bandgap of the active layer is sufficiently small with respect to the bandgap of the cladding layers. The double heterojunction also assures the optical confinement, i.e. it provides the waveguiding mechanism. This waveguiding is usually of the index-guiding type; the refractive index of the active layer is larger than the index of both cladding layers. We will therefore restrict ourselves to such structures in the following. For more information on different waveguiding used in laser diodes, we refer to the existing literature ([2], [5]).

Fig. I.2 shows a typical longitudinal view of a laser diode. The most general case is considered here. The laser can consist of several sections, whereby each section can be pumped independently and can have a different lateral/ transverse geometry. A grating, causing distributed reflections in the longitudinal direction, can be present in the cladding layers or in the active layer. Furthermore, discrete reflections can occur at the interfaces between two sections or at the front and rear facet.

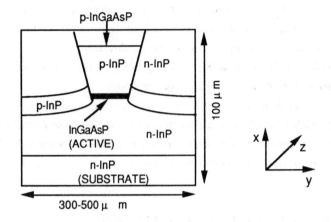

Fig. I.1: schematic cross-section of an etched-mesa buried heterostructure laser.

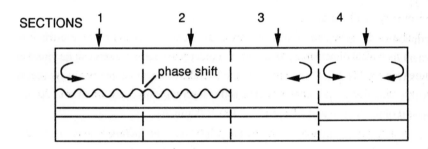

Fig. I.2: Longitudinal structure of a general laser diode

I.1.2 The physical processes and their approximation

A detailed description of a laser diode would require the determination of the carrier density N, the electrical potential V and the optical field E in each point (x,y,z) of the laser. As all these quantities depend on each other, a 3-dimensional description is practically impossible. Furthermore, the physical insight and hence the derivation of easy design rules would not really benefit from such a complicated description. In the following, we will show how some plausible assumptions allow a simplification.

-The rate equation for the carrier density

We assume that all sections are current controlled and that, due to the leakage, only a fraction η of the injected carriers is captured by the active layer, where the carriers distribute uniformly in the lateral and transverse direction. No doping is assumed and hence it follows from the neutrality condition that the hole density equals the electron density. We can then write down the following rate equation for the carrier density $N(z,t)$ in the active layer at the longitudinal position z and at a time t:

$$\frac{\partial N(z,t)}{\partial t} = \frac{\eta J}{qd} - \frac{N}{\tau} - B_0 N^2 - C_0 N^3 - R_{st} \tag{I.1}$$

The first term on the right hand side of (I.1) represents the current injection, with J being the current density, q the electron charge and d the active layer thickness. N/τ represents spontaneous carrier recombination via traps and at surfaces. $B_0 N^2$ represents the bimolecular recombination (or spontaneous emission) rate and the $C_0 N^3$-term stands for the Auger recombination rate.

It must be noticed that carrier diffusion, in the lateral/ transverse direction as well as in the longitudinal direction is neglected here. Ignorance of the longitudinal diffusion is justified by the fact that the corresponding diffusion length is generally small with respect to the typical distance on which the carrier density varies due to the other terms in (I.1).

-Description of the optical field

The lateral/transverse cross section forms an optical waveguide, of which we assume that it only sustains the lowest order TE-mode. The small dimensions of the active layer justify this assumption. The forward (+) and backward (-) propagating parts of the lateral electrical laser field can therefore in each waveguide section be expanded as:

$$E_y^+(x,y,z,t) = \text{Re}\left\{ \phi_0(x,y) \sum_m R_m^+(z,t)\, e^{j(\omega_m t - \beta_g z)} \right\}$$

$$E_y^-(x,y,z,t) = \text{Re}\left\{ \phi_0(x,y) \sum_m R_m^-(z,t)\, e^{j(\omega_m t + \beta_g z)} \right\} \tag{I.2}$$

where ϕ_0 is the eigenmode of the unperturbed waveguide (i.e. the waveguide without grating and in the absence of carrier injection). In reality, the eigenmode is different for different frequencies ω_m and is varying in time and with axial position due to the presence of the grating and the variation with time or axial position of the refractive index of the active layer. These variations are taken into account for the calculation of the propagation constant but their influence on the eigenmode itself is marginal. Furthermore, the different longitudinal modes are located very closely one to another and as such they correspond with almost identical eigenmodes.

The complex amplitudes $R_m^{\pm}$ in (I.2) represent the slowly varying (with time and axial position) parts of the fields. All rapid variations are included in the exponentials, although the choice of ω_m and β_g is not unique. In the following, we will always choose ω_m so that it coincides with the optical frequency of the m-th longitudinal mode when only static current is injected. For β_g, we choose the Bragg-number $\beta_g = n\pi/\Lambda$ if the section contains a grating of order n. Λ denotes the period of the grating in this case and is, in practical devices, usually chosen so that β_g nearly equals the propagation constant of the waveguide. When a section has no grating, β_g is set equal to the propagation constant β_m of the mode ϕ_0 of the unperturbed waveguide at the frequency ω_m.

-The electron-field interaction

A part of the optical field intensity, propagating inside the laser cavity, is lost through absorption or through the front and rear facets. In lasers, this loss is compensated by the stimulated emission in the active region, which amplifies the intensity during propagation. The amplification is expressed mathematically by the gain g, which, in semiconductors, depends on the carrier density and on the wavelength λ (or the frequency $\omega = 2\pi c/\lambda$) of the light:

$$\frac{\partial(|E(\lambda)|^2)}{\partial s} = g(N,\lambda) \, |E(\lambda)|^2 \qquad (I.3)$$

$E(\lambda)$ represents the field component with wavelength λ and propagating in the s direction. From equation (I.3), it can be readily seen how g corresponds with an imaginary refractive index, given by:

$$n = \frac{jg}{2\,(\omega/c)} \qquad (I.4)$$

g can be calculated quantummechanically from the Fermi-functions of electrons and holes, from the densities of states and from the transition probability for a transition from conduction to valence band [6]. From such a calculation, one readily finds that the gain is approximately linear in the carrier density N and that the gain can be positive only for wavelengths in the neighbourhood of the bandgap wavelength λ_g (=1.24μm/E_g, E_g in eV).

The expression (I.3) is, strictly speaking, not longer valid when large optical intensities exist in the active layer. The stimulated emission becomes non-linear in this case and the gain becomes intensity dependent. The non-linearity can be taken into account in gain calculations based on the density matrix formalism, as was first proposed by Asada and Suematsu [7]. From curve-fitting, applied on the numerical results obtained with such a gain model, it follows that the gain can be approximated analytically as:

$$g(\lambda) = [a(\lambda)\ N - b(\lambda)] \left\{ 1 - \sum_m \varepsilon(\lambda, \lambda_m)\ P_m \right\} \tag{I.5}$$

in which P_m denotes the power density of the mode m at wavelength λ_m.

The carriers not only cause gain, but also absorption. One can distinct between intervalence band absorption, which has its origin in transitions within the valence band and is dominant in 1.55µm semiconductors, and free carrier absorption, which is a consequence of the plasma behaviour of the electron-hole gas. Additional absorption also occurs in the passive layer, where the photons can be scattered by e.g. impurities or phonons. We will just include all absorption in one constant absorption coefficient α_{int}.

So far, we have described only the stimulated emission and the absorption (which both correspond with an imaginary part of the refractive index). The real part of the refractive index, to which the carrier density also contributes, is of equal importance for the wave propagation. The carrier-induced refractive index contributions are connected with the gain and absorption through the well-known Kramers-Kronig relations [8]. From these relations, it can be concluded that the refractive index Δn_r corresponding with the gain and the absorption is again a linear function of the carrier density.

I.2 The Longitudinal Field Equations

A rigourous derivation of the coupled wave equations, which describe the longitudinal wave propagation in a section with a uniform waveguide structure, can be found in many textbooks on optoelectronics. Contributions from spontaneous emission however are usually omitted in these derivations and therefore we present a modified theory here. Some steps in the derivation may seem rather intuitive, in which case we refer to the literature (e.g. [9] and [10]) for more details.

I.2.1 The three dimensional wave equation
We start from the vectorial wave equation which holds for non-magnetic and isotropic media:

$$\nabla^2\ \mathbf{E} = \mu\ \varepsilon_0\ \frac{\partial^2 \mathbf{E}}{\partial t^2} + \mu\ \frac{\partial^2 \mathbf{P}}{\partial t^2} \tag{I.6}$$

The polarisation $\mathbf{P}$ in this equation consists of three distinctive parts: the polarisation $\mathbf{P}_0$ of the structure without its grating, the polarisation $\mathbf{P}_{pert}$ induced by a possible grating and the polarisation $\mathbf{P}_{spont}$ corresponding with the fluctuating spontaneous emission. The polarisations P_0 and P_{pert} can be expressed with the help of the refractive index for numerical calculations:

$$P_0 = \varepsilon_0 \, n^2(x,y,z) \, E = \varepsilon_0 \, (n_0(x,y) + \Delta n(x,y,z))^2 E$$

$$P_{pert} = \varepsilon_0 \, n_p^2(x,y,z) \, E \qquad\qquad (I.7)$$

The perturbation in the refractive index Δn, represents the influence of the carrier density on the complex refractive index and includes contributions such as gain, absorption and carrier induced refractive index. Δn is assumed to be small and the derivative of n with respect to the time is neglected. Even under high speed (e.g. 10 GHz) modulation, the variation of n is much slower than the field variation (with a frequency of $\pm 10^5$ GHz).

The wave equation (I.6) can now be written as:

$$\nabla^2 E - \mu\varepsilon_0\{n_0^2(x,y)+2n_0(x,y).\Delta n(x,y,z)\} \frac{\partial^2 E}{\partial t^2} = \mu \frac{\partial^2(P_{pert} + P_{spont})}{\partial t^2} \qquad (I.8)$$

Since only the lowest order TE-mode, corresponding with a lateral electrical field, can reach the threshold in most laser diodes (see e.g. [11]), we only need to consider the lateral component ($E = E1_y$) of the fields and the wave equation becomes scalar:

$$\nabla^2 E - \mu\varepsilon_0\{n_0^2+2n_0.\Delta n\} \frac{\partial^2 E}{\partial t^2} = F(x,y,z,t)+\mu\varepsilon_0 n_p^2 \frac{\partial^2 E}{\partial t^2} \qquad (I.9)$$

The spontaneous emission term is now, for the sake of a simple notation, denoted by F. The stochastic nature of spontaneous emission implies that F represents a stochastic driving force or a so-called Langevin function ([12]) for the wave equation.

I.2.2 The Langevin force

It can be assumed that the Langevin force has a negligible spatial and temporal correlation, at least if we denote by F the spontaneous emission, averaged over a small volume ΔV and during a short time Δt. Indeed, spontaneous emissions can only be correlated if they originate from the same carrier and if that carrier has not been scattered in the time between the successive emissions. ΔV and Δt are thus determined by the typical scattering time ($\sim 10^{-13}$s) and the scattering distance ($\sim 10^{-2}$ µm).

The averaging over ΔV and Δt also implies that F can be approximated as a gaussian process since it is the result of many processes. F is therefore completely characterised by the first and second order moments, which can be calculated quantummechanically or semiclassical. A simple, semiclassical calculation has been reported by Henry [13], who derived the moments of F by requiring the spontaneously emitted light to be in equilibrium with the semiconductor. The moments are then found to be:

$$< F(x,y,z,t) > = 0$$

$$< F(x,y,z,t) \, F(x',y',z',t') > = 2 \, D_{FF} \, \delta(t-t') \, \delta(x-x') \, \delta(y-y') \, \delta(z-z')$$

$$2 \, D_{FF} = \frac{4\omega^3 \hbar}{\varepsilon_0 c^3} \, g \, n \, n_{sp} \tag{I.10}$$

with g being the gain, n the refractive index, ω the optical frequency and

$$n_{sp} = \frac{1}{1 - \exp\left(\dfrac{\hbar\omega - eV}{kT}\right)} \tag{I.11}$$

the so-called inversion factor. eV is the difference between the Fermi-levels of conduction and valence band. We will further consider n_{sp} as a constant (with a numerical value of 2). D_{FF} is called the diffusion constant of the stochastic process .

I.2.3 Reduction in the lateral/transverse direction

The (x,y)-dependence in (I.8) can be eliminated by substitution of the field expansions (I.2). The eigenmode $\phi(x,y)$ of the actual waveguide obeys, for the longitudinal mode m at the frequency ω_m, the 2-dimensional Helmholtz equation:

$$\nabla_{xy}^2 \phi + \omega_m^2 \mu_0 \varepsilon_0 \, n^2(x,y,z) \, \phi = \beta_{c,m}^2 \, \phi \tag{I.12}$$

We approximate ϕ by ϕ_0, the real eigenmode of the unperturbed waveguide which obeys the equation:

$$\nabla_{xy}^2 \phi_0 + \omega_m^2 \mu_0 \varepsilon_0 \, n_0^2(x,y) \, \phi_0 = \beta_m^2 \, \phi_0 = \left(\frac{2\pi}{\lambda_m} n_{eff}\right)^2 \phi_0 \tag{I.13}$$

n_{eff} is called the effective refractive index and it is considered as constant in our approximation and β_m is the propagation constant of the unperturbed waveguide. From (I.12) and (I.13), it follows for $\beta_{c,m}$ [10]:

$$\beta_{c,m} = \frac{2\pi}{\lambda_m} \{ n_{eff} + \Gamma \, \Delta n_a + (1-\Gamma) \, \Delta n_{cl} \} \tag{I.14}$$

with Γ being the fraction of the modal power that is confined to the active layer (the confinement or power filling factor) and Δn_a, resp.Δn_{cl} the perturbations in the complex refractive index of the active layer, resp. the cladding layers:

$$\Delta n_a = \frac{j(g(\lambda) - \alpha_{ac})}{2(\omega/c)} + \Delta n_r \,, \; \Delta n_{cl} = \frac{-j\alpha_{cl}}{2(\omega/c)} \tag{I.15}$$

α_{ac}, resp. α_{cl} represents the absorption in the active layer, resp. the cladding layers.

The complex propagation number $\beta_{c,m}$ finally becomes:

$$\beta_{c,m} = \frac{2\pi}{\lambda_m}(n_{eff} + \Delta n_r \Gamma) + j0.5 \; \Gamma \; g(\lambda_m) - j0.5 \; \alpha_{int}$$

$$\alpha_{int} = \Gamma \alpha_{ac} + (1 - \Gamma) \; \alpha_{cl} \tag{I.16}$$

Substitution of the expansions (I.2), which we rewrite as:

$$E(x,y,z,t) = \mathrm{Re}\left\{ \phi_0(x,y) \sum_m g_m(z,t) \, e^{j\omega_m t} \right\} \tag{I.17}$$

and taking into account that g_m is a slowly varying function of the time, gives:

$$\phi_0 \sum_m \left\{ \frac{d^2 g_m}{dz^2} + \beta_{c,m}^2 \, g_m - j\omega_m \frac{2n^2}{c^2} \frac{dg_m}{dt} \right\} e^{j\omega_m t} = F' - \phi_0 \frac{n_p^2}{c^2} \sum_m \omega_m^2 \, g_m e^{j\omega_m t} \tag{I.18}$$

It must be noticed here that F' in (I.18) is slightly different from F in (I.9). Indeed, F' is the analytical signal derived from F and its second order moments are twice as large as the 2nd order moments of F. Multiplication of the equation (I.18) with ϕ_0 and integration over the x,y-plane eliminates the x,y-dependence:

$$\sum_m \left\{ \frac{d^2 g_m}{dz^2} + \beta_{c,m}^2 \, g_m - 2j\frac{\beta_m}{v_g} \frac{dg_m}{dt} \right\} e^{j\omega_m t} = f(z,t) - \frac{n_{p,eff}^2}{c^2} \sum_m \omega_m^2 \, g_m e^{j\omega_m t} \tag{I.19}$$

v_g represents the group velocity and is assumed to be a constant. f and $n_{p,eff}$ are given by:

$$n_{p,eff}^2 = \frac{\displaystyle\iint_{(x,y)-\text{plane}} n_p^2 \, \phi_0^2 \, dxdy}{\displaystyle\iint_{(x,y)-\text{plane}} \phi_0^2 \, dxdy} \,, \; f(z,t) = \frac{\displaystyle\iint_{(x,y)-\text{plane}} F'(x,y,z,t) \, \phi_0 \, dxdy}{\displaystyle\iint_{(x,y)-\text{plane}} \phi_0^2 \, dxdy} \tag{I.20}$$

Equation (I.20) can be decomposed into equations for each longitudinal mode separately by integrating over a few periods T, with T being defined by the mode spacing $\Delta\omega_m$: $T = 2\pi/\Delta\omega_m$. $\Delta\omega_m$ is typically of the order of 100 GHz, and hence the resulting equations are only valid for modulation frequencies up to a few times 10 GHz. The field quantities then denote the average values over 10 psec. The longitudinal equation for each mode m can be written as:

$$\frac{d^2g_m}{dz^2} + \beta_{c,m}^2\, g_m - 2j\frac{\beta_m}{v_g}\frac{dg_m}{dt} = \frac{1}{nT}\int_{nT} f\, e^{-j\omega_m t}\, dt - \frac{n_{p,eff}^2}{c^2}\,\omega_m^2\, g_m \qquad (I.21)$$

The correlation function of the resulting Langevin forces can easily be derived from the correlation function of F'.

I.2.4 Derivation of the coupled wave equations

The second term on the r.h.s. of (I.21) can be transformed by remarking that, for a grating of period Λ, $n_{p,eff}$ is a periodic function of z and can be expanded in a Fourier series:

$$n_{p,eff}^2 = \sum_{q=-\infty}^{+\infty} a_q\, e^{j\frac{2q\pi}{\Lambda}z} \qquad (I.22)$$

Substitution of (I.22) in (I.21), together with

$$g_m(z,t) = R_m^+(z,t)\, e^{-j\beta_g z} + R_m^-(z,t)\, e^{j\beta_g z} \qquad (I.23)$$

results in the equation:

$$\left\{-2j\beta_g\frac{\partial R_m^+}{\partial z} - 2j\frac{\beta_m}{v_g}\frac{\partial R_m^+}{\partial t} + (\beta_{c,m}^2 - \beta_g^2)\, R_m^+\right\}e^{-j\beta_g z} +$$

$$\left\{2j\beta_g\frac{\partial R_m^-}{\partial z} - 2j\frac{\beta_m}{v_g}\frac{\partial R_m^-}{\partial t} + (\beta_{c,m}^2 - \beta_g^2)\, R_m^-\right\}e^{j\beta_g z} =$$

$$f_m - \frac{\omega_m^2}{c^2}\,(R_m^+ e^{-j\beta_g z} + R_m^- e^{j\beta_g z})\sum_{q=-\infty}^{+\infty} a_q\, e^{j\frac{2q\pi}{\Lambda}z} \qquad (I.24)$$

f_m is the time-averaged Langevin force, given by the integral in (I.21). Averaging (I.24) over a few grating periods and noticing that $\beta_g = n\pi/\Lambda$ results in the coupled wave equations, which are usually written in the following form:

$$\frac{\partial R_m^+}{\partial z} + \frac{1}{v_g}\frac{\partial R_m^+}{\partial t} + j\Delta\beta_m\, R_m^+ = F_m^+ + \kappa_{FB}\, R_m^-$$

$$-\frac{\partial R_m^-}{\partial z} + \frac{1}{v_g}\frac{\partial R_m^-}{\partial t} + j\Delta\beta_m\, R_m^- = F_m^- + \kappa_{BF}\, R_m^+ \qquad (I.25)$$

In the derivation of (I.25), it has been taken into account that $\beta_{c,m} \approx \beta_g$. In the absence of a grating, the averaging can occur over a few wavelengths. The a-coefficients vanish in this case. The coupling coefficients κ_{FB} and κ_{BF}, the complex Bragg-deviation $\Delta\beta_m$ and F_m^+ and F_m^- are defined by:

$$\kappa_{FB} = -\frac{j\omega_m^2 a_{-n}}{2\beta_g c^2}, \; \kappa_{BF} = -\frac{j\omega_m^2 a_n}{2\beta_g c^2}, \; \Delta\beta_m = \beta_{c,m} - \beta_g, \; F_m^\pm = \frac{j}{n\Lambda 2\beta_g}\int_{n\Lambda} f_m e^{\pm j\beta_g z} dz \quad (I.26)$$

The equations (I.25) are the standard equations used in the analysis of laser diodes with or without a grating.

I.2.5 Moments of the new Langevin forces

The Langevin forces $F_m^\pm$ are obtained from the original Langevin function F' by linear transformations. Their first order moments are therefore zero and the second order moments can easily be calculated. Starting from the relation (I.20), it follows:

$$<f(z,t)f^*(z',t')> = \frac{8\omega^3 \hbar n_{sp} n_{eff} \Gamma g}{\varepsilon_0 c^3 \iint_{(x,y)} dxdy \, \phi_0^2(x,y)} \delta(t-t') \, \delta(z-z') \quad (I.27)$$

The correlation functions of the Langevin forces f_m are calculated as:

$$<f_l(z,t)f_m^*(z',t')> = \frac{1}{(nT)^2}\int_{nT} d\tau_1 \int_{nT} d\tau_2 <f(z,\tau_1)f^*(z',\tau_2)> e^{j[\omega_m \tau_1 - \omega_l \tau_2]}$$

$$= \frac{8\omega_m^3 \hbar n_{sp} n_{eff} \Gamma g}{\varepsilon_0 c^3 \iint_{(x,y)} dxdy \, \phi_0^2} \delta(z-z') \frac{1}{(nT)^2} \delta_{lm} \int_{nT} d\tau_1 \int_{nT} d\tau_2 \, \delta(\tau_1 - \tau_2) \quad (I.28)$$

with δ_{lm} being the Kronecker delta and being a result of the averaging of the exponential function. The double integration in the last expression of (I.28) vanishes when $|t-t'|>nT$. More exactly, one has:

$$\frac{1}{(nT)^2}\int_t^{t+nT} d\tau_1 \int_{t'}^{t'+nT} d\tau_2 \, \delta(\tau_1 - \tau_2) = \frac{1}{nT}\left(1 - \frac{|t-t'|}{nT}\right), \; |t-t'| < nT \quad (I.29)$$

This function approaches the Dirac δ-function for small nT, which implies that on our timescale (of the order of 0.1 nsec.) the moments of f_m can be expressed as:

$$<f_l(z,t)f_m^*(z',t')> = \frac{8\omega_m^3 \hbar n_{sp} n_{eff} \Gamma g}{\varepsilon_0 c^3 \iint_{(x,y)} dxdy \, \phi_0^2} \delta(z-z') \, \delta_{lm} \, \delta(t-t') \quad (I.30)$$

At first sight, it looks remarkable that all f_l, which are the components of f, have an energy equal to that of f. In reality however, we have, by approximating the function (I.29) by a δ-function, reduced the bandwidth of f_l with respect to that of f. The time averaging filters the white spectrum. The moments of $F_m^\pm$ can now be derived in a similar way from the moments of f_m. We finally find:

$$\langle F_l^\pm(z,t)F_m^{\pm*}(z',t')\rangle = 2\,\frac{\omega_m^3\hbar n_{sp}n_{eff}\Gamma g\,\delta(z-z')\,\delta(t-t')}{\varepsilon_0 c^3\iint_{(x,y)}dxdy\,\phi_0^2\,|\beta_g|^2}\,\delta_{lm}\,\delta_{+-} \tag{I.31}$$

The presence of δ_{+-} in (I.31) indicates that Langevin forces corresponding with forward propagating waves are uncorrelated with the ones corresponding with backward propagating waves. The averaging over a few periods now implies a reduction of the spatial bandwidth (i.e. the bandwidth of the spatial Fourier transform). The spatial Fourier transform of $F_l^\pm$ is thereby located in a small band around $\pm\beta_g$.

I.3 The Coupled Wave Equations II

This section explores the full physical meaning of the coupled wave equations and discusses alternative forms of the equations. We first describe a proper choice for the field normalisation. This is followed by the transformation of the equations (I.25) into longitudinal rate equations. Both steps make the physical interpretation, which is given subsequently, considerably more simple.

I.3.1 The field normalisation
The normalisation of ϕ_0 will further be chosen so that the average optical power in the + or - direction can be expressed as:

$$P^+ = \sum_m P_m^+ = \sum_m |R_m^+|^2\ , P^- = \sum_m P_m^- = \sum_m |R_m^-|^2 \tag{I.32}$$

This optical power flux can be calculated by integration of the Poynting vector [14] over the lateral/transverse cross-section:

$$P_z = P^+ - P^- = \frac{1}{2}\,\text{Re}\iint_{x,y}(\mathbf{E}\times\mathbf{H}^*)_z\,dxdy = \frac{1}{2}\,\text{Re}\iint_{x,y}(-E_yH_x^*)\,dxdy \tag{I.33}$$

where E_y is given by (I.2) , while H_x can be derived with the help of :

$$H_x = \frac{-j}{\mu_0\omega}\frac{\partial E_y}{\partial z} = -\frac{\beta_g\phi_0}{\mu_0\omega}\sum_m(R_m^+e^{-j\beta_g z} - R_m^-e^{j\beta_g z})\,e^{j\omega_m t} \tag{I.34}$$

When calculating the product $E_y H_x^*$, it can be taken into account that cross products of different mode fields will vanish after averaging over a time of e.g. 10 psec. The power P_z then reduces to:

$$P_z = \frac{\beta_g}{2\mu_0\omega} \iint_{(x,y)-plane} |\phi_0(x,y)|^2 dxdy \sum_m (|R_m^+|^2 - |R_m^-|^2) \qquad (I.35)$$

Consistency with (I.32) thus requires the normalisation:

$$\iint_{(x,y)-plane} |\phi_0(x,y)|^2 dxdy = \frac{2\mu_0\omega}{\beta_g} \approx \frac{2\mu_0 c}{n_{eff}} \qquad (I.36)$$

The energy density (per unit distance in the longitudinal direction) inside the laser cavity on the other hand can be calculated as [14]:

$$u = \frac{1}{v_g} \sum_m (|R_m^+|^2 + |R_m^-|^2) \qquad (I.37)$$

This energy is related with the photon density inside the cavity. From (I.32) and (I.37), it can be concluded that the number of photons $i_m^\pm$ in mode m per unit length in the longitudinal direction and propagating in the +, - direction is given by:

$$i_m^\pm = \frac{|R_m^\pm|^2}{v_g \hbar \omega_m} \qquad (I.38)$$

The relation between the field amplitudes and the photon densities makes a transformation of the coupled wave equations (I.25) into equations for the field amplitudes and phases attractive for physical interpretation. The normalisation discussed here gives rise to new expressions for the second order moments of these Langevin forces:

$$\langle F_l^\pm(z,t) F_m^{\pm *}(z',t')\rangle = \hbar\omega_m n_{sp} \, \Gamma g(\omega_m) \, \delta(z-z') \, \delta(t-t') \, \delta_{lm} \, \delta_{+-} \qquad (I.39)$$

I.3.2 Longitudinal rate equations

Complex quantities such as $R_m^\pm$, $F_m^\pm$ are often characterised by their amplitude and phase for practical interpretations. We therefore write:

$$R_m^\pm(z,t) = r_m^\pm(z,t) \, e^{j\phi_m^\pm(z,t)}, \quad F_m^\pm(z,t) = |F_m^\pm(z,t)| \, e^{j\phi_{F,m}^\pm(z,t)}$$

$$\Delta\beta_m(z,t) = \Delta\beta_{m,r}(z,t) + j\Delta\beta_{m,i}(z,t)), \quad \kappa_{FB} = |\kappa| \, e^{j\phi_\kappa} \qquad (I.40)$$

where all quantities on the r.h.s. are now real functions. Substitution of (I.40) in (I.25) results in the longitudinal rate equations:

$$\pm\frac{\partial r_m^\pm}{\partial z} + \frac{1}{v_g}\frac{\partial r_m^\pm}{\partial t} - \Delta\beta_{m,i} r_m^\pm = |F_m^\pm|\cos(\varphi_{F,m}^\pm - \varphi_m^\pm) \pm |\kappa|\, r_m^\mp \cos(\varphi_\kappa + \varphi_m^- - \varphi_m^+)$$

and

$$\pm\frac{\partial\varphi_m^\pm}{\partial z} + \frac{1}{v_g}\frac{\partial\varphi_m^\pm}{\partial t} + \Delta\beta_{m,r} = \frac{|F_m^\pm|}{r_m^\pm}\sin(\varphi_{F,m}^\pm - \varphi_m^\pm) + |\kappa|\frac{r_m^\mp}{r_m^\pm}\sin(\varphi_\kappa + \varphi_m^- - \varphi_m^+) \qquad (I.41)$$

Three processes contribute to the time variation of the field amplitudes and hence to the time variation of the power or of the photon density: the absorption and the stimulated emission (expressed by $\Delta\beta_{m,i}$), the spontaneous emission (expressed by $F_m^\pm$ and $\varphi_m^\pm$) and the distributed reflections (caused by a grating and expressed by $|\kappa|$)

The effect of stimulated emission and absorption is more or less obvious. The increase of the amplitude (or of the photon density) is proportional with the amplitude. The Langevin functions on the right hand side of (I.41) represent the spontaneously emitted photons that couple into the mode. The spontaneous emissions add up to the fields $r_m^\pm$, but only in an incoherent way. The presence of the Langevin functions causes fluctuations in the modal photon densities, but it also contributes to the average photon densities. The latter effect however is not easily seen from (I.41) and we will therefore treat the spontaneous emission in more detail in section (I.3.3). The last term on the r.h.s. of (I.41) corresponds with reflected power of the backward (forward) propagating wave that couples into the forward (backward) propagating wave. The reflected backward (forward) propagating fields are not necessarily in phase with the forward (backward) propagating fields. The phase mismatch, expressed by $\varphi_\kappa + \varphi^- - \varphi^+$ in (I.41), results in an imperfect energy (photon) transfer from the reflected waves. It is included in the coupling term in (I.41) through the cosine factor.

The interference effect also affects the phases of forward and backward propagating waves, as can be seen from equations (I.41). The influence of the interference obviously depends on the relative strength of reflected and transmitted waves; e.g. a weak reflected wave r^- will only have a small impact (expressed by r^-/r^+) on the phase of the resulting field r^+. Other contributions to the phase variation have their origin in spontaneous emission and the Bragg deviation. The first contribution is related to the character of the spontaneously emitted photons, which are not coherent but which interfere with the laser photons. The random phase of the spontaneously emitted photons results in random fluctuations in the phases of the laser fields. The phase variation due to the Bragg deviation accounts for the discrepancy between actual wave vector and the as reference wave vector chosen Bragg vector. It is the result of our choice for the field representation (I.2) and has only a physical meaning when a grating is present, i.e. when κ is non-zero.

The Bragg deviation determines the phase mismatch in this case and hence the efficiency of the distributed feedback.

I.3.3 The spontaneous emission

Before discussing the new Langevin functions, we first remark that equation (I.39) can also be written as:

$$< F_l^\pm(z,t)\, F_m^{\pm*}(z',t') > = \frac{\hbar\omega_m}{v_g}\, S_m\, \delta(z-z')\, \delta(t-t')\, \delta_{lm}\, \delta_{+-} \qquad (I.42)$$

with $S_m = \Gamma g n_{sp} v_g$ being the rate of spontaneous emission that couples into the mode m, as it can be derived from the famous Einstein relations [6]. S_m is a fraction of the total spontaneous emission rate $B_0 N^2 V_a$ (see I.1.2).

By taking the statistical average of equations (I.41), it can be shown that, if $F_m^\pm$ has zero mean, the spontaneous emission doesn't affect the field amplitudes. On the other hand however, there is a contribution from the spontaneous emission to the average power (or photon number) in the mode. This can be proven by integration of (I.25), in which the time dependence can be removed by Fourier transformation. We will illustrate this for the forward propagating wave. After integration, we find for the fields:

$$R_m^+(z,t) = \frac{1}{2\pi}\int_{-\infty}^{+\infty} d\Omega\, e^{j\Omega t}\left\{ \frac{1}{a_m}\int^z \sin[a_m(z-z')]\, G_m^+(z',\Omega)dz' \right.$$

$$\text{with: } a_m(\Omega) = \left[\left(\frac{\Omega}{v_g} + \Delta\beta_m \right)^2 - |\kappa|^2 \right]^{1/2}$$

$$G_m^+(z,\Omega) = \int_{-\infty}^{+\infty} dt\, e^{-j\Omega t}\left\{ \frac{\partial F_m^+}{\partial z} - \kappa_{FB}F_m^- - j\Delta\beta_m F_m^+ - \frac{1}{v_g}\frac{\partial F_m^+}{\partial t} \right\} \qquad (I.43)$$

Multiplication of the first equation of (I.25) with $R_m^{\pm*}$, adding the complex conjugate of the resulting expression and taking the statistical average of the expression, results in:

$$\frac{\partial <(r_m^+)^2>}{\partial z} + \frac{1}{v_g}\frac{\partial <(r_m^+)^2>}{\partial t} - 2\Delta\beta_{m,i}<(r_m^+)^2> = 2|\kappa| <r_m^+ r_m^- \cos(\varphi_\kappa + \varphi_m^- - \varphi_m^+)>$$

$$+ < R_m^{+*}(z,t)\, F_m^+(z,t) > + < R_m^+(z,t)\, F_m^{+*}(z,t) > \qquad (I.44)$$

Replacing $R_m^\pm$ by the expression (I.43) in the last two terms of (I.44) and taking into account the $\delta(z-z')$-function in (I.39) shows that R_m^+ in (I.44) can be replaced by:

$$R_m^+(z,t) = \frac{1}{2\pi} \int dt' \int d\Omega \, e^{\Omega(t-t')} \left\{ \frac{1}{a_m} \int^z \sin[a_m(z-z')] \frac{\partial F_m^+(z',t')}{\partial z'} \right\} dz'$$

$$= \frac{1}{2\pi} \int dt' \int d\Omega \, e^{\Omega(t-t')} \int^z dz' \, F_m^+(z',t) \cos[a_m(z-z')] \tag{I.45}$$

And hence, taking into account the $\delta(z-z')$ and $\delta(t-t')$ in (I.42):

$$<R_m^+(z,t)F_m^{+*}(z,t)> = \int^z dz' \, <F_m^+(z,t)F_m^{+*}(z',t)> \tag{I.46}$$

The last term, for which the integrand is only non-zero at the integration boundary, cannot be calculated unambiguously. However, the problem can be overcome by using a more exact function (I.29 in the time as well as in the axial variable) for the correlation. This gives:

$$<R_m^{+*}(z,t)F_m^+(z,t)> + c.c. = \frac{2\hbar\omega_m}{v_g T} S_m \int_{z-n\Lambda}^z \frac{1}{n\Lambda} \left(1 - \frac{z-z'}{n\Lambda}\right) dz' = \frac{\hbar\omega_m}{2L} S_m \tag{I.47}$$

The value of T in this formula is defined by the bandwidth of the Langevin function and hence by the modal spacing ($\Delta f = v_g/2L$). We have chosen the value n=1 in (I.29) to include the complete spectral content of the mode; a higher value for n would imply a filtering of this spectrum.

Equation (I.44) can be transformed into an equation for the average photon density per unit distance propagating in the $\pm$ direction ($d/dt_{f,b} = \partial/\partial t \pm v_g \partial/\partial z$):

$$\frac{d<i_m^+>}{dt_f} - 2\Delta\beta_{m,i} v_g <i_m^+> = \frac{S_m}{2L} + \frac{2|\kappa|}{\hbar\omega_m} <r_m^+ r_m^- \cos(\varphi_\kappa + \varphi_m^- - \varphi_m^+)> \tag{I.48}$$

A similar equation also holds for $<i_m^->$. Equation (I.48) indicates that the spontaneous emission rate (S_m/L per unit distance in the longitudinal direction) is divided equally between the forward and backward propagating waves (or more generally that the spontaneously emitted photons have a uniformly in space distributed propagation direction). It also implies that the field amplitudes include contributions from the Langevin forces.

The equation for $i_m^\pm$ can now be written as:

$$\pm \frac{d<i_m^\pm>}{dt_{f,b}} - 2\Delta\beta_{m,i} v_g <i_m^\pm> = \frac{S_m}{2L} \pm 2|\kappa| <i_m^+ i_m^- \cos(\varphi_\kappa + \varphi_m^- - \varphi_m^+)> + F_{i,m}^\pm(z,t)$$

$$\text{with: } F_{i,m}^\pm = \frac{R_m^{+*} F_m^+ + R_m^+ F_m^{+*} - <R_m^{+*} F_m^+ + R_m^+ F_m^{+*}>}{\hbar\omega_m} \tag{I.49}$$

The new Langevin functions $F_{i,m}$ now have a zero mean. Under the assumption that the energy of the spontaneous emission is relatively small when compared with the energy of the coherent fields, one can approximate the 2nd order moment of $F_{i,m}$ by:

$$< F_{i,m}^{\pm}(z,t) F_{i,l}^{\pm}(z',t') > = 2 \, S_m \, <i_m^{\pm}> \, \delta(z-z') \, \delta(t-t') \, \delta_{lm} \, \delta_{+-} \qquad (I.50)$$

The Langevin functions for the phase equations can be characterized in a similar way. Multiplying e.g. the first equation of (I.25) with $R_m^{\pm *}$ and subtracting the complex conjugate of the resulting equation, gives:

$$\pm \frac{1}{v_g} \frac{d\varphi_m^{\pm}}{dt_{f,b}} + \Delta\beta_{m,r} = |\kappa| \frac{r_m^{\pm}}{r_m^{\pm}} \sin(\varphi_\kappa + \varphi_m^- - \varphi_m^+) - \frac{j}{2} \left\{ \frac{F_m^{\pm}}{R_m^{\pm}} - \frac{F_m^{\pm *}}{R_m^{\pm *}} \right\} \qquad (I.51)$$

and one finds:

$$< F_{\varphi,m}^{\pm}(z,t) F_{\varphi,l}^{\pm}(z',t') > = \frac{S_m}{2(v_g)^2 <i_m^{\pm}>} \, \delta(z-z') \, \delta(t-t') \, \delta_{lm} \, \delta_{+-} \qquad (I.52)$$

I.4 The Rate Equation For The Carrier Density

The carrier rate equation given in I.1.2 can now be written as:

$$\frac{\partial N}{\partial t} = \frac{\eta J}{qd} - \frac{N}{\tau} - B_0 N^2 - C_0 N^3 \frac{\Gamma}{wd} \sum_m \frac{g(\omega_m)}{\hbar\omega_m} [(r_m^+)^2 + (r_m^+)^2] + F_N(z,t) \qquad (I.53)$$

in which the rapidly varying terms of $|E|^2$ have been neglected (in reality their influence on the carrier density is damped by the diffusion). We have added the Langevin force F to account for the discrete nature of the carrier density and of the recombination and creation processes. We can assume that $F_N(z,t)$ and $F_N(z',t')$ are gaussian and uncorrelated for $|t-t'| > 10^{-13}$s and $|z-z'| > 10^{-2} \mu$m. Again, F_N has zero average.

A part of F_N has its origin in the spontaneous emission and is thus related to the Langevin functions which we encountered in the previous section. However, spontaneous emissions induce opposite changes in photon number and in carrier number (one spontaneous emission increases the photon number by one, but it decreases the carrier number by one). We can therefore write:

$$wd \, F_N(z,t) = wd \, F_S(z,t) - \sum_m \{ F_{i,m}^+(z,t) + F_{i,m}^-(z,t) \} \qquad (I.54)$$

The multiplication with wd is necessary to obtain the number of carriers per unit length in the longitudinal direction, a quantity similar to i_m.

wdF$_S$ represents the shot noise related to the spontaneous carrier recombination. This shot noise is not correlated with the spontaneous emission and its second order moment can be derived from the standard formula for shot noise. One finds:

$$< F_S(z,t) \, F_S(z',t') > = \frac{2}{wd} \left(\frac{N}{\tau} + B_0 N^2 + C_0 N^3 \right) \delta(t-t') \, \delta(z-z') \tag{I.55}$$

I.5 Boundary Conditions

The coupled wave equations and the carrier rate equation are valid only in a section with a uniform waveguide structure. In the case of lasers, consisting of different sections with different waveguiding properties, we can use the coupled wave and the carrier rate equations μ each section separately. The fields in two different sections A and B are thereby connected by the boundary conditions at the sections' interface. These boundary conditions can be expressed as usual in terms of a reflection (ρ_j) and a transmission (t_j) coefficient. They impose the following relations between the fields in two neighbouring sections A and B:

$$\begin{pmatrix} E_m^+ \\ E_m^- \end{pmatrix}_B = \frac{1}{t_j} \begin{pmatrix} 1 & -\rho_j \\ -\rho_j & 1 \end{pmatrix} \begin{pmatrix} E_m^+ \\ E_m^- \end{pmatrix}_A \tag{I.56}$$

The reflection and transmission coefficient are determined by the difference in lateral/transverse mode or in effective refractive index between the different sections. In the following, we will treat them as given constants. The relations (2.2.75) also apply at the output facets, i.e. the boundary between the semiconductor and the air. One of the fields ((E_m^+)$_A$ for z=0 and (E_m^-)$_B$ for z=L, L: laser length) vanishes in this case.

The continuity of the carrier density however can not be taken into account unless carrier diffusion is included. The diffusion length in the longitudinal direction on the other hand is so small that it allows large carrier density variations over a small distance. We can therefore, by denoting by N the average carrier density over a certain distance (e.g. a few μm), translate the boundary condition so that discontinuities in the average carrier density are permitted.

It is convenient in DFB laser theory to use complex field reflectivities. The phase of the reflectivity can then account for the random variations in the phase of the grating at z=0 and z=L. This phase of the grating, which cannot be controlled with the current technological means and which practically varies from chip to chip (fig. I.3), could also be included in the coupling coefficient and by considering small random variations in the laser length. However, one usually assumes a constant phase at z=0 (phase zero in our case) and a constant laser length.

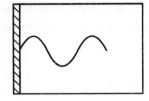 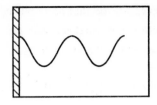

Fig. I.3:   Variation in the phase of the grating at the facets
for different lasers of the same wafer.

II A NUMERICAL, LONGITUDINAL MODEL: 'CLADISS'

Translation of the longitudinal equations and the boundary conditions into an algorithm
for numerical solutions results in a very powerful laser diode simulator. Such a laser
simulator has been developped at the Laboratory for Electromagnetism and Acoustics in
Gent and it is called CLADISS ('Compound Laser Diode Simulation Software') [15],
[16].

CLADISS can handle the analysis of Fabry-Perot lasers, C^3-lasers (Cleaved Coupled
Cavity lasers), DFB lasers , DBR lasers and external cavity lasers. It consists of three
separate modules: for the analysis of the static, the dynamic and the stochastic behaviour.
Especially the fact that longitudinal spatial hole burning is included in a detailed way and
the multi mode analysis in each module have made CLADISS to one of the most
powerful and unique laser models that have been developed in the past years.

II.1 The Analysis Of The Static Behaviour

This analysis aims at calculating quantities such as the total output power and the
central wavelength of each line in the optical spectrum, and this for a given static current
injection. The laser is assumed to have reached a static (time-independent) regime in this
case and all derivatives with respect to the time are thus neglected. Langevin forces,
which determine the fluctuations in the output power or the lineform of each line, are not
included either.

It is convenient to distinguish between two parts in the analysis: a threshold analysis
and an above-threshold DC analysis. The threshold analysis serves to investigate at
which current levels (the threshold currents) a transition from amplifier operation to
oscillator operation of the laser occurs. At the same time, the wavelength of the main
longitudinal modes and an estimate of the side mode suppression follow from the
analysis.

Different approximations and numerical techniques are introduced in both analysises and a separate description of each analysis will be given.

II.1.1 The threshold analysis

The transition from amplifier to oscillator operation takes place when the population inversion in the active layer provides an amount of stimulated emission, sufficient to compensate for the absorption and mirror loss. The current at which this transition takes place, but were the power is still very small is the threshold current.

In the theoretical threshold analysis, one must neglect the stimulated emission rate in the carrier rate equation, the power dependence of the gain and the spontaneous emission in the coupled wave equations. The carrier rate equation (I.53) in a section with a uniform waveguide geometry and a uniform injection then reduces to a simple relation between current and (uniform) carrier density and the 3rd order equation can be solved analytically. Equations (I.25), in which the Langevin forces and the time dependences are ignored, can be solved for this uniform carrier density:

$$\begin{pmatrix} R_m^+(z) \\ R_m^-(z) \end{pmatrix} = \begin{pmatrix} a_{11}^m(z-z_0) & a_{12}^m(z-z_0) \\ a_{21}^m(z-z_0) & a_{22}^m(z-z_0) \end{pmatrix} \begin{pmatrix} R_m^+(z_0) \\ R_m^-(z_0) \end{pmatrix}$$

with: $a_{11}^m(z) = \cosh(\Delta_m z) - j\dfrac{\Delta\beta_m}{\Delta_m} \sinh(\Delta_m z)$

$\qquad a_{22}^m(z) = \cosh(\Delta_m z) + j\dfrac{\Delta\beta_m}{\Delta_m} \sinh(\Delta_m z)$

$\qquad a_{12}^m(z) = \dfrac{\kappa_{FB}}{\Delta_m} \sinh(\Delta_m z) \; ; \; a_{21}^m(z) = -\dfrac{\kappa_{BF}}{\Delta_m} \sinh(\Delta_m z)$

and: $(\Delta_m)^2 = -[\ (\Delta\beta_m)^2 + \kappa_{FB}\kappa_{BF}\ ]$ 　　　　　　　　　　　　(II.1)

$\Delta\beta_m$ is a function of the carrier density and of the wavelength.

To develop a numerical threshold algorithm, we consider a general multi-section laser as shown in fig. II.1. For such a multi-section laser, it is assumed that the currents injected into the different sections depend linearly on an independent current I_v, injected into the section v. For a given current I_v and a given wavelength, it is then possible to calculate the Bragg deviation $\Delta\beta$ and the a_{ij}-coefficients in each section.

In CLADISS, a point z_v in the middle of the section v is chosen and the field reflections ρ_L and ρ_R of the left and the right part of the laser cavity are calculated at z_v for a given injection and a given wavelength λ. Laser oscillation then occurs if the product $\rho_L\rho_R$ (which is called the roundtrip gain as it denotes the field gain after one roundtrip in the cavity) equals one. The field reflectivities ρ_L and ρ_R can be derived from the propagator matrices (II.1), which must be transformed into relations between the fields $E^\pm$, and the matrices (I.56), expressing the boundary conditions. One can write:

$$\rho_L = \frac{\rho_f (F_L)_{11} + (F_L)_{12}}{\rho_f (F_L)_{21} + (F_L)_{22}} \quad , \rho_R = - \frac{\rho_b (F_R)_{11} - (F_R)_{21}}{\rho_b (F_R)_{12} - (F_R)_{22}} \tag{II.2}$$

where the propagator matrices F_L and F_R are defined by:

$$\begin{pmatrix} E^+(L) \\ E^-(L) \end{pmatrix} = F_R \begin{pmatrix} E^+(z_v) \\ E^-(z_v) \end{pmatrix} , \begin{pmatrix} E^+(z_v) \\ E^-(z_v) \end{pmatrix} = F_L \begin{pmatrix} E^+(0) \\ E^-(0) \end{pmatrix} \tag{II.3}$$

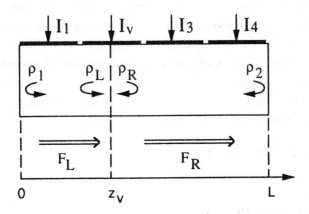

Fig. II.1: Division of a laser in 2 parts for the calculation of the roundtrip gain.

The determination of the threshold current and of the wavelength of the main modes proceeds as follows. Initially, an estimate of the threshold current I_v and of the wavelength λ_0 of the main mode must be chosen. For this value of I_v, the roundtrip gain is calculated for a wide range of wavelengths around λ_0. The values of the roundtrip gain are subsequently scanned to check whether a resonance occurs. If this is not the case, the current is gradually increased (if all points with zero phase, i.e. if all longitudinal modes, have an amplitude below one) or decreased (if a point with zero phase and an amplitude above one exists) until at least one wavelength gives a zero phase and an amplitude of one. This defines the threshold current, as well as the wavelength λ_m of the most important longitudinal modes.

As an illustration, the complex roundtrip gain at threshold is shown in fig. II.2 for a 300 μm long DFB laser with $\kappa L = 1.5$ and with facet reflectivities $\rho_f = 0.566\, e^{j\pi}$ and $\rho_b = 0.224\, e^{j3\pi/2}$. Other parameters are listed in table 1. The threshold current has the value 21.5 mA in this case. Both the main (lasing) mode and the most important side mode are indicated by an arrow in fig. II.2.

The roundtrip gain of the side mode could be considered as a measure for the side mode suppression. A more conventional quantity however is given by the difference in threshold gain ΔgL between main and side mode. This is the difference between the normalized gain g_1L at the threshold current and the gain g_2L for which the side mode starts lasing. For the example, this quantity ΔgL has the value 0.25, a value which is generally regarded as guaranteeing a large side mode suppression [17], [18].

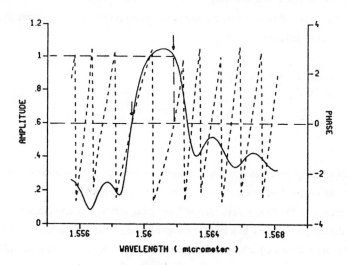

Fig. II.2: roundtrip gain at threshold for a DFB laser (see text).

Table 1

Parameter		Typical value	
w	[μm]	1.5	Stripe width
d	[μm]	0.12	Thickness active layer
G		0.5	Confinement factor
L	[μm]	300	Laser length
Λ	[μm]	0.2413	Grating period
n_e		3.25	Effective index of the unperturbed waveguide
η		0.8	Injection efficiency
τ	[s]	$5. \, 10^{-9}$	Carrier lifetime
B_0	[μm³/s]	100	Bimoleculaire recombination
C_0	[μm⁶/s]	$20 \, 10^{-5}$	Auger recombination
n_{sp}		2.	Inversion factor
α_{int}	[μm⁻¹]	$50. \, 10^{-4}$	Internal absorption loss

II.1.2 The DC analysis

Above threshold, one can no longer neglect the stimulated emission rate in the carrier rate equation or the power dependence of the gain. Since the optical power is generally not uniform in the longitudinal direction, it follows that a non-uniform carrier density (and hence a non-uniform Bragg deviation) might exist in the cavity. Furthermore, one must also include the static contribution from the spontaneous emission. The equations (I.25) should therefore be changed into:

$$\frac{\partial R_m^+}{\partial z} + \left\{ j\Delta\beta_m - \frac{S_m}{4L\,|R_m^+|^2} \right\} R_m^+ = \kappa_{FB}\, R_m^-$$

$$-\frac{\partial R_m^-}{\partial z} + \left\{ j\Delta\beta_m - \frac{S_m}{4L\,|R_m^-|^2} \right\} R_m^- = \kappa_{BF}\, R_m^-$$

(II.4)

It can be seen that the spontaneous emission manifests itself as a non-linearity and an exact solution of the coupled wave equations is now no longer possible.

The z-dependence of the Bragg deviation and the spontaneous emission can be taken into account by dividing the cavity in a large number of small segments (with length l_v in a section v). The carrier rate equation and the coupled wave equations can be solved approximately in such small segments (which typically must be a few μm long).

-Solution in a small section

For small segments, we can approximate both the carrier density N and the optical powers by constants with a value equal to their value in the middle of the segment (at z'_i). The optical powers can be obtained as the average of the power levels at z_i and z_{i+1}. For a uniform carrier density and uniform powers, the approximate solution of (II.4) reduces to:

$$\begin{pmatrix} R_m^+(z_{i+1}) \\ R_m^-(z_{i+1}) \end{pmatrix} = s_c(l_i) \begin{pmatrix} a_{11}'(l_i) & a_{12}'(l_i) \\ a_{21}'(l_i) & a_{22}'(l_i) \end{pmatrix} \begin{pmatrix} R_m^+(z_i) \\ R_m^-(z_i) \end{pmatrix}$$

$$s_c(l_i) = \exp\left\{ \frac{S_m l_i}{8L} \left(\frac{1}{P_m^+(z_i')} - \frac{1}{P_m^-(z_i')} \right) \right\}$$

$a_{ij}' = a_{ij}$ with $j\Delta\beta_m$ replaced by : $j\Delta\beta_m - \dfrac{S_m}{8L}\left(\dfrac{1}{P_m^+(z_i')} + \dfrac{1}{P_m^-(z_i')} \right)$ (II.5)

The propagator matrix now obviously depends on the resulting fields through $P_m^{\pm}(z'_i)$ and an accurate solution can be achieved only after iteration. E.g. as a first approximation, one can replace the carrier density and the optical powers at z'_i by their value at z_i. Substitution of these values in (II.5) yields better estimates for the fields at z_i.

This then allows the determination of more accurate values for the carrier density and the optical powers at z_i, which in turn lead to more accurate values for the fields at z_i. This iteration can be repeated until further iterations no longer result in significant changes (typically, only about 5 iterations are needed).

-Self-consistent determination of the laser state

The field quantities and the carrier density in each point a-long the longitudinal axis can be determined completely if proper values for the fields at the front facet and for the wavelengths are chosen. More specificly, a value for λ_m and $R_m^-(z=0)$ (or $R_m^+(z=0)$, both quantities are related by the boundary condition at the facet) must be chosen for each longitudinal mode m under consideration. We notice that all quantities $R_m^-(z=0)$ can be considered as real numbers. This is justified by the absence of any phase relation between the different modes (as a result of the time averaging (I.21)) and by the fact that, at least for the static regime, the time origin can be chosen arbitrarely.

We assume that q longitudinal modes are included in the calculations. The 2q values for λ_m and $R_m^-(z=0)$ then allow to determine the fields at the right facet z=L, where the boundary condition must be fulfilled. This boundary condition actually constitutes a complex equation which, for the case of q modes, gives 2q real equations, which the 2q choices of λ_m and $R_m^-(z=0)$ (m=1,q) must obey. These non-linear equations can be solved by the Newton-Raphson method [3.6].

The complete algorithm can be explained as follows. One chooses initial estimates for λ_m and $R_m^-(z=0)$. With these values, the fields are propagated subsequently along the small segments, until to the right laser facet is reached. At the same time, the derivatives of the fields with respect to λ_m and $R_m^-(z=0)$ are propagated as well. The resulting system of equations at the right facet and its jacobian can thus be calculated numerically and, if necessary, better approximations for λ_m and $R_m^-(z=0)$ can be derived by means of the Newton-Raphson algorithm. The iteration can continue until sufficient accuracy has been reached. It must be noticed here that the propagation of the field derivatives can proceed in a very similar way as the field propagation; mainly as a result of the analytical expression for the field propagation over a small segment.

Output of the DC-model includes the variation of the output power or the wavelength of several longitudinal modes as a function of the injected currents, as well as the longitudinal variation of the optical power, of the carrier density, the refractive index or the Bragg deviation at a certain bias level. As an illustration, we have shown the longitudinal variation of the power in the main mode (fig. II.3) and of the carrier density (fig. II.4) for the laser chosen as example. The injected current is chosen so as to obtain an output power of 1 mW.

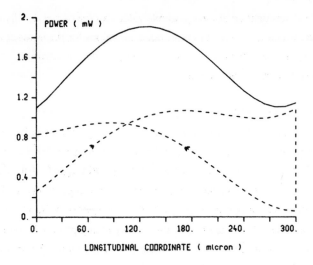

Fig.II.3: Longitudinal variation of the power in the main mode at 1 mW output power.

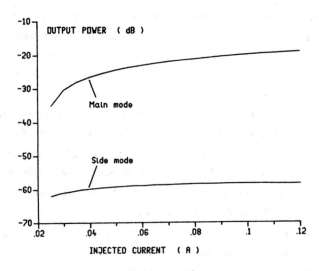

Fig.II.4: Output power in main and side mode vs. injected current.

The DC-analysis furthermore allows to control whether the side modes start lasing at a certain power levels. Two methods can be followed. A first method is to include several modes in the analysis and to calculate the power in each mode for increasing injection. The lasing of a side mode can be observed as a kink in the power-current relation in this case. The rise of a side mode can even be seen if only the main mode is included in the calculations. In this case, one can calculate the roundtrip gain vs. the wavelength at different bias levels and check whether the roundtrip gain of the side mode reaches one. The roundtrip gain above threshold is calculated in an identical way as the roundtrip gain at threshold, although the gain suppression and the non-uniform carrier density must be taken into account. This implies that many small sections ought to be used now.

II.2 Analysis Of The Dynamic Behaviour

The noise sources are still ignored in the analysis of the dynamic behaviour. However, the analysis now is based on solution of the equations (I.41) and (I.53), in which the derivatives with respect to time are taken into account as well. We thereby recall that the steady state spontaneous emission should be included in the amplitude equations. Langevin functions are still ignored here.

II.2.1 Assumptions

The analysis is restricted to sinusoidal regimes, i.e. the injected current densities are expressed as:

$$J_v(t) = J_{v0} + \text{Re}\{ J_{v1} e^{j\Omega t} \} \tag{II.6}$$

with J_{v0} being the static current density and J_{v1} the sinusoidal current density, injected into the section v. The periodic excitation makes the assumption of a periodic response plausible and we express the field quantities and the carrier densities as:

$$r_m^\pm(z,t) = r_{m0}^\pm(z) + \text{Re}\left\{ \sum_{k=1}^{\infty} r_{mk}^\pm(z,k\Omega) e^{jk\Omega t} \right\}$$

$$\varphi_m^\pm(z,t) = \varphi_{m0}^\pm(z) + \text{Re}\left\{ \sum_{k=1}^{\infty} \varphi_{mk}^\pm(z,k\Omega) e^{jk\Omega t} \right\}$$

$$N(z,t) = N_0(z) + \text{Re}\left\{ \sum_{k=1}^{\infty} N_k(z,k\Omega) e^{jk\Omega t} \right\}$$

$$\Delta\omega_m(t) = \text{Re}\left\{ \sum_{k=1}^{\infty} \Delta\omega_{mk}(k\Omega) e^{jk\Omega t} \right\} \tag{II.7}$$

The terms with subscript 0 obviously represent the static solution, whereas the other terms denote deviations from the static solution, caused by the sinusoidal currents.

It can be remarked that (II.7) does not include alll possible solutions of the dynamic laser equations. Indeed, due to the non-linearity of these equations, a periodic excitation does not necessarily imply a periodic respons with the same period. We restrict ourselves to relatively small modulation depths, for which the expansion (II.7) can be used. The terms with subscript 1 then represent the linear or small-signal modulation responses, while the terms with higher subscript are an indication for the harmonic distortion.

It must finally be noted that the time dependent parts of the field phases and of the optical frequencies are not uniquely defined. This ambiguity is removed in CLADISS by requiring $\varphi_{mk}^-(z=0,kW)$ to be zero for all m and k. The frequency then denotes the frequency of the light that leaves the laser at the front facet. Up to frequencies of several tens of GHz, this frequency can safely be regarded as 'the' optical frequency.

II.2.2 Approximations and algorithm

We restrict the sums in (II.7) to the first three terms, which is acceptable for sufficiently small modulation depths. In fact, it follows both from experiments [19] and theory [20] that the terms with index k are proportional with the k-th power of the modulation depth.

The laser equations are solved by substitution of the expansions (II.7), whereby the $\exp(jk\Omega t)$-time variation removes the derivatives with respect to the time. Multiplication of the resulting equations with $\exp(-jk\Omega t)$ and integration over the period corresponding with Ω leads to separate equations for the terms with index k. The separate equations for the terms with index 1 can be written under the form of a matrix equation (i.e. X_k, B_k and A are matrices) as:

$$\frac{dX_1(z,\Omega)}{dz} = A(z,\Omega)\, X_1(z,\Omega) + B_1(J_{v1},z,\Omega)$$

with $X_1^T(z,\Omega) = [\ (\Delta\omega_{mk}, r_{mk}^+, r_{mk}^-, \varphi_{mk}^+, \varphi_{mk}^-),\ m=1,q]$ $\hspace{2cm}$ (II.8)

with q being the total number of modes taken into account. For the terms with index 2 or 3, the equations have a similar form and identical homogeneous parts. The equations for $d\Delta\omega_{mk}/dz$, included in (II.8) are the trivial identities $d\Delta\omega_{mk}/dz=0$.

We will not give the equations (II.8) in any further detail because they are too extended. Their derivation is nonetheless straightforward. By applying finite differences with the same longitudinal discretisation scheme as in the DC analysis, the equations (II.8) can be transformed into a set of linear algebraic equations. The discretised equations, together with the boundary equations can now be solved by standard techniques. The simulator CLADISS uses the ACM software package COLROW, which decomposes the system matrix into triangular matrices [21]

The AC analysis allows to calculate the small signal FM- (Frequency Modulation) and IM- (Intensity Modulation) response as a function of the modulation frequency, as well as the 2nd and 3rd order harmonic distortion in the FM- and IM-response. An example of the AC-analysis is given in fig. II.5, which depicts the amplitude and the phase of the FM-response vs. the modulation frequency for the example at different bias points and for a modulation current of 1mA.

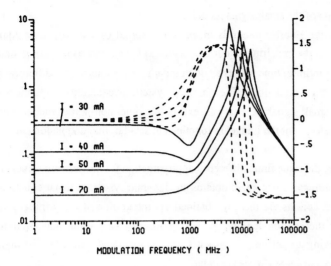

Fig. II.5: Amplitude (——) and phase (--) of the FM-response of the example.

II.3 Analysis Of The Noise Behaviour

We restrict the analysis here to situations with static injection only. The fluctuations, mainly caused by spontaneous emission, are nevertheless time dependent and we must use the time-dependent equations (I.49), (I.51) and (I.53). The Langevin forces, which represent the noise sources, can thereby be regarded as time dependent excitations with a small amplitude.

The small amplitude of these Langevin forces will in general result in small fluctuations of the field amplitudes and phases and of the carrier density. Hence, a linearisation of the laser equations into these small fluctuations is justified. Anyway, the linearisation can be more generally justified (also for the phase equations) by the small distortion in the FM- and IM-responses, which is usually found. In order to eliminate the derivatives with respect to the time, we now use the Fourier transform of the fluctuations, e.g. for the frequency:

$$\Delta\omega_m(t) = \frac{1}{2\pi}\int_{-\infty}^{+\infty} d\Omega\, \Delta\omega_m(\Omega)\, e^{j\Omega t} \tag{II.9}$$

The Langevin functions can be Fourier transformed in a similar way. Since they can be considered as stationary (i.e. for time differences of the order of 0.01 nsec. or more), it follows that their Fourier transforms at different Fourier frequencies are also uncorrelated.

II.3.1 Application of the small signal model

Substitution of the Fourier integrals in the small signal (linearised) laser equations, multiplication of the resulting equations with $\exp(-j\Omega t)$ and averaging over a long (theoretically an infinite) time results again in separate equations for the different spectral components of the fluctuations. The obtained system of equations is identical to that obtained in the small signal AC analysis, except that other non-homogeneous terms (i.e. the Fourier transforms of the Langevin functions in stead of the modulation currents) are now present.

By applying the same finite difference approximation for the derivatives with respect to the axial coordinate z, one again obtains an algebraic system of linear equations. The new Langevin functions are thereby obtained via integration of the original Langevin functions over the section lengths. It can also easily be proven that the Langevin functions referring to different segments are uncorrelated (the $\delta(z-z')$-function is transformed into a discrete Kronecker delta).

The linear relation between the different Langevin functions and the fluctuations of the field quantities in the different discretisation points allows to calculate the 2nd order moments of these fluctuations in a straightforward way. The spectrum of the FM-noise or the relative intensity noise (RIN) can be extracted from these 2nd order moments.

II.3.2 Power spectrum and linewidth

It has been shown before that the power spectrum and its width (the linewidth) are mainly due to the fluctuations in the optical frequency of each mode [22]. We can therefore assume the fields, emitted at each laser facet, to be of the form:

$$E_m(t) = E_{0m} \exp\left\{ j\omega_{m0}t + j\int^t \Delta\omega_m(t')dt' \right\} \tag{II.10}$$

$\Delta\omega_m$ has a gaussian distribution and the relative power spectrum is thereby given as:

$$S_{\Delta P,m}(\omega) = F\left\{ e^{j\omega_{m0}t} e^{-0.5 \left< \left(\int_0^\tau \Delta\omega_m(t')dt'\right)^2 \right>} \right\} \tag{II.11}$$

in which F denotes the Fourier transform. The exponent can easily be calculated by inverse Fourier transform of the spectral density of the FM-noise:

$$\left< \left(\int_0^\tau \Delta\omega_m(t')dt'\right)^2 \right> = \frac{1}{2\pi}\int_0^\tau dt_1 \int_0^\tau dt_2 \int_{-\infty}^{+\infty} d\Omega\, S_{\Delta\omega,m}(\Omega)\, e^{j\Omega(t_1-t_2)} \tag{II.12}$$

In CLADISS, the integration is performed numerically and the Fourier transforms are calculated with the help of FFT-routines. In this way, one readily finds the detailed power spectrum, including possible relaxation oscillation peaks. However, another option, based on a more simple approximation for the linewidth is also offered by CLADISS. The simple approximation assumes a white FM-noise, which, at not too low a power level, is justified. The spectrum of the FM-noise is usually constant up to ±1 GHz, while the linewidth is usually far below 1 GHz. The linewidth in this approximation is given by:

$$\Delta \nu_m = \frac{S_{\Delta \omega,m}(\Omega=0)}{2\pi} \tag{II.13}$$

and one only needs to calculate the low frequency value of the FM-noise in this case.

As an example, we have shown the power spectrum of the main mode of laser A at 1 mW output power in fig. II.6. The linewidth is 43.5 MHz.

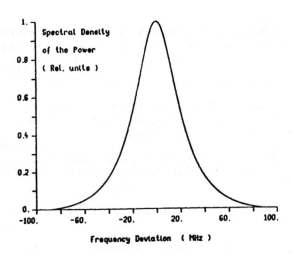

Fig. II.6: Spectrum for the main mode of the example at 1 mW.

References

[1] K. Petermann, "Laser Diode Modulation and Noise", KTK Publishers, Tokyo, 1988.

[2] G. Agrawal, N. Dutta, "Long-wavelength semiconductor lasers", Van Nostrand Reinhold, New York, 1986.

[3] M. Amann, "New stripe-geometry laser with simplified fabrication process", El. Lett., Vol. 15, pp.441-442, July, 1979.

[4] K. Saito, R. Ito, "Buried-heterostructure AlGaAs lasers", IEEE Journ. Quant. El., Vol. 16, pp. 205-215, February, 1980.

[5] D. Cook, F. Nash, "Gain-induced guiding and astigmatic output beam of GaAs lasers", Journ. Appl. Phys., Vol. 46, p. 1660, 1975.

[6] H. Casey, M. Panish, "Heterostructure Lasers, part A: Fundamental Principles", Academic Press, New York, 1978.

[7] M. Asada, Y. Suematsu, "Density-matrix theory of semiconductor lasers with relaxation broadening model: gain and gain-suppression in semiconductor lasers", IEEE Journ. Quant. El., Vol. 21, pp. 434-442, May, 1985.

[8] A. Yariv, "Quantum Electronics", 2nd Ed., Wiley, New York, 1980.

[9] D. Marcuse, "Principles of Quantum Electronics", Academic Press, New York, 1980.

[10] P. Vankwikelberge, "Theoretische studie van statische en dynamische longitudinale effecten in Fabry-Perot en DFB-diodelasers", Ph. D. thesis (in Dutch), university of Gent, 1990.

[11] G. Thompson, "Physics of semiconductor laser devices", Wiley, New York, 1980.

[12] M Lax, "Classical Noise IV: Langevin Methods", Rev. Mod. Phys., Vol. 38, pp. 541-566, July, 1966.

[13] C. Henry, "Theory of Spontaneous Emission Noise in Open Resonators and its Application to Lasers and Optical Amplifiers", Journ. Lightw. Techn., Vol. 4, pp. 288-297, March, 1986.

[14] J. D. Jackson, "Classical Electrodynamics", Wiley, New York, 1962.

[15] P. Vankwikelberge, G. Morthier, R. Baets, "CLADISS, a longitudinal, multi mode model for the analysis of the static, dynamic and stochastic behaviour of diode lasers with distributed feedback", IEEE Journ. Quant. El., October 1990.

[16] P. Vankwikelberge, G. Morthier, K. David, R. Baets, "CLADISS, a new diode laser simulator", Technical Digest Topical Meeting on Integrated Photonics Research, Hilton Head, March, 1990.

[17] J. Buus, "Mode selectivity in DFB lasers with cleaved facets", Electron. Lett., Vol. 21, pp. 179-180, 1985.

[18] P. Mols, P. Kuindersma, W. Van Es, I. Baele, "Yield and device charac-teristics of DFB lasers: statistics and novel coating design in theory and experiment", IEEE Journ. Quant. El., Vol.25, June, 1989.

[19] G. Morthier, F. Libbrecht, K. David, P. Vankwikelberge, R. Baets, "Theoretical investigation of the 2nd order harmonic distortion in the AM-response of 1.55 μm F-P and DFB lasers", IEEE Journ. Quant. El., Vol. 27, pp. 1990 - 2002, August, 1991.

[20] K. Lau, A. Yariv, "Intermodulation distortion in semiconductor injection lasers", Appl. Phys. Lett., Vol. 45, pp. 1034-1036, 1984.

[21] 'Subroutine COLROW, Algorithm 603, ACM-Trans. Math. Software, Vol. 9, pp. 376-380, September, 1983.

[22] K. Vahala, A. Yariv, "Semiclassical Theory of Noise in Semiconductor Lasers - Part I", IEEE Journ. Quant. El., Vol. 19, pp. 1096-1101, June, 1983.

Equivalent Circuit Modelling

Stavros Iezekiel
University of Leeds

1. INTRODUCTION

During the last decade, there has been a shift in emphasis in device modelling research away from equivalent circuits and in favour of numerical device simulation. This has been driven by a rapid increase in the computing power to cost ratio and the assertion that the solution of the fundamental device equations is required to adequately model modern devices. Physical models have now progressed to the point where their use in the design and optimisation of existing and future device structures is feasible. Nevertheless, numerical device simulations have yet to displace equivalent circuits in the armoury of circuit design engineers, and equivalent circuit modelling techniques still appear frequently in the literature. It is commonly stated that as computing capabilities increase, numerical device simulations will play a more significant part in circuit simulation. Quite often, however, the complexity of physical models simply increases to use the resources at hand. Moreover, circuit engineers are familiar with circuit concepts and are likely to choose to work with equivalent circuits if they are sufficiently accurate for design purposes. It appears safe to assume that equivalent circuits will remain a useful tool for the foreseeable future. A reappraisal of equivalent circuit modelling is therefore highly appropriate.

Numerous papers have been written about equivalent circuit modelling of compound semiconductor devices. Those who are unfamiliar with the subject will probably experience difficulty in selecting a suitable equivalent circuit model from amongst the many varieties available for a given device. Rather than merely catalogue various equivalent circuits, this chapter aims to examine the basic principles involved in their derivation and use. Equivalent circuits for many of the devices that are the subject of other chapters will act as vehicles for reviewing these principles. Models for the metal-semiconductor field effect transistor (MESFET), high electron mobility transistor (HEMT), heterojunction bipolar transistor (HBT) and laser diode will be presented. The strengths and weaknesses of the equivalent circuit model will be discussed, as will its relationship with models based on numerical simulation of the semiconductor equations. Much of the discussion will revolve around practical topics such as parameter extraction and the application of models in circuit design. However, before looking at these issues, it is pertinent to establish what is meant by an equivalent circuit and what role it is intended for.

2. THE ROLE OF DEVICE MODELS

Three avenues are available for improving the performance of microwave and optoelectronic systems, namely passive circuit design, device optimisation, and simultaneous optimisation of the device and embedding circuit. Both circuit design and device optimisation cannot be conducted satisfactorily on an empirical basis, using measurements alone. The reasons are twofold. The first is the need to curb development costs and reduce the time to market in order to gain competitive advantage. With an empirical approach, many time-consuming and costly fabrication and characterisation cycles would be needed to obtain the optimum device-circuit interaction. The second reason is the interest in monolithic integration of microwave and optoelectronic devices. Monolithic microwave integrated circuits (MMICs) and optoelectronic integrated circuits (OEICs) offer potential rewards which mirror those of silicon ICs, and include low cost mass production, improved reliability and smaller dimensions when compared with hybrid integration. Unlike hybrid integration, however, MMICs and OEICs offer very limited scope for post-production tuning. Once again, an empirical approach to MMIC and OEIC design would lead to numerous fabrication and characterisation cycles.

The successful design of circuits which meet the specification in as few fabrication-characterisation cycles as possible depends on accurate modelling of semiconductor devices. Ideally, a device model should predict the correct performance under small-signal and large-signal conditions and over a wide frequency range. It is also vital that the model be computationally efficient so that its incorporation in circuit simulation and optimisation is feasible, while the ability to relate the physical structure of the device to the performance is highly desirable. In many cases, it will be necessary to use a model for yield predictions and examining the effects of process variations on device performance. Even though computing facilities are available, it is also desirable to have closed-form expressions that link the material properties and device structure to the terminal performance. Finally, it is essential that the model have parameters which can be readily extracted. The ultimate goal of modelling is to produce models with these properties. Unfortunately, many aspects of the modelling process, including the complexity of semiconductor device physics, dictate that a less ambitious and more realistic course be taken. Depending on the demands of the end user, this approach includes the consideration of equivalent circuit models.

3. DEFINITION OF EQUIVALENT CIRCUIT MODELLING

Strictly speaking, it can be argued that there is no such thing as an equivalent circuit model (Chua, 1980). The circuit model of a semiconductor device is not an equivalent circuit because it has terminal characteristics which approximate, rather than equate, to the actual performance. In the majority of cases, a semiconductor device will have many distinct circuit models that correspond to different operating modes of the device, such as small-signal, large-signal and noise models. It is more accurate to speak of a circuit model instead of an equivalent circuit. However, this chapter is not intended to be heretical, and the widely accepted definition of equivalent circuits will be adhered to.

In order to appreciate what is meant by an equivalent circuit, one has to consider the schemes which are used to classify device models. An outstanding treatise on the principles of semiconductor device modelling was given by Chua (1980). Although no general theory of device modelling is presently available, he suggests that most existing models of devices have been derived (sometimes in an *ad hoc* manner) via two basic routes: the physical approach and the black-box approach. Chua also lays down criteria for a truly realistic model, showing that in addition to being quantitatively accurate, it should:

1. display the same qualitative behaviour as the actual device;
2. be capable of predicting previously unknown operating modes;
3. have qualitative properties that do not change under "small" perturbations of the model parameters;
4. be well-posed in the sense that when connected with other well-posed models, no non-physical situation arises.

The black-box approach to modelling is usually adopted when the device physics are poorly understood. According to Chua (1980), four steps are involved: measurements, mathematical modelling, model validation and network synthesis. When all four steps are carried out, the resulting model is often called an (empirically derived) equivalent circuit.

Although equivalent circuits usually require measurements for obtaining element values, their topology can be derived from the device physics in many cases. Indeed, the most logical route to a reliable device model is through a detailed study of the physical and operating mechanisms of the device. Chua's understanding of the physical approach to modelling also involves four basic steps: study of the salient features of the device physics, physical equation formulation, equation simplification and network synthesis. This definition embraces two specific descriptions of device modelling. For many researchers, only the first two steps belong to the realm of physical modelling. The inclusion of the last two steps then leads to a physically-based equivalent circuit model. In order to eliminate possible confusion, the term "physical model" will only apply here to a model that has progressed through the first two stages of Chua's definition.

At this stage of the discussion, it can be seen that the definition of equivalent circuit modelling is far from being straightforward and unequivocal. One has to qualify the term "equivalent circuit" by the method of derivation (black-box, physical or a combination), by the type of performance that is predicted (small-signal, large-signal or noise), and by the kind of calculations needed to predict device behaviour (closed-form analytical expressions or numerical solutions).

4. EQUIVALENT CIRCUIT MODELS VERSUS PHYSICAL MODELS

Before embarking on a derivation of an equivalent circuit model, it is instructive to consider when such a model would be preferred over a physical model, and vice-versa. Device models serve two groups of engineers: device and circuit designers. Equivalent circuits have traditionally been geared towards engineers who are well versed in circuit analysis and design, whereas physical models were until recently the sole domain of the device physicist.

152

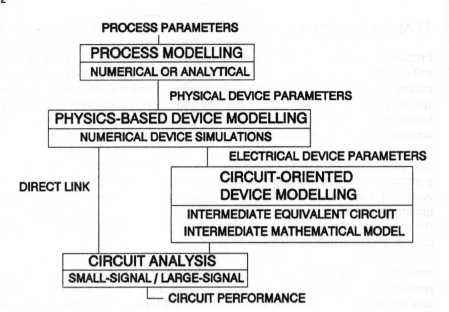

Fig.1. Relationship between physics-based equivalent circuits and physical models, process modelling and circuit analysis. (Filicori *et al*. 1992).

The internal workings of the device are usually of little interest to the circuit designer (who is concerned about how the device behaves as a black-box) while they are of paramount importance to the device engineer. This demarcation is gradually being eroded as MMIC foundries start to allow the circuit designer to have limited control over processing steps so that active devices can be tailored to improve the overall circuit performance. Moreover, it has been realised that design centring and yield analysis are best dealt with by physical models. (Improving MMIC yield is an important factor in reducing costs). On the other hand, most physical models are computationally expensive and cannot be used as computer-aided design (CAD) tools for circuit analysis, let alone optimisation.

When properly formulated, physical models have the potential to accurately describe the terminal or black-box characteristics of a device, even under highly nonlinear conditions. Furthermore, the ability to examine the effect of physical parameter and geometry variations on these characteristics has made physical models extremely attractive to device designers. In contrast, it is difficult to relate equivalent circuit models to the geometry and doping profiles in modern compound semiconductor devices, especially in the sub-micron range. This suggests that physical modelling (based on numerical simulations) will be the method that is increasingly chosen for the development of new devices. However, once a device has been developed, it is still the case that any circuit incorporating it will usually be designed with equivalent circuit techniques. It remains to be seen for how long equivalent circuits can maintain their supremacy over physical models in this area.

4.1 Application of Physical Models to Circuit CAD

Proponents of physical modelling point out that microwave IC design technology will come to rely on integrated CAD tools which link geometry, layout, physical parameter and process parameter descriptions with performance, yield and system specifications. A block diagram of this process is shown in Fig.1, where physically-based equivalent circuits are included as an optional intermediate step. This scenario assumes that fast and predictable physical modelling will be available for the simulation of MMICs (and OEICs eventually). Unfortunately, a model that can provide complete device performance prediction in terms of physical parameters alone requires the solution of fundamental semiconductor equations. As device dimensions continue to shrink, this inevitably entails consideration of quantum effects. However, even the relatively simple drift diffusion model involves the numerical solution of sets of partial differential equations over two- or three-dimensional space. The computational burden imposed by such a model in a circuit analysis procedure would be prohibitively large. This is borne out by the work of Licqurish *et al.* (1989) who found that using a full two-dimensional model required over 7000 CPU seconds on an Amdahl 5860 mainframe to simulate the first two cycles of an intermediate frequency of 5 GHz for a dual-gate MESFET mixer circuit. Contemporary large-signal circuit design makes use of harmonic balance techniques. These require the analysis of the circuit at each iterative step, and therefore demand computationally efficient models.

It is inevitable that simplifying assumptions have to be employed in the derivation of computationally and analytically tractable models. In doing so, the predictive powers of the model should not be jeopardised. A good example of reliable model simplification is the quasi-two-dimensional model as implemented by Pantoja *et al.* (1989). This approach takes advantage of the specific structure of microwave MESFETs to reduce the semiconductor equations to a consistent one-dimensional approximation that reduces computing times by a factor of 1000. A quasi-static large-signal model can then be developed and applied in a harmonic balance simulation. This is an example of indirect linking of physical models via equivalent circuit models in nonlinear circuit simulators. However, Filicori *et al.* (1992) point out that although quasi-two-dimensional models can be used for simple large-signal analyses, only analytical models are directly compatible with optimisation-driven circuit analysis techniques such as harmonic balance at present.

A further problem with physically-based models is the difficulty in obtaining proprietary information from manufacturers about the physical structure of the device. Although some physical parameter values can be extracted from measured data in the same way that equivalent circuit element values are obtained, this somewhat defeats the purpose of physical modelling. The ability to use the model for yield analysis is subsequently impaired. Yield optimisation is a major reason for using physical models in circuit design. In yield-driven MMIC design, for example, the electrical device parameters must be characterised statistically. It turns out that physical parameters (such as doping profile and recess depth) resulting from the manufacturing process are either virtually uncorrelated or subject to simple correlations. In contrast, the device physics contrive to produce complex correlations in the electrical device parameter statistics.

4.2 Application of Equivalent Circuits to Circuit CAD

The major advantage which equivalent circuit models have over physical models is that they are easily implemented in circuit analysis procedures. Most commercial circuit design packages can only accept circuit models. Another beneficial property of equivalent circuits is their computational efficiency. This is an important factor not only in harmonic balance simulations which repetitively analyse the model in the quest for a large-signal solution, but also in bias-dependent small-signal circuit optimisation. Unfortunately, equivalent circuits require measurements-based parameter extraction procedures whether they are physically- or empirically-based. There are examples, such as the small-signal equivalent circuit for the MESFET, where this can be carried out analytically. However, equivalent circuits increase in complexity at higher frequencies to account for parasitics. A small-signal FET model may require as many as twenty elements to describe its behaviour at 10 GHz. In many cases, therefore, parameter extraction is based on multi-variable optimisation which means there are no unique solutions and there are even models with non-physical element values. There is no reason why an equivalent circuit model should not have negative element values, but it is unlikely that it will have a strong physical basis. Nevertheless, equivalent circuits have been shown to accurately match measurements, especially in the small-signal regime. Moreover, they are ideally suited to analysing large-scale integrated circuits, although it should be noted that most microwave and optoelectronic circuits have a low device count.

The main obstacle to equivalent circuit modelling is the fact that it is difficult to relate element values to physical parameters, and bias conditions, signal levels and temperature are all known to have an effect on element values. Consequently, whenever a new device is chosen the parameter extraction process must be repeated. This makes the use of equivalent circuits in device design virtually impossible and also limits their usefulness in yield analyses. Despite these problems, equivalent circuits are still popular with circuit designers, if only because physical models are computationally expensive. In addition, it has also been seen that physical models are not without their own drawbacks.

To summarise, it appears likely that physical models will dominate device design while equivalent circuit models will continue to be used for circuit design. The optimal all-purpose device model has alluded engineers, and will probably continue to do so for some time to come. However, the distinction between the two basic model types is becoming increasingly blurred as MMIC designers begin to have some control over device parameters. There is also a noticeable trend in the use of physically-based circuit models in circuit design. Often, the only difference between physical models and non-empirical equivalent circuit models is that the latter interpret the physical equations of the former as network equations. Even if a physical model can be found to simultaneously satisfy the requirements for accuracy and computational speed, it is almost certain that circuit designers will still use an equivalent circuit representation of that model largely because of their familiarity with this type.

The stage is now set to examine how many of the above considerations relate to equivalent circuit models for microwave and optoelectronic devices. The following models will be discussed: small-signal MESFET and HEMT, large-signal MESFET and HBT, and small-signal and large-signal laser diode.

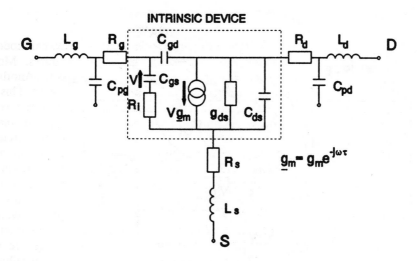

Fig.2. Small-signal MESFET equivalent circuit.

5. EXAMPLES OF EQUIVALENT CIRCUIT MODELS

5.1 Small-Signal Model of the MESFET and HEMT

A small-signal equivalent circuit model of the MESFET is a useful tool in the design of microwave active circuits. Such a model must be representative of an actual device. This is usually achieved by carefully choosing the topology so that each element provides a lumped approximation to some aspect of the device physics. Not only does this provide a physically-based model, it also makes the task of matching element values to measured S-parameters relatively easy. If a rigorous extraction procedure is employed to determine the element values, the model can be used to extrapolate S-parameters to frequencies outside the measurement range. Furthermore, some of the element values can be scaled with gate width, allowing the designer to predict the S-parameters of devices with different dimensions (Golio 1990). However, as Ladbrooke (1989) has mentioned, element values cannot be varied independently without distorting the physical correctness of the model. All element values are interdependent, and changing one physical parameter affects all of them. As mentioned earlier, this renders an equivalent circuit virtually useless for yield analysis. Ladbrooke has taken this fact as the basis of his reverse modelling philosophy, in which the physical parameters of the device are deduced from the equivalent circuit element values. In this case, the equivalent circuit must adequately represent all phenomena which have a bearing on device performance, including signal delay, charge and energy storage, current modulation and power loss.

A typical small-signal equivalent circuit that appears in many texts is shown in Fig.2 and an illustration of the physical origin of the elements is given in Fig.3. R_g

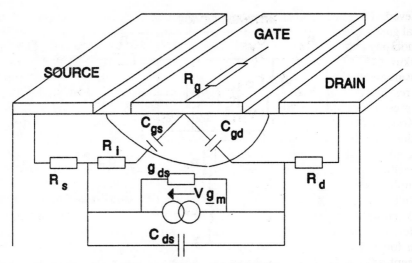

Fig.3. Physical origin of the small-signal MESFET equivalent circuit elements.

is the ohmic resistance of the gate and R_s and R_d are the source and drain ohmic contact resistances respectively. R_i is the resistance of the semiconductor region under the gate. C_{ds} is the drain-source capacitance and C_{gs} and C_{gd} are the gate channel capacitances. If the gate leakage current (which requires a conductance in parallel with C_{gs}) is neglected, this equivalent circuit can also model the HEMT.

It may be disconcerting to find that a microwave device is modelled by lumped elements rather than distributed elements. Heinrich (1987) has developed a field-theoretical analysis to model the FET by structures of growing complexity. Unfortunately, this approach required vast amounts of computing time. This was later mitigated (Heinrich 1989) by deriving an equivalent distributed circuit model that approximated the full wave analysis. The importance of distributed (travelling wave) effects can be assessed by considering the FET structure as a set of coupled transmission lines (source, gate and drain contacts) and evaluating the influence of all modes of propagation on device performance. It has been found that distributed effects can be neglected if the gate line wavelength is more than an order of magnitude greater than the device width. Consequently, the lumped approximation topology has provided an excellent match to measured S-parameters at frequencies up to 26 GHz.

Of course, there is little value to the circuit designer in having a highly detailed equivalent circuit model if the element values cannot be found easily. Extracting the element values from measured S-parameters has received much attention in the literature. Two basic techniques can be identified. The first method involves a manipulation of the circuit equations so that the element values can be related to one another and at least one of them can be directly related to the measured S-parameters. This is the method to aim for, since it produces an accurate fit between the model and measurements. The less attractive method involves an optimisation routine that varies the element values until an acceptable fit is achieved. The success of this method is highly dependent on the quality of the initial guesses, otherwise it is possible that the optimum solution that is found will be non-physical. This dilutes the physical significance of the equivalent circuit. Vaitkus (1983) analysed the effect that errors in broadband S-parameter measurements can have on extracted element values. Measurement uncertainties and residual dependence on initial guesses can affect final element values through the presence of multiple local minima in the objective function.

Kondoh (1986) has sought to improve the behaviour of optimisation-based routines by an *a priori* partitioning of the problem. Groups of elements are selected from the equivalent circuit and these are optimised one by one against a specific S-parameter over the whole frequency range or the upper half. For example, the elements g_{ds} and C_{ds} are optimised first by being fitted to S_{22} over the full frequency range. The choice of the order of elements to be optimised and the S-parameter to be fitted to is based on a sensitivity analysis performed on the model's S-parameters as a function of element value changes. This technique has been found to perform well and to be relatively immune to poor guesses for element values. However, an accurate broadband S-parameter measurement facility is required in order to obtain meaningful element values.

Despite the success of Kondoh's method, it is preferable to be able to uniquely extract the element values from the measured S-parameters. One of the advantages of the model in Fig.2 over more complex equivalent circuits is the availability of exact element value extraction procedures. The method was first described by Minasian (1977) and later developed by Dambrine *et al.* (1988) and Berroth and Bosch (1990). It has since been applied to an extended model which includes gate current and resistive feedback (Berroth and Bosch 1991), while Vickes (1991) has extracted element values for a model that includes charge accumulation in the conducting channel. Hughes and Tasker (1989) have also developed a similar procedure for HEMTs while Costa *et al.* (1991) have extracted element values for an HBT equivalent circuit.

The first step in the parameter extraction procedure is to extract the parasitic elements and convert the S-parameters of the resulting intrinsic circuit to y-parameters. These are given by:

$$y_{11} = \frac{R_i C_{gs}^2 \omega^2}{D} + j \omega \left(\frac{C_{gs}}{D} + C_{gd}\right) \tag{1}$$

$$y_{12} = -j \omega C_{gd} \tag{2}$$

$$y_{21} = \frac{g_m e^{-j\omega\tau}}{1 + j R_i C_{gs}\omega} - j \omega C_{gd} \tag{3}$$

$$y_{22} = g_d + j \omega(C_{ds} + C_{gd}) \tag{4}$$

where $D = 1 + \omega^2 C_{gs}^2 R_i^2$. Dambrine et al. (1988) restrict their technique to frequencies below 5 GHz by assuming that $D \approx 1$, even though this was later shown to be unnecessary by Berroth and Bosch (1990). Separating the y-parameters into their real and imaginary parts, the equivalent circuit elements can be found analytically:

$$g_{ds} = Re\ (y_{22}) \tag{5}$$

$$C_{gd} = -\frac{Im(y_{12})}{\omega} \tag{6}$$

$$C_{gs} = \frac{Im(y_{11}) - \omega C_{gd}}{\omega}\left[1 + \frac{(Re(y_{11}))^2}{(Im(y_{11}) - \omega C_{gd})^2}\right] \tag{7}$$

$$R_i = \frac{Re(y_{11})}{(Im(y_{11}) - \omega C_{gd})^2 + (Re(y_{11}))^2} \tag{8}$$

$$g_m = \sqrt{((Re(y_{21}))^2 + (Im(y_{21}) + \omega C_{gd})^2)\ (1 + \omega^2 C_{gs}^2 R_i^2)} \tag{9}$$

$$\tau = \frac{1}{\omega}\sin^{-1}\left(\frac{-\omega C_{gd} - Im(y_{21}) + \omega C_{gs} R_i\ Re(y_{21})}{g_m}\right) \tag{10}$$

$$C_{ds} = \frac{Im(y_{22}) - \omega C_{gd}}{\omega} \tag{11}$$

Provided the parasitic element values are known, the intrinsic equivalent circuit can be extracted through a series of transformations to and from z- and y-parameters. The procedure comprises the following sequence.

1. Measurement of the extrinsic device parameters.
2. Transformation of the S-parameters to z-parameters and extraction of L_g and L_d.
3. Transformation of the z-parameters to y-parameters and extraction of C_{pg} and C_{pd}.

4. Transformation of the y-parameters to z-parameters and extraction of R_g, R_s, L_s and R_d.

5. Transformation of the z-parameters to the y-parameters of the intrinsic circuit.

6. Extraction of the intrinsic element values using the above equations.

The principal technique used to determine the parasitic element values is the "cold modelling" developed by Diamont and Laviron (1982). This involves the measurement of the S-parameters at zero drain bias with a strongly forward biased gate and thus introduces an extra characterisation step. The parasitic inductances can be found from the imaginary parts of the corresponding z-parameters while the resistances are derived from the real parts. The parasitic capacitances can then be deduced from S-parameters taken under pinch-off condition and a zero drain-to-source voltage. Curtice and Camisa (1984) have used the biasing condition of Diamont and Laviron to optimise the device parasitics. Dambrine *et al.* (1988) however, have proposed an alternative approach where all the parasitics are directly deduced from measurements performed at $V_{ds} = 0$.

One of the implicit assumptions of "cold modelling" is that the parasitic elements are independent of bias. Anholt and Swirhun (1991) have made a detailed study of the possible bias-dependencies of FET parasitic inductances and capacitances at forward bias. They found that many of the elements are largely unaffected by changes in bias. It was established that for recess etched FETs R_s increases slowly as the gate bias becomes more negative, but the sensitivity of g_m decreases faster than R_s increases. Although they could not estimate the increase in R_d with drain bias, the sensitivities of all intrinsic elements to changes in R_d are small except for τ.

The fact that "cold modelling" requires S-parameter characterisation at zero drain bias to extract the parasitic inductances is a hindrance and can also lead to inaccuracies. This is especially true of MMIC devices where the parasitic gate inductance is proportional to depletion depth under the gate. Arnold *et al.* (1990) have eliminated the need for additional microwave characterisation by employing the dc measurement technique of Fukui (1979) to determine parasitic resistances and then determining the inductances and capacitances from the S-parameter data at the operating bias level.

While this section is ostensibly concerned with MESFETs and HEMTs, it is worth noting that many of the concepts are applicable to other devices. Conversely, parameter extraction techniques for other devices may also be relevant for FETs. A case in point is the work of Bilbro *et al.* (1990) who have solved the parameter extraction problem for small-signal physically-based HBT equivalent circuits with a new formulation of simulated annealing. Their algorithm allows equivalent circuits to be derived that are as physically meaningful as those found from conventional optimisation, but without the need for expert knowledge of good starting values.

Although the equivalent circuit model of the MESFET is physically-based, it has no direct relation to numerical device simulations. This does not mean, however, that numerical solutions cannot be used to derive equivalent circuits. Curtice (1989), for example, has used fully two-dimensional simulations to extract element values for the MESFET equivalent circuit. This is carried out by Fourier analysis of the gate and drain current response to voltage steps on these electrodes which yield complex y-parameters. These can then be employed instead of S-parameter measurements in the extraction techniques described above.

It was stated previously that one of the limitations with equivalent circuit modelling is the difficulty in ascertaining the temperature dependence of element values. The usefulness of equivalent circuits to designers and the importance of temperature effects has motivated a recent study by Anholt and Swirhun (1992). Although this is one of the most comprehensive treatments of the subject, the expressions for temperature dependence are empirical. One of the problems in modelling temperature dependence of equivalent circuits is the inability to distinguish between ambient and self-heating effects.

5.2 Large-Signal MESFET Model

Whenever a MESFET is operating under nonlinear conditions, as in power amplifier applications, a large-signal model is a crucial design aid. Most CAD-oriented large-signal equivalent circuits are characterised by dc and S-parameter measurements.

A commonly adopted approach is to derive the nonlinear behaviour of the device from the bias dependence of the small-signal equivalent circuit. Considering the equivalent circuit in Fig.2, the strongest bias dependence (i.e. nonlinearity) is exhibited by g_m, g_{ds}, C_{ds} and C_{gs}. All other elements are essentially linear. Look-up tables are then generated which relate the bias-dependent elements to the relevant controlling voltages, and these can be used in a harmonic balance simulator. A similar method has been adopted by Pantoja et al. (1989). Here, a quasi-static large-signal MESFET equivalent circuit was derived from a quasi-two-dimensional simulation (Fig. 4). The nonlinearly controlled elements of the equivalent circuit were directly estimated from the charge and internal potential distribution. For each simulated point of the device characteristics (I_{DS} versus V_{DS}), small perturbations of the gate voltage and/or source current provide the bias-dependent equivalent circuit element values G_m, G_{DS}, C_{GS}, C_{DG} and C_{DOM}. The suitability of this model for nonlinear circuit CAD has been proven by comparing the results of a two-tone harmonic balance analysis with measured load-pull contours for a power MESFET.

Although the potential of physical models for nonlinear CAD has been demonstrated, most CAD software packages make use of empirically-based equivalent circuits. The equations in these models are typically derived by fitting mathematical approximation functions such as polynomials and splines to measured data. A well known example of this is the Curtice model (Golio 1990), which describes the drain current with respect to the drain-source and gate-source voltages by means of a polynomial and hyperbolic function:

$$I_{ds}(V_{gs}, V_{ds}) = \beta\, (V_{gs} - V_{T0})^2\, (1 + \lambda V_{ds})\, \tanh(\alpha\, V_{ds}) \qquad (12)$$

where V_{gs} and V_{ds} are the intrinsic terminal voltages and β, V_{T0}, λ and α are the model parameters. This equation has no physical basis, and the coefficients have no physical significance. Furthermore, it relies on parameter extraction procedures which do not have unique solutions and can provide models that will not converge in harmonic balance simulations. Even so, this equation is extremely flexible in that it can be made to fit the drain current characteristics of MESFETs with

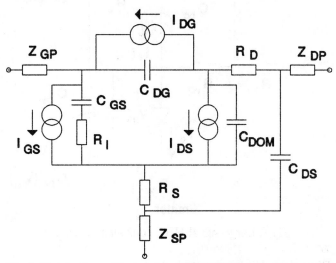

Fig.4. Physics-based large-signal MESFET equivalent circuit.

different doping profiles. The same cannot always be said of physical models such as the Shockley model which uses simplifying assumptions (including the gradual channel approximation and field-independent mobility) to obtain a closed-form expression for the channel current.

Finally, it is worth reviewing the Root model (Root *et al.* 1991). It would be correct to classify this model as a behavioural black-box rather than an equivalent circuit, although it can be related to a generalised topology that is very similar to Fig.4. The aim of this type of model is to provide an accurate technology-independent large-signal representation which can be automatically and unambiguously derived either from measurements or numerical device simulations. In the case of the Root model, this is achieved by explicitly obtaining the constitutive relations for the terminal characteristics from solutions of the algebraic and /or differential equations involving the measured data. No optimisation is needed to construct the model. Moreover, it provides good large-signal predictions even though it is derived from small-signal measurements. The generalised nature of the equivalent circuit topology makes it applicable to many types of FET structure.

5.3 Large-Signal HBT Equivalent Circuit Model

The GaAs/AlGaAs heterojunction bipolar transistor (HBT) has aroused considerable interest amongst microwave circuit designers. Unlike the MESFET and HEMT, the HBT is a vertically-structured device. As such, it is not well suited to planar processing or for low white noise applications. Despite these problems, it has many advantages which still warrant considering it for circuit applications. For example, the HBT offers useful gain at millimetre-wave frequencies, and has a transconductance that is as much as 100 times that of the MESFET. In addition,

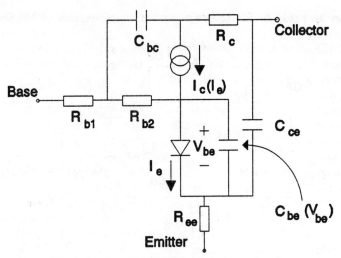

Fig.5. Large-signal HBT equivalent circuit.

this device has better matching properties and lower $1/f$ noise when compared with either the MESFET or HEMT.

One of the most noticeable characteristics of the HBT is the strong nonlinearity in the base-emitter voltage to collector current transfer characteristic. However, the device can exhibit surprisingly linear behaviour if it is operated at relatively low bias levels. In a recent paper, Maas *et al.* (1992) tried to explain this unusually high linearity by investigating small-signal intermodulation distortion. They showed that the intermodulation current generated in the exponential junction is partly cancelled by the intermodulation current generated in the junction capacitance, and that this effect results in the comparatively low levels of intermodulation distortion. It is vital that an accurate nonlinear model should accurately model both forms of nonlinearity.

The equivalent circuit in Fig.5 has been proposed as a nonlinear model suitable for intermodulation calculations. This circuit includes three nonlinearities: the resistive junction (which is modelled as an ideal diode), the capacitive junction and the nonlinear current source $I_c(I_e)$. Parasitic elements are also included, the most important being the emitter resistance R_{ee}.

Maas *et al.* (1992) then proceed to extract the model parameters from a combination of dc and S-parameter measurements. For example, the junction I/V characteristic can be found from Gummel plots, in which $\log(I_e)$ is plotted as a function of V_{be} at low current levels. Provided the junction temperature remains constant over an adequate range of bias voltages, the exponential characteristic relating the base-emitter voltage to the collector current can be determined.

The model in Fig.5 has been validated by being used in a commercial harmonic balance simulator to predict the gain and second- and third-order intercept points. Measured and simulated results differed by no more than 2 dB, indicating the potential utility of physically-based equivalent circuit models in HBT circuit design.

Yet another harmonic balance simulation of HBT equivalent circuits was undertaken by Frankel and Pavlidis (1992). In this case, a bias-dependent small-signal equivalent circuit with distributed base elements was used. As with the MESFET-HEMT model, the parasitic elements were taken to be bias-independent. The element values were extracted from measured S-parameters by optimisation-based fitting routines. In order to maintain the physical significance of the circuit, the optimisation was constrained by using dc determined transconductance values. Bias-dependent equations were then obtained for the intrinsic equivalent circuit elements by employing approximate HBT physical equations. Again, this approach ensured that the equivalent circuit retained physical significance. It also resulted in a more efficient harmonic balance simulation of the device output power and gain characteristics as a function of input power. The predictions of the model were found to be in close agreement with measurements under both class A and class AB conditions.

An alternative approach to HBT large-signal equivalent circuit modelling was adopted by Grossman and Choma, Jr. (1992). A SPICE macro model was developed to approximate the HBT operation. This study represents another example of circuit engineers adapting physically-based circuit models for use in standard circuit CAD programs. As such, it reflects the current inability of most commercial CAD tools to directly handle physical models. The resulting model accurately simulates circuits that operate below 3 GHz but above this frequency the limitations of the time-dependent charge reduce simulation accuracy for large-signal inputs.

5.4 Small-Signal and Large-Signal Models of the Laser Diode

The modelling of microwave optoelectronic devices in general, and laser diodes in particular, represents a difficult challenge that has commanded the time of many researchers for over three decades. This effort has received added impetus within recent years due to the immense advantages that fibre-optic and integrated optic techniques can proffer to microwave and telecommunications systems.

The complexity in dynamic laser diode modelling is introduced by the need to account for both electronic and optical phenomena within the same device. Consequently, most laser diode modellers have been forced to strive for qualitatively, rather than quantitatively correct models. It would be unwise to incorporate all known physical properties of the laser diode into one model if only one specific aspect of the operation is of interest. The success of a physically-based model therefore hinges on the modeller's skill in simplifying the describing equations whilst maintaining adequate simulation accuracy and efficiency. In this section, models of the direct modulation of laser diodes at microwave frequencies will be discussed.

Future microwave optoelectronic systems and OEICs will require similar tools as those that have been developed for MMICs. One of these tools is computer-aided circuit design that can incorporate models for optoelectronic devices and optical active and passive networks. Current physical models of microwave devices have proved inadequate for circuit CAD, so it is not surprising that sophisticated physical models for the laser diode, such as the travelling-wave rate equations, are also inappropriate for CAD. The travelling-wave rate equations are of the partial differential variety, and require extensive computing facilities.

In order to reach the stage where a physically-based equivalent circuit can be derived, apposite simplifications must be made so that the travelling-wave rate equations can be reduced to the more manageable set of rate equations. The single-mode rate equations are given by:

$$\frac{dN}{dt} = \frac{I_A}{\alpha} - \frac{N}{\tau_n} - G(N) \ (1 - \epsilon S) \ S \tag{13}$$

$$\frac{dS}{dt} = \Gamma G(N) \ (1 - \epsilon S) \ S - \frac{S}{\tau_p} + \Gamma \beta \frac{N}{\tau_n} \tag{14}$$

where S and N are the photon and carrier densities, τ_p and τ_n are the photon and carrier lifetimes, α is the product of the electron charge and active region volume, $G(N)$ is the optical gain, I_A is the current entering the active region, β is the spontaneous emission coefficient, Γ is the ratio of the active region volume to the modal volume, and ϵ is a gain compression factor.

Being ordinary differential equations, these are particularly easy to solve numerically. They can also be used to analytically obtain the small-signal intensity modulation response of the intrinsic device. However, the advantage in reducing the complexity of the laser diode physical model to the single-mode rate equations is that small-signal and large-signal circuit models may be derived. There are significant benefits in choosing a circuit modelling technique. Firstly, an equivalent circuit of the intrinsic device includes information on the I-V characteristics of the forward biased heterojunction. This is achieved via the classical Shockley relationship. Secondly, provision can be made for the inclusion of the parasitic network into the model. Finally, commercially available circuit analysis programs can be used to determine the response of relatively complex laser diode circuits.

The subject of small-signal equivalent circuits of the laser diode is well developed, and such models are regularly used in conjunction with microwave CAD tools for matching network design. The small-signal impedance characteristics of an intrinsic laser diode were first studied by Morishita et al. (1979). Katz et al. (1981) then extended the model to account for spontaneous emission into the lasing mode and saturation losses. Harder et al. (1982) included the effects of Langevin noise sources and multiple cavity modes in a later model. Tucker and Pope (1983 (a)) presented a model which included parasitics, and they have shown that a gain compression term can be used to model diffusion damping (Tucker and Pope 1983 (b)). The quasi-static assumption in the latter model is only valid for narrow-stripe devices, and was removed by Ng and Sovero (1984).

A widely-used small-signal equivalent circuit (Tucker and Pope 1983 (b)) is shown in Fig. 6. The input terminals represent the electrical contacts on the device chip. Contact capacitance and series resistance between the contacts and the active layer are modelled by C_s and R_s respectively. The small-signal photon density is proportional to the current in the inductive branch of the model. Alternatively, the output voltage across R_{s2} can be used to represent optical intensity. It is interesting to note that circuit elements are used to represent optical effects. This is analogous to the use of resistances in thermal impedance calculations.

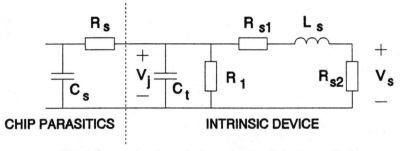

Fig.6. Small-signal equivalent circuit of the laser diode.

In common with small-signal models for microwave devices, the extraction of equivalent circuit element values is an important task. All the intrinsic device elements are related to physical parameters, some of which can be determined from the device dimensions, material parameters and measurements of the dc, small-signal and optical spectrum characteristics. Unfortunately, laser diode manufacturers are reluctant to release information about the physical structure of their devices. Even if this information was available, the modeller would still have to contend with finding values for the parasitic elements.

The parameter extraction tools available to the laser diode modeller are similar to those for the MESFET, namely dc and S-parameter characterisation. However, the presence of an optical port tends to complicate matters. The relative immaturity of existing lightwave network analysis equipment means that many engineers rely on measurements of the input impedance (S_{11}) and modulation response, whereas the MESFET modeller can call on all four S-parameters. Suggestions for improving lightwave characterisation capabilities have been made by Iezekiel *et al.* (1991). Nevertheless, it is still possible to obtain parasitic element values from electrical input impedance measurements. Further dc and small-signal intensity modulation response measurements suffice to give the values of the intrinsic elements. At present, an analytically-based extraction procedure similar to that employed for MESFETs has not been derived. Use is made of optimisation and fitting routines instead. In this situation, care must be exercised in the initial values chosen for the element values. As in MESFET modelling, many sets of element values can satisfy the small-signal response.

Compared with small-signal equivalent circuits, the synthesis of a large-signal model is trivial. The rate equations can be simply rearranged into the form of nodal circuit equations, and the various physical terms are assigned to circuit elements or circuit voltages and currents. The resulting model is shown in Fig.7 (Tucker and Kaminow 1984). Tucker (1981) first derived a large-signal equivalent circuit from the single-mode rate equations and used this in a SPICE simulation. (Tucker's model was preceded by a more complex multimode model derived by Habermayer (1981), but this has not been adopted by many workers). The nonlinear resistor $\bar{R}$ models the spontaneous recombination process, while I_{SP} and I_{ST} model spontaneous and stimulated emission respectively. The lumped elements R_{PH} and C_{PH} represent photon loss and storage respectively. It should be stressed

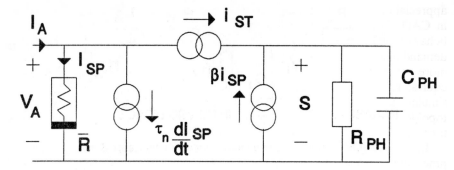

Fig.7. Large-signal equivalent circuit of the laser diode.

that the representation of photon density as the output voltage is not physically meaningful but is necessary if the model is to be used in circuit analysis and design.

The problem of parameter extraction for the large-signal case is even more demanding than for the small-signal model. All element values depend on physical model parameters. Although the measurements used to extract the small-signal model can also be used to obtain some physical parameters, virtually all extraction procedures require a knowledge of the active device dimensions. A common technique is the use of optimisation to indirectly fit physical parameters, via the expressions for the small-signal equivalent circuit elements, to the small-signal response. To date, there have been no reports of extracting physical parameters directly from the element values of the small-signal equivalent circuit. If reverse modelling for the laser diode could be performed, it would be a significant improvement to large-signal modelling of the device.

Having derived a large-signal model, it can be used for the analysis and design of optoelectronic circuits in the nonlinear regime. Until recently, investigations of harmonic and intermodulation distortion in directly modulated laser diodes either neglected the device-circuit interaction or were restricted to small-signal inputs. This approach will be inadequate in the design of future systems, where the interaction of the device with its embedding electrical and optical circuit must be accounted for. To this end, Iezekiel *et al.* (1990) have shown that it is feasible to apply harmonic balance techniques to the analysis (and potential optimisation) of laser diodes and their driver circuits.

6. CONCLUSIONS

This chapter has outlined the strengths and weaknesses of the equivalent circuit modelling technique. The emphasis has been placed on circuit design applications given the established superiority of physical models for device design. Although physical models are now starting to have an impact on circuit CAD applications, most engineers still work with equivalent circuits. One reason for the continued success of circuit modelling is that it provides a conceptual tool that electronic engineers readily understand, whereas physical models require at least some appreciation of device physics. However, practicality also has a strong role to play in CAD. Physical models may indeed be able to predict all modes of device behaviour, but they require data that can be difficult to obtain and they normally demand unrealistic computing times for CAD. Those models that have been made more efficient usually sacrifice some of their accuracy. On the other hand, equivalent circuits are readily implemented in small-signal and large-signal simulators. They cannot claim to be completely physically correct, but when sound topology formulation and parameter extraction techniques are used, they can be formidably accurate for most CAD applications.

In future, it may be possible to discard equivalent circuits as both the practicality and efficiency of physical models is improved. Quasi-two-dimensional models, for example, have already proven their worth in CAD applications. Nevertheless, it has been found useful to derive an equivalent circuit from this model for use in nonlinear CAD (Pantoja *et al.* 1989). A similar example is provided by the large-signal laser diode equivalent circuit, which is simply a network representation of a simplified physical model (Tucker and Kaminow 1984). Hence it can be seen that equivalent circuits have retained their appeal even when their "status" has been reduced to that of a circuit analogy of physical models. It may be that they will survive as the intermediate modelling step depicted in Fig.1. Perhaps a reason for this is that passive embedding networks tend to be modelled as circuits, in which case the equivalent circuit should enjoy a secure future in system design.

REFERENCES

Anholt, R. and Swirhun, S. (1991) Equivalent-circuit parameter extraction for cold GaAs MESFET's, IEEE Trans. Microwave Theory Tech., Vol. MTT-39, pp. 1243-1247.

Anholt, R.E. and Swirhun, S.E. (1992) Experimental investigation of the temperature dependence of GaAs FET equivalent circuits, IEEE Trans. Electron Devices, Vol. ED-39, pp. 2029-2036.

Arnold, E., Golio, M., Miller,M. and Beckwith, B. (1990) Direct extraction of GaAs MESFET intrinsic element and parasitic inductance values, IEEE MTT-S Int. Microwave Symp. Dig., pp. 359-362.

Berroth, M. and Bosch, R. (1990) Broad-band determination of the FET small-signal equivalent circuit, IEEE Trans. Microwave Theory Tech., Vol. MTT-38, pp. 891-895.

Berroth, M. and Bosch, R. (1991) High-frequency equivalent circuit of GaAs FET's for large-signal applications, IEEE Trans. Microwave Theory Tech., Vol. MTT-39, pp. 224-229.

Bilbro, G.L., Steer, M.B., Trew, R.J., Chang, C.-R., and Skaggs, S.G. (1990) Extraction of the parameters of equivalent circuits of microwave transistors using tree annealing, IEEE Trans. Microwave Theory Tech., Vol. MTT-38, pp. 1711-1718.

Chua, L.O. (1980) Device modeling via basic nonlinear circuit elements, IEEE Trans. Circuits Syst., Vol. CAS-27, pp. 1014-1044.

Costa, D., Liu, W.U., and Harris, Jr., J.S. (1991) Direct extraction of the AlGaAs/GaAs heterojunction bipolar transistor small-signal equivalent circuit, IEEE Trans. Electron Devices, Vol. ED-38, pp. 2018-2024.

Curtice, W.R. and Camisa, R.L. (1984) Self-consistent GaAs FET models for amplifier design and device diagnostics, IEEE Trans. Microwave Theory Tech., Vol. MTT-32, pp.1573-1578.

Curtice, W.R. (1989) Intrinsic GaAs MESFET equivalent circuit models generated from two-dimensional simulations, IEEE Trans. Computer Aided Design, Vol. CAD-8, pp. 395-402.

Dambrine, G., Cappy, A., Heliodore, F., and Playez, E. (1988) "A new method for determining the FET small-signal equivalent circuit, IEEE Trans. Microwave Theory Tech., Vol. MTT-36, pp. 1151-1159.

Diamont, F. and Laviron, M. (1982) Measurement of the extrinsic series elements of a microwave MESFET under zero current condition, Proc. 12th European Microwave Conf., pp. 451-456.

Filicori, F., Ghione, G., and Naldi, C.U. (1992) Physics-based electron device modeling and computer-aided MMIC design, IEEE Trans. Microwave Theory Tech., Vol. MTT-40, pp. 1333-1352.

Frankel, M.Y. and Pavlidis, D. (1992) An analysis of the large-signal characteristics of AlGaAs/GaAs heterojunction bipolar transistors, IEEE Trans. Microwave Theory Tech., Vol. MTT-40, pp. 465-474.

Fukui, H. (1979) Determination of the basic device parameters of a GaAs MESFET, Bell Syst. Tech. J., Vol. 58, pp. 771-797.

Golio, J.M. (1990) Microwave MESFETs and HEMTs, (Artech House,Norwood MA)

Grossman, P.C. and Choma,Jr., J. (1992) Large signal modeling of HBT's including self-heating and transit time effects, IEEE Trans. Microwave Theory Tech., Vol. MTT-40, pp. 449-464.

Habermayer, I. (1981) Nonlinear circuit model for semiconductor lasers, Opt. Quantum Electron., Vol.13, pp. 461-468.

Harder, C.H., Katz, J., Margalit, S., Shacham, J., and Yariv, A. (1982) Noise equivalent circuit of a semiconductor laser diode, IEEE J. Quantum Electron., Vol. QE-18, pp. 333-337.

Heinrich, W. (1987) Distributed equivalent-circuit model for traveling-wave FET design, IEEE Trans. Microwave Theory Tech., Vol. MTT-35, pp. 487-491.

Heinrich, W. (1989) High-frequency MESFET noise modeling including distributed effects, IEEE Trans. Microwave Theory Tech., Vol. MTT-37, pp. 836-842.

Hughes, B. and Tasker, P.J. (1989) Bias dependence of MODFET intrinsic model element values at microwave frequencies, IEEE Trans. Electron Devices, Vol. ED-36, pp. 2267-2273.

Iezekiel, S., Snowden, C.M., and Howes, M.J. (1990) Nonlinear circuit analysis of harmonic and intermodulation distortions in laser diodes under microwave direct modulation, IEEE Trans. Microwave Theory Tech., Vol. MTT-38, pp. 1906-1915.

Iezekiel, S., Snowden, C.M., and Howes, M.J. (1991) Scattering parameter characterization of microwave optoelectronic devices and fiber-optic networks, IEEE Microwave Guided Wave Lett., Vol. MGWL-1, pp. 233-235 and p. 399.

Katz, J., Margalit, S., Harder, C., Wilt, D., and Yariv, A. (1981), The intrinsic electrical equivalent circuit of a laser diode, IEEE J. Quantum Electron., Vol. QE-17, pp. 4-7.

Kondoh, H. (1986) An accurate FET modeling from measured S-parameters, Microwave J., Vol. 29, p. 86.

Ladbrooke, P.H. (1989) MMIC Design: GaAs FETs and HEMTs (Artech House, Norwood MA).

Licqurish, C., Howes, M.J., and Snowden, C.M. (1989) A new model for the dual-gate GaAs MESFET, IEEE Trans. Microwave Theory Tech., Vol. MTT-37, pp. 1497-1505.

Maas, S.A., Nelson, B.L., and Tait, D.L. (1992) Intermodulation in heterojunction bipolar transistors, IEEE Trans. Microwave Theory Tech., Vol. MTT-40, pp. 442-448.

Minasian, R.A. (1977) Simplified GaAs MESFET model to 10 GHz, Electron. Lett., Vol. 13, pp. 549-551.

Morishita, M., Ohmi, T., and Nishizawa, J. (1979) Impedance characteristics of double-heterostructure laser diodes, Solid-State Electron., Vol. 22, pp. 951-962.

Ng, W.W. and Sovero, E.A. (1984) An analytic model for the modulation response of buried heterostructure lasers, IEEE J. Quantum Electron., Vol. QE-20, pp. 1008-1015.

Pantoja, R.R., Howes, M.J., Richardson, J.R., and Snowden, C.M. (1989) A large-signal physical MESFET model for computer-aided design and its applications, IEEE Trans. Microwave Theory Tech., Vol. MTT-37, pp. 2039-2045.

Root, D.E., Fan, S., and Meyer, J. (1991) Technology independent large signal non quasi-static FET models by direct construction from automatically characterized device data, Proc. 21st European Microwave Conf., pp. 927-932.

Tucker, R.S. (1981) Large-signal circuit model for simulation of injection-laser modulation dynamics, Proc. Inst. Elec. Eng., Vol. 128, Pt.I, pp. 180-184.

Tucker, R.S. and Kaminow, I.P. (1984) High-frequency characteristics of directly modulated InGaAsP ridge waveguide and buried heterostructure lasers, J. Lightwave Technol., Vol. LT-2, pp. 385-393.

Tucker, R.S. and Pope, D.J. (1983 (a)) Microwave circuit models of semiconductor injection lasers, IEEE Trans. Microwave Theory Tech., Vol. MTT-31, pp. 289-294.

Tucker, R.S. and Pope, D.J. (1983 (b)) Circuit modeling of the effect of diffusion on damping in a narrow-stripe semiconductor laser, IEEE J. Quantum Electron., Vol. QE-19, pp. 1179-1183.

Vaitkus, R.L. (1983) Uncertainty in the values of GaAs MESFET equivalent circuit elements extracted from measured two-port scattering parameters, Proc. IEEE/Cornell Conf. High-Speed Semiconductor Devices and Circuits, pp. 301-308.

Vickes, H.-O. (1991) Determination of intrinsic FET parameters using circuit partitioning approach, IEEE Trans. Microwave Theory Tech., Vol. MTT-39, pp. 363-366.

Large-Signal Models

Robert J. Trew
North Carolina State University

Many electronic devices, such as power amplifiers, oscillators, mixers, switches, etc., operate in the large-signal regime. The term *large-signal* refers to the magnitude of the RF signal relative to the operating bias. Therefore, the magnitude of the RF signal that produces large-signal effects can vary widely, depending upon the type of device and the bias conditions under which the device operates. For example, semiconductor power diodes may be operated with 100's or 1000's of volts applied bias, sustaining 10's to 100's of volts of RF or operating signal before large-signal effects become significant; whereas, a submicron quantum well device may be in the large-signal regime with only microvolt or millivolt RF voltage applied.

Certain types of electronic components operate only in the large-signal regime. For example, a self-starting oscillator must operate in a saturation mode that only occurs under large-signal operating conditions. If large-signal saturation did not occur the oscillations would build up without bound until the oscillator self destructed. In general, the large-signal operation of an electronic component requires analyses of both the active device and the dc and RF circuit in which the active device is embedded. Large-signal operation conditions are established by interactions between the active device, operating in a nonlinear mode, and the linear circuit. The technique used to mathematically interface the active device with the circuit is an important consideration in development of a large-signal model. In recent years the Harmonic Balance method has been extensively developed for this purpose.

In this chapter various techniques for formulating large-signal models of semiconductor microwave devices are discussed. Computational difficulties in interfacing nonlinear device models, which are most efficiently formulated in the time domain, with linear circuit models, which are best formulated in the frequency domain, have historically limited the use of physically based device models in comprehensive, frequency domain nonlinear simulators. For this reason, most simulators that are available make use of equivalent circuit techniques. Recently, however, physically based models that can be used in these simulators have been developed. Both the equivalent circuit based approach and the physically based device models are discussed. The discussion will concentrate upon models that have been developed for the microwave MESFET (*ME*tal *S*emiconductor *F*ield-*E*ffect *T*ransistor). An analytic large-signal model will be described in detail, including comparison of model performance with experimental data.

LARGE-SIGNAL MODEL CONSIDERATIONS

Large-signal models have been developed for semiconductor devices using a variety of techniques, including empirically based approaches that establish an electrical network for the device from measured data, and theoretical approaches in which the model is developed from fundamental principles according to physical describing equations. The former approach is often used in practical applications and yields simple models that are useful in design. The latter approach is more rigorous and yields models that provide a physical understanding of device performance. These models provide significant information concerning physical phenomena involved in device operation and are, therefore, very useful in investigations of optimum device structures and operational limitations. This type of information is, in general, not available from the empirically based models.

Basic Semiconductor Device Equations

The fundamental set of semiconductor device equations provides the basis for development of a device model, either small-signal or large-signal. These equations consist of the current-density equations, the continuity equations, Poisson's equation, and Faraday's law. These equations must be solved simultaneously with the appropriate boundary conditions in order to obtain an accurate representation of the device operation. For devices such as bipolar transistors, where both electrons and holes are important to the operation of the device, two sets of equations are required, one for each type of charge carrier. These equations form the drift-diffusion approximation that is applicable to conditions where the mobile charge carriers can be considered to be in thermal equilibrium with the crystal lattice. These conditions are generally valid when the device dimensions are large compared to the wavelength of the operation frequency, or when the RF period is long compared to the charge carrier relaxation time. When these conditions are not applicable, non equilibrium (i.e., hot-electron) effects can be significant and the drift-diffusion approximation is not valid. Under these conditions alternate modeling approaches must be employed. The MESFET models discussed in this chapter will only consider conditions consistent with drift-diffusion operation.

The drift-diffusion approximation allows the current density equation to be separated into terms that describe current conduction by drift and diffusion mechanisms. For electrons the current density equation can be written as

$$\vec{J}_n = q\mu_n n\vec{E} + qD_n \nabla n \qquad (1)$$

where the first term indicates that electrons drift in response to an applied electric field, with current density proportional to the electric field according to the electron mobility μ_n (units of $cm^2/V\text{-}sec$). The second term indicates that electron conduction also occurs by a diffusion mechanism, where current density flows in response to a gradient in the electron density. For the latter case the proportionality constant is the electron diffusion coefficient D_n (units of cm^2/sec). For nondegenerate semiconductors the diffusion coefficient and the mobility are related by the Einstein relation

$$D_n = \mu_n \frac{kT}{q} \qquad (2)$$

The current density equation, as expressed in (1) applies to relatively low electric field operation where electron flow is linear. In this region the electron velocity is directly related to the magnitude of the electric field according to the relation

$$v = \mu_n E \qquad (3)$$

For large electric fields, as are encountered under large-signal operating conditions, the electron velocity saturates and becomes a nonlinear function of electric field. For example, the electron velocity versus electric field characteristics for several common semiconductors are shown in Fig. 1. As shown, the details of the v-E characteristic depend upon the particular semiconductor, and a variety of nonlinear behaviors are obtained. The characteristic for GaAs is particularly interesting due to the region of negative differential mobility where the electron velocity decreases with increasing electric field. This behavior can produce significant complexities in device operation. The highly nonlinear characteristic can also produce complications in the development of device models. It is primarily for this reason that GaAs device models are more difficult to formulate that those for Si devices.

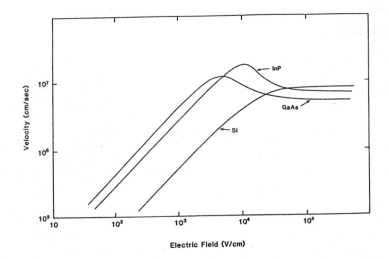

Figure 1   Electron Velocity versus Electric Field for Several Semiconductors

The continuity equation for electrons is

$$\frac{\partial n}{\partial t} = G_n + \frac{1}{q}\nabla \cdot \vec{J}_n \qquad (4)$$

This equation states that the time rate of change of charge within a volume is equal to the rate of flow of charge out of the volume. Charge within the volume may be generated by some mechanism (G_n), such as optical excitation or avalanche ionization. Poisson's equation and Faraday's law complete the basic set of equations required for device analysis. Poisson's equation is

$$\nabla \cdot \vec{E} = \frac{q}{\varepsilon}(N_d - n) \qquad (5)$$

where N_d represents the density of ionized donor impurities and n is the free electron density (both have units of cm^{-3}). Faraday's law is

$$\nabla \times \vec{E} = -\frac{\partial \vec{B}}{\partial t} \qquad (6)$$

Magnetic effects (B) can usually be ignored so that the right hand side of (6) is zero. This, in turn, allows the electric field to be determined from the spatial gradient of the electric potential according to the expression

$$\vec{E} = -\nabla V \qquad (7)$$

Large-Signal Models, RF Circuit Simulators, and the Harmonic Balance Method

A large-signal device model based upon the semiconductor device equations is highly nonlinear and is most efficiently solved in the time domain. Both analytic and numerical solution techniques can be applied. The analytic approach is attractive since it yields a deterministic set of analytic equations that can describe large-signal operation with good accuracy, for certain conditions. The approach requires approximations regarding device geometry and operation, which can prove limiting for certain device structures. The numerical approach is more robust and is suitable for investigation of most device structures. The technique generally requires significant execution time.

Large-signal models that have good accuracy in comparison with experimental data can quickly become complicated due to the necessity of explicitly including additional nonlinear phenomena. For example, large-signal analysis requires that many device parameters be known functions of variables, such as electric field, temperature, current density, etc. In addition, operational nonlinearities due to phenomena such as forward and reverse conduction of contact electrodes can complicate model formulation. Often it is possible to determine explicit analytic expressions for the parameter variations. However, when these expressions are included in the set of semiconductor equations the resulting device model becomes quite complicated. Although the device model can generally be solved in the time domain, it is not unusual to encounter significant model execution time. The simulation time can prove to be a major limitation to utilization of the model in frequency domain simulators.

Microwave engineers generally work in the frequency domain. The interfacing of time domain device models and frequency domain circuit simulators can be a difficult process. The harmonic balance method has been extensively developed for this purpose. The harmonic balance method is conceptually simple. In this method the circuit is divided into linear and nonlinear sub-circuits as shown in Fig. 2.

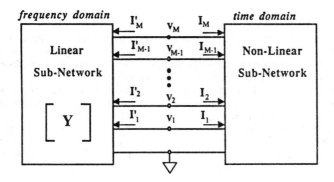

Figure 2   Block Diagram for the Harmonic Balance Method

The linear sub-circuit is analyzed in the frequency domain, whereas the nonlinear circuit is characterized in the time domain. It is assumed that the excitations present in the circuit are sinusoidal and harmonically related. In advanced harmonic balance formulations it is possible to remove the latter restriction. Also, it is assumed that there are $(M+1)$ nodes between the linear and nonlinear sub-circuits including one selected as the reference (or ground) node. The time domain voltage at node m relative to the reference node can be approximated by a truncated Fourier Series

$$v_m(t) = \text{Re}\left\{\sum_{n=0}^{N} V_{mn} e^{jn\omega_1 t}\right\} \qquad (8)$$

where V_{mn} is the voltage phasor at the nth harmonic and mth node. The lower index of the sum corresponds to the dc value of $v_m(t)$. Similarly, the current flowing through node m into the nonlinear sub-circuit can be expressed as

$$i_m(t) = \text{Re}\left\{\sum_{n=0}^{N} I_{mn} e^{jn\omega_1 t}\right\} \tag{9}$$

where I_{mn} is the corresponding current phasor at the nth harmonic and mth node. Also $i'_m(t)$ is defined as the current flowing through the mth node into the linear sub-circuit with the associated current phasors I'_{mn}. Applying Kirchoff's Current Law to the M nodes of the circuit (excluding the ground node) yields

$$i_m(t) + i_m'(t) = 0, \qquad\qquad m = 0, 1, \cdots, M \tag{10}$$

In practice, due to the truncation of the Fourier Series representation of the currents, equation (10) will not hold and the solution requires that the currents be adjusted until the net current flowing out of all nodes is minimized. This is equivalent to the minimization of an objective function of the form

$$E = \sum_{m=1}^{M}\sum_{n=0}^{N} |I_{mn} + I_{mn}'|^2 \tag{11}$$

in which the unknowns of the least-square minimization are the interface voltage phasors V_{mn}. Both I_{mn} and I_{mn}' are functions of V_{mn}. I_{mn}' is related to V_{mn} by the admittance matrix of the linear sub-circuit. Therefore,

$$[I_{mn}']_{M\times 1} = [Y(n\omega_1)]_{M\times M} \cdot [V_{mn}]_{M\times 1} \qquad n = 0, 1, \cdots, N \tag{12}$$

where the first term on the right-hand side denotes the admittance matrix calculated at the nth harmonic. The phasor I_{mn} is a nonlinear function of V_{mn} and it is necessary to transform between the time and frequency domains using the Fourier Transform. A flow diagram for a Harmonic Balance algorithm is shown in Fig. 3.

A difficulty encountered with the harmonic balance method results from the Fourier transformation required by the algorithm. The Nyquist Sampling Theorem states that in order to recover all of the harmonics of a band-limited waveform such as $i_m(t)$ from its samples (i_{mk}), the sampling frequency $f_s = 1/T_s$ must be greater than or equal to twice the bandwidth of the original waveform. For the harmonic balance method, this translates into the condition that the number of time samples per cycle must be at least $(2N+1)$ where N is the highest harmonic number present in the circuit. Therefore,

$$N_s \geq (2N + 1) \tag{13}$$

This restriction is severe in the case of intermodulation or mixing analysis where the signal bandwidths are several orders of magnitude larger than the frequency separation generally of interest.

The most significant fraction of the simulation time for the harmonic balance method is used in the error minimization. Various types of root finding algorithms have been investigated in an effort to reduce execution time, and significant progress has been achieved. Simulation time for highly nonlinear circuits, however, can still be unacceptably long. Time delays, as are often encountered in microwave devices, introduce significant complexity into Harmonic Balance algorithms, and when these delays are significant the technique often fails to converge.

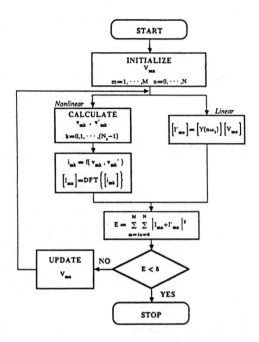

Figure 3   Harmonic Balance Method Algorithm

Although much progress has been achieved in harmonic balance simulation procedures, it is still difficult to employ the method using highly nonlinear device models. Also, it is difficult to use numerically based physical device models in these simulators. The long execution time normally required by the device model, coupled with the harmonic balance simulation time can lead to unsatisfactory performance. For this reason, physically based models are not often used in frequency domain simulators. This has resulted in the wide application of equivalent circuit models in generally available microwave simulators, especially those with large-signal, nonlinear device models.

LARGE-SIGNAL EQUIVALENT CIRCUIT MODELS

Historically, equivalent circuit modeling techniques have been widely utilized by device and circuit engineers. The equivalent circuit approach has proved to be effective and has permitted complex physical phenomena associated with electronic device operation to be simplified so that engineering descriptions useful in practical applications are available. The technique permits device description in terms of a network of circuit elements that can be related to physical phenomena by means of simplified mathematical expressions. The network topology and the circuit elements are selected so that the terminal RF performance of the equivalent circuit accurately simulates that measured for the actual device. Such techniques generally are only valid over a specified range of operating conditions and frequency.

The technique is implemented for a given transistor by expanding the two-port equations representing the RF operation of the device about a dc operating point. The resulting equations are then modeled as *tee* or *pi* equivalent circuits. Equivalent circuit topology can take on various configurations, but is generally selected so that the network reflects the physical operation of the device. In this manner, the equivalent circuit for a transistor can be either a *tee* or *pi* circuit since either will accurately simulate the

176

performance defined by the network equations. In practice, however, the *tee* and *pi* circuits are commonly selected for MESFET's and bipolar transistors, respectively. The input and output ports of the bipolar transistor consist of *pn* junction diodes which are accurately modeled as parallel *RC* circuits, which are easily implemented in the *pi* configuration. The output port of the MESFET is accurately modeled as a parallel *RC* network, but the input is modeled as a series *RC* network. The MESFET is, therefore, more suitable for the *tee* circuit configuration. Once the basic circuit topology has been determined the element values are defined either from device physics or by comparison to experimental measurements. The former technique is commonly employed in device design and operational investigations. The latter procedure dominates in practice and is commonly used in defining the equivalent circuit for circuit design applications as well as device optimization studies. Definition of the equivalent circuit element values is termed *parameter extraction* and is accomplished by determining element values that result in network performance in agreement with measured data. Two basic procedures are used. The first involves adjustment of the element values by means of a mathematical optimization routine until RF terminal characteristics (generally small-signal S-parameters) calculated from the network model are in agreement with the measured data [1,2]. This procedure, although, robust and accurate, can be time intensive and subject to uncertainties due to inherent difficulties in defining unique equivalent circuit element values from terminal measurements [3]. The technique can be effective and is used in many commercially available software packages. The second approach uses direct de-embedding of the circuit without the need to use optimization routines [4]. This approach is computationally faster, but less accurate due to uncertainties in determining embedding impedances.

The basic equivalent circuit model for the microwave MESFET is shown in Fig. 4. This circuit contains elements that describe the fundamental, linear operation of the device and the model reproduces the small-signal RF terminal characteristics of the device with excellent accuracy [5]. For many applications the equivalent circuit is often simplified to the form shown in Fig. 5. This circuit has also proven to be very useful for characterizing the linear, small-signal RF operation of the device. The circuit is widely used to extrapolate device performance (e.g., network S-parameters) to frequencies for which experimental data are not available. The main difference between the circuits in Fig. 5 and Fig. 4 is that several of the resistances and the domain capacitance known to be of importance to device operation are not included due to increased model complexity and difficulty in determining element values. The circuit in Fig. 5 is widely used for characterization purposes where the element values are determined by parameter extraction. The simplified circuit is desirable due to the difficulty in determining element values for a complex equivalent circuit from terminal measurements.

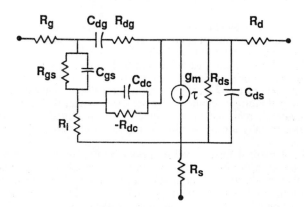

Figure 4   Basic Small-Signal MESFET Equivalent Circuit Model

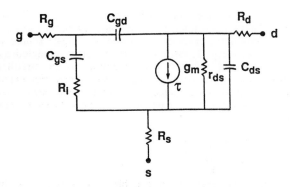

Figure 5   Simplified Small-Signal MESFET Equivalent Circuit Model

The equivalent circuit technique has proved very useful for a variety of device and circuit design applications. As large-signal applications become of interest the natural tendency is to extend the equivalent circuit technique to nonlinear operation. This approach has been extensively pursued and, while good results have been obtained in some applications, only limited success has been achieved in others. For example, large-signal equivalent circuit models have been applied to microwave power amplifier design and analysis with good results, especially for single frequency operation. However, under multi-frequency operation the equivalent circuit models do not yield results in good agreement with experimental data. Problems result from the simplifications employed in the model development, especially the techniques used to characterize the nonlinear elements.

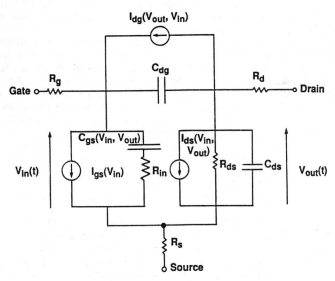

Figure 6   Large-Signal MESFET Equivalent Circuit Model

A large-signal, nonlinear equivalent circuit model often employed can be derived from the circuit shown in Fig. 5 by allowing certain elements to be nonlinear functions of applied bias. The resulting equivalent circuit is shown in Fig. 6. The main non linearity is the drain-source current generator, I_{ds}. The nonlinear current sources I_{gs} and I_{dg} are used to represent forward and reverse conduction of the gate-source and gate-drain Schottky diodes, respectively. These elements are often modeled as simple nonlinear switches. The remaining resistors and capacitors in the circuit are also nonlinear elements and can be modeled using simple expressions. For example, the capacitors can be modeled using the inverse square root voltage dependence expression that results from the Schottky diode mode. For example, at low voltages the gate-source capacitance can be expressed as

$$C_{gs} = \frac{C_{gs0}}{\sqrt{1 - \dfrac{V}{V_{bi}}}} \tag{14}$$

where C_{gs0} is the gate-source capacitance at zero gate bias and V_{bi} is the built-in voltage of the junction. The gate-source and gate-drain capacitances are easily modeled in this manner, with the capacitances assumed to saturate to a constant value at high gate bias. The other elements are often assumed to be constants.

Many nonlinear models have been presented. All of these models produce dc and RF performance that is in qualitative agreement with experimental data. Significant differences occur, however, in the quantitative performance of the models, both in comparison to each other and in comparison with experimental data. The primary difference in the various models is the expression used to characterize the drain current generator. Some of the most significant formulations [6-13] are listed in Table 1. The different expressions are all based upon the basic Shockley model [6], suitably modified to improve accuracy. Inclusion of the tanh function [8-10] is the generally accepted technique used to extend operation to the saturation region. In practice the various coefficients are used to calibrate the model to experimental data, usually dc I-V characteristics for a device. The use of more coefficients allows better agreement between the simulated and measured dc I-V characteristics, but the techniques required to define the coefficients become more complicated.

The main advantage of the equivalent circuit models is the ease with which they can be integrated into RF circuit simulators. For linear operation (i.e., small-signal operation) the interface is direct since the entire device and circuit model are simulated in the frequency domain. For nonlinear applications the device models are formulated in the time domain and are interfaced with the frequency domain linear circuit simulators by means of the harmonic balance method. The RF performance obtained from these simulators can be satisfactory to good for a well defined circuit, especially for mildly nonlinear applications such as class A power amplifiers not operating in hard saturation. The large-signal equivalent circuit models generally do not scale well with varying operational conditions such as frequency or bias. As the circuit becomes increasingly nonlinear, simulator performance degraded.

The main disadvantage of the equivalent circuit models is inherent inaccuracy resulting from simplifications in the model formulation, such as neglect of domain capacitance and the interdependencies of the nonlinear elements. In the actual device all of the nonlinear elements are interdependent. For example, in a MESFET it is not possible to change the device transconductance without also changing elements such as the gate-source capacitance. Such interdependence is generally ignored in the models, entering only when attempting to use such models for yield analysis. In such applications correlation matrices between the various elements are established, based upon experimental characterization procedures. Perhaps the most significant limitation of the equivalent circuit models, however, is the need to experimentally characterize every device that is to be used. The devices must be designed, fabricated, and characterized before computer-aided design models can be defined. Although useful data bases of commonly used devices can be established, each device type must be individually characterized. A simple change in any design parameter (such as gate length, width, or channel impurity concentration) requires a complete re-characterization since scaling techniques are difficult to apply. This limits the designer's flexibility in obtaining optimum performance from the device in integrated circuit applications, where tailoring the device design for special applications would be desirable.

Table 1   Large-Signal Drain Current Generator Models for MESFETs

Shockley (1952) [6]

$$I_{ds} = I_p \left[1 - \frac{(V_{gs} - V_{bi})}{V_p} \right]^n = \beta (V_{gs} - V_T)^2 \tag{15}$$

SPICE2 - Shichman and Hodges (1968) [7]

Linear Region: $I_{ds} = \beta V_{ds} [2(V_{gs} - V_p) - V_{ds}](1 + \lambda V_{ds})$ $\tag{16}$

Saturation Region: $I_{ds} = \beta (V_{gs} - V_p)^2 (1 + \lambda V_{ds})$ $\tag{17}$

Van Tuyl and Liechti (1974) [8] / Taki (1978) [9] / Curtice (1980) [10]

$$I_{ds} = \beta (V_{gs} - V_p)^2 (1 + \lambda V_{ds}) \tanh(\alpha V_{ds}) \tag{18}$$

Time delay can be included [10]: $I_{ds} = f[V_{gs}(t - \tau), V_{ds}]$ $\tag{19}$

Materka and Kacprzak (1985) [11]

$$I_{ds} = I_{dss} \left(1 - \frac{V_g}{V_p} \right)^2 \tanh \left(\frac{\alpha V_d}{V_g - V_p} \right) \tag{20}$$

$$V_p = V_{p0} + \gamma V_d \tag{21}$$

Curtice and Ettenberg (1985) [12]

$$I_{ds} = (A_0 + A_1 V_1 + A_2 V_1^2 + A_3 V_1^3) \tanh(\alpha V_{out}) \tag{22}$$

$$V_1 = V_{in}(t - \tau)[1 + \beta(V_{ds0} - V_{out})] \tag{23}$$

Statz et. al. (1987) [13]

$$I_{ds} = \frac{\beta (V_{gs} - V_T)^2}{1 + b(V_{gs} - V_T)} (1 + \lambda V_{ds}) \tanh(\alpha V_{ds}) \tag{24}$$

PHYSICAL LARGE-SIGNAL MESFET MODELS

Physical large-signal MESFET models are derived from solutions to a set of the basic semiconductor device equations previously presented, subject to appropriate boundary conditions determined by device geometry, channel doping details, and applied terminal potentials. The main difficulty in applying physical device models to harmonic balance microwave computer-aided design simulators is the trade-off between model accuracy and the large execution times generally required. Most of the physical models that have been reported solve the semiconductor device equations using some form of numerical technique such as finite differences or finite elements. The models can be formulated with varying complexity, depending upon the operational detail included. For example, the drift-diffusion approximation, where the charge carriers are assumed to be in thermal equilibrium with the crystal lattice is often used, and produces good results for MESFET's with gate dimensions in the range of 0.5-1 μm and operating to about X-band (i.e., 8-12 GHz). This type of model has proved very valuable in illuminating operational physical phenomena. More advanced formulations include energy-momentum relaxation phenomena to account for non equilibrium effects that become significant for submicron gate dimensions and very high frequency operation. In these formulations the free electrons are not in thermal equilibrium with the crystal lattice over at least a portion of the conducting channel. Non equilibrium effects tend to be most important in semiconductor materials such as GaAs at low bias where the majority of the carriers are in the low effective mass conduction band central valley (i.e., the Γ valley). As bias or RF terminal voltage is increased the carriers transfer to high conduction band valleys (i.e., the X or L valleys) where higher effective mass tends to dampen the hot electron behavior. As a result, hot electron phenomena is reduced for large-signal and power devices operated in saturation where the terminal voltages have large magnitudes.

Quantum effects can be included in the models for devices such as *High Electron Mobility Transistors (HEMT's)* by including Schroedinger's equation in the set of semiconductor equations. Solutions of this type tend to be complex and require extensive computer execution time to obtain solutions. The models have not yet been developed to the point where they can be efficiently used to simulate the RF performance of a device. In general, non equilibrium phenomena can be simulated using hydrodynamic [14], numerical [15], or Monte Carlo [16] solution techniques. These models are very useful for investigating in detail the physical operation of the device. However, simulation time increases rapidly with model complexity and practical operation of the models is usually limited to dc solutions. Large-signal operation, such as transient switching performance, is usually investigated by performing a series of dc simulations for varying electrode potential. There have been a few attempts at using the numerical formulations to investigate the RF operation of a device [17,18]. Conceptually, there is no fundamental reason that these models could not be applied to RF simulations, given sufficient computer resources. Such simulations are, however, beyond the current state-of-the-art.

Solution of the semiconductor device equations completely by analytic methods is possible and provides an engineering compromise between model accuracy and execution time. In this approach the device structure is simplified so that only the active region in the neighborhood of the gate electrode is explicitly modeled. The remainder of the device structure is modeled as a network of impedances that interface with the active portion of the device. This approach yields a large-signal, physical MESFET model that can be efficiently used in harmonic balance simulators. Since the model is based upon solutions of the semiconductor device equations it retains the advantages of the more elaborate physical models. However, the analytic formulation permits very efficient computation so that the model can be operated in harmonic balance simulators with the same order of magnitude execution time obtained with the large-signal equivalent circuit models. However, since device nonlinearities are inherent in the semiconductor device equations, it is not necessary to make *a priori* assumptions as to the form or identity of model nonlinearities. The analytic model is well suited for both device and RF circuit optimization investigations.

ANALYTIC LARGE-SIGNAL PHYSICAL MESFET MODELS

The first significant analytic model for the field-effect transistor was presented by Shockley [6]. In this work Shockley presented a simple dc and ac incremental analysis of the *Junction Field-Effect Transistor (JFET)*. The model was based upon three major assumptions: constant mobility in the material, a one-dimensional electric field in the conducting channel (i.e., the *Gradual Channel Approximation (GCA)*), and an abrupt transition between the depletion region and the conducting channel. The model was only applicable for devices that had a long gate length relative to the conducting channel depth, but demonstrated qualitatively the correct *I-V* characteristics for the device. The magnitude of the drain current that was calculated with this model, however, was significantly larger than measured in experimental devices. The inclusion of velocity saturation, as shown in Fig. 1, was demonstrated by Dacey and Ross [19] to improve the quantitative accuracy of the model and much improved agreement between measured and modeled performance was obtained.

Many models that offer improvements over the basic Shockley model have been presented for both MESFET's and HEMT's. These models generally are dc and ac small-signal formulations that include effects such as velocity saturation and non uniform channel doping details. These models are usually one dimensional or quasi two-dimensional and are applicable over a limited range of gate-length to channel depth ratio. Large-signal models that attempt to solve the device equations with a minimum number of simplifying assumptions have been presented. In order to obtain more accurate *I-V* results in the saturation region Yamaguchi and Kodera [20] introduced the concept of *velocity vector rotation* in which the electric field in the channel under the gate electrode becomes increasingly more horizontally directed (i.e., in the direction of current flow) in the direction of the drain electrode. This effect has been confirmed in many detailed numerical simulations and is the basis for the quasi two-dimensional models proposed. Yamaguchi and Kodera also removed the abrupt depletion approximation and allowed for a linear decrease as a function of distance in the channel carrier density. This phenomenon has been confirmed in many numerical simulations. Consideration of the effect allows a self-consistent solution for the potential in the saturation region. Using the approach of Yamaguchi and Kodera, Madjar and Rosenbaum [21] demonstrated that the introduction of excess channel charge at the drain electrode allows for a continuous potential solution. These models demonstrated that deviation from channel space-charge neutrality is fundamental to the operation of the field-effect transistor. This point was clearly demonstrated in the numerical model presented by Kennedy and O'Brien [22]. Using this model they provided a simple argument for the formation of a charge domain dipole in the conducting channel of the device whenever velocity saturation occurs. Since all modern field-effect transistors (including HEMT's) operate in velocity saturation the formation of the charge dipole domain is fundamental to their operation. This fact has been subsequently verified in many numerical simulations. The charge dipole results in the domain capacitance (e.g., C_{dc} in Fig. 4). An analytic formulation of the charge dipole domain has been presented by Khatibzadeh and Trew [23,24]. This formulation has, in turn, resulted in the derivation of a self-consistent large-signal analytic MESFET model that can easily be implemented in microwave harmonic balance simulators. This model will be explained in detail in the next section.

An Analytic Large-Signal MESFET Model

The derivation of an analytic, large-signal MESFET model [23,24] requires several simplifying assumptions due to difficulties imposed by the MESFET boundary conditions. The primary simplifications consist of restricting the solution to the channel region under the gate and assuming equilibrium conditions. The model is quasi two-dimensional in that a two-dimensional Poisson solution is obtained for the channel potential, while the current flow is assumed one-dimensional. Detailed numerical simulations indicate that this assumption is justified for modern MESFET's. The model is quasi-static and does not account for non equilibrium effects such as velocity overshoot, etc.

Model Formulation.   The region of the MESFET that is modeled consists of the region under the gate electrode as shown in Fig. 7. The model is formulated in this region. All other areas of the device are modeled phenomenologically using external linear elements. The semiconductor device equations listed in (1)-(7) are applied to the device and written in the form

$$\nabla^2 \Psi = -\frac{q}{\varepsilon}[N(y) - n(x,y)] \tag{25}$$

$$\vec{J} = -qn\vec{v} + qD_n \nabla n \tag{26}$$

$$\nabla \cdot \vec{J} = q\frac{\partial n}{\partial t} \qquad (27)$$

and

$$\vec{J}_t = \vec{J} + \varepsilon\frac{\partial \vec{E}}{\partial t} \qquad (28)$$

where

$$\vec{E} = -\nabla\Psi \qquad (29)$$

is the electric field, Ψ is the electrostatic potential, N is the arbitrary donor density in the channel, n is the free electron density, v is the electron velocity, D_n is the diffusion coefficient, J is the total conduction current density, J_t is the total current density (conduction and displacement). It is assumed that v and E are in the same direction. The model permits the use of an arbitrary electron velocity-field characteristic, but for example, the saturating v-E characteristic shown in Fig. 8 will be used.

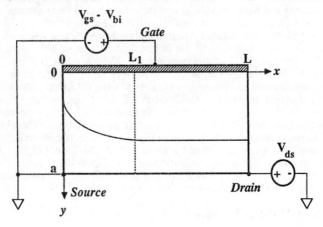

Figure 7   Active Region of a MESFET

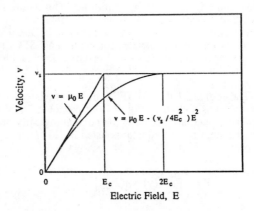

Figure 8   Simplified Electron Velocity versus Electric Field Characteristic

Based upon the magnitude of the electric field in the conducting channel of the MESFET, specifically at the $y=a$ boundary (i.e., $E(x,a)$), the device can operate in three different modes. These modes can be represented as

Mode A $\qquad E(0,a) < E(L,a) < E_c$ $\hfill$ (30)

Mode B $\qquad E(0,a) < E(L_1,a) = E_c < E(L,a)$ $\hfill$ (31)

Mode C $\qquad E_c < E(0,a) < E(L,a)$ $\hfill$ (32)

The plane $x=L_1$ indicated in Fig. 7 separates the saturation and linear regions of the device. The source and drain currents are calculated by integrating J over the entire $x=0$ and $x=L$ boundary planes, respectively. The gate contact $(y=0)$ is modeled as an ideal constant potential boundary, while the substrate interface $(y=a)$ is modeled as an ideal reflecting boundary. Mode A indicates that the device is operating with the channel electric field below the critical field E_c necessary to produce saturation. For Mode B operation the critical field for saturation occurs in the gate region so that the channel is operating partly linear and partly in saturation. For Mode C the entire channel region under the gate is in velocity saturation..

Since equations (25)-(27) are coupled together a functional form for $n(x,y)$ is assumed *a priori* in order to solve for Ψ analytically. In general, n must vary continuously from approximately zero near the gate $(y\sim0)$ to a value almost equal to the background doping density in the channel $(y\sim a)$. In order to produce a smooth transition for an arbitrary doping profile, the following functional form for $n(xy)$ is assumed:

$$n(x,y) = \left[1 + \gamma(x - L_1)\right]T(d(x),y)N(y) \tag{33}$$

where $T(d(x),y)$ is a transition function and is defined as

$$T(d(x),y) \equiv 1 - \frac{1}{1 + \exp\left[\dfrac{y - d(x)}{\lambda}\right]} \tag{34}$$

The term $[1+\gamma(x-L_1)]$ allows for charge accumulation or depletion in the channel. The excess charge in the channel allows for a self-consistent solution of the Poisson equation in the saturation region (i.e., x between L_1 and L). In the linear region (i.e., x less than L_1) the term $\gamma=0$. Whether accumulation or depletion occurs in the channel will depend upon the sign of γ, which is determined from the solution of Ψ in the saturation region.

This formulation for free electron density in the conducting channel ignores out-diffusion of carriers from the channel region into the substrate. A transition region can be defined as the region where n varies from 5 percent to 95 percent of the impurity doping N. From (33) and (34) it follows that this corresponds to the region between $d-3\lambda$ and $d+3\lambda$ so that the width of the transition region is 6λ. Numerical simulations indicate that this region usually extends to about $6\lambda_D$, where λ_D is the Debye length. It follows that λ is on the order of the Debye length.

Solution for the Potential Ψ. The solution to equation (25) can be represented by a linear superposition of two components $\Psi = \Psi_0 + \Psi_1$ where Ψ_0 is the Laplacian potential due to the impressed voltages on the electrodes and Ψ_1 is due to the space charge in the channel and satisfies Poisson's equation. In mathematical terms

$$\nabla^2 \Psi_0 = 0 \tag{35}$$

with boundary conditions

$$\Psi_0(0,a) = 0 \tag{36a}$$

$$\Psi_0(L,a) = V_0 \tag{36b}$$

$$\frac{\partial \Psi_0}{\partial y}(x,a) = 0 \tag{36c}$$

$$\Psi_0(x,0) = 0 \tag{36d}$$

and

$$\nabla^2 \Psi_1 = -\frac{q}{\varepsilon}(N-n) \tag{37}$$

with boundary conditions

$$\Psi_1(0,a) = 0 \tag{38a}$$

$$\Psi_1(L,a) = V_1 \tag{38b}$$

$$\frac{\partial \Psi_1}{\partial y}(x,a) = 0 \tag{38c}$$

$$\Psi_1(x,0) = V_{gs} - V_{bi} \tag{38d}$$

where V_{bi} is the built-in voltage of the gate Schottky contact, and V_{gs} and $V_{ds}=V_1+V_0$ are the applied gate-source and drain-source voltages across the gate region of the MESFET, respectively. The solution to (35) with the stated boundary conditions is

$$\Psi_0(x,y) = \sum_{j=0}^{\infty} A_j \sinh\left[\frac{(2j+1)\pi}{2a}x\right]\sin\left[\frac{(2j+1)\pi}{2a}y\right] \tag{39}$$

where

$$A_j = \frac{4V_0}{(2j+1)\pi\sinh\left[\dfrac{(2j+1)\pi L}{2a}\right]} \qquad j = 0,1,\cdots \tag{40}$$

For typical microwave MESFET's $L/a>1$ and A_j converges rapidly so that only the first term need be considered. It follows, however, that Ψ_0 will no longer satisfy boundary condition (36b). In order to satisfy the boundary condition, (36b) is reapplied to the first term in (39) to get

$$\Psi_0(x,y) = \frac{V_0}{\sinh\left(\dfrac{\pi L}{2a}\right)}\sinh\left(\frac{\pi x}{2a}\right)\sin\left(\frac{\pi y}{2a}\right) \tag{41}$$

The solution for Ψ_1 depends upon the carrier concentration n. Using definitions (33) and (34), Ψ_1 is determined assuming

$$\left|\frac{\partial^2 \Psi_1}{\partial y^2}\right| >> \left|\frac{\partial^2 \Psi_1}{\partial x^2}\right| \tag{42}$$

For $0 \leq x \leq L_1$ the solution is

$$\Psi_1(x,y) = -\frac{q}{\varepsilon}\int_y^a\int_{y'}^a [1 - T(d(x),y'')]N(y'')dy''dy' + \frac{V_1}{L}x \tag{43}$$

For $L_1 < x \leq L(d(x) \equiv d_1)$, the solution is

$$\Psi_1(x,y) = -\frac{q}{\varepsilon}\int_y^a\int_{y'}^a [1 - T(d_1,y'')]N(y'')dy''dy' + \frac{V_1}{L}x$$

$$+ \frac{q}{\varepsilon}\gamma(x - L_1)\int_y^a\int_{y'}^a T(d_1,y'')N(y'')dy''dy' \tag{44}$$

The last term in (44) is the potential due to the accumulation or depletion charge. For $x \leq L_1$ the depletion depth d varies as a function of x. To solve for $d(x)$, apply boundary condition (38d) to (43) to get for the region $x \leq L_1$

$$-\frac{q}{\varepsilon}F_1(d(x)) + \frac{V_1}{L}x = V_{gs} - V_{bi} \tag{45}$$

where

$$F_1(d) \equiv \int_0^a\int_{y'}^a [1 - T(d,y'')]N(y'')dy''dy' \tag{46}$$

For the case of uniform donor density (i.e., $N(y) \cong N_d$) and using the abrupt depletion approximation (i.e., $\lambda \to 0$), $F_1(d)$ reduces to $N_d d^2$ and $d(x)$ can be solved to give the result

$$d(x) = \sqrt{\left(\frac{2\varepsilon}{qN_d}\right)\left(\frac{V_1}{L}x - V_{gs} + V_{bi}\right)} \tag{47}$$

In general, however, $F_1(d)$ depends upon the channel donor doping profile and cannot be calculated analytically. For a specified donor profile it can be tabulated numerically as a function of d in the interval $0 \leq y \leq 2a$. For a given value of V_{gs} and x, the corresponding value of $F_1(d)$ can be calculated and $d(x)$ determined by interpolation. The procedure is computationally efficient and accurate.

The pinch-off voltage V_{p0} is defined as the gate-source voltage for which the channel is completely pinched off at the source side. The pinch-off voltage can be determined from the expression

$$V_{p0} = V_{bi} - \frac{q}{\varepsilon} \int_0^a \int_{y'}^a N(y'') dy'' dy' \tag{48}$$

For $L_1 \le x \le L$, $d(x) \equiv d_1$, where d_1 is found by substituting $x=L_1$ in equation (45). Applying conditions (38d) to (44) yields for the region $L_1 \le x \le L$

$$-\frac{q}{\varepsilon} F_1(d_1) + \frac{V_1}{L} x + \frac{q}{\varepsilon} \gamma(x - L_1) F_2(d_1) = V_{gs} - V_{bi} \tag{49}$$

where

$$F_2(d) \equiv \int_0^a \int_{y'}^a T(d, y'') N(y'') dy'' dy' \tag{50}$$

From the definitions of V_{p0} and $F_1(d)$ it follows that

$$F_2(d) = -\frac{\varepsilon}{q}(V_{p0} - V_{bi}) - F_1(d) \tag{51}$$

The additional integration is not required for the calculation of $F_2(d)$. Equation (49) must hold for all $L_1 \le x \le L$, therefore

$$\gamma = \frac{-\varepsilon V_1}{qL F_2(d_1)} \tag{52}$$

Equation (52) gives the solution for the slope of the carrier concentration in the saturation region. For $\gamma < 0$ $(V_1 > 0)$, charge depletion in the channel occurs, whereas for $\gamma > 0$ $(V_1 < 0)$ accumulation results. The solution for V_1 will arise from the current continuity equation. The effect of the extra charge in the channel is to enable Ψ to satisfy the potential boundary condition (38d) while the depletion layer width is held constant in the saturation region.

Drain Current. In order to solve for the current, the electric field distribution in the device is calculated. The electric field has two components: one due to Ψ_0 and the other due to Ψ_1. The x component of the electric field in both the linear and saturation regions can be approximated by

$$E_x(x,y) = -\frac{V_1}{L} \left[1 - \left(\frac{y}{a} - 1 \right)^2 \right] - \xi \cosh\left(\frac{\pi x}{2a} \right) \sin\left(\frac{\pi y}{2a} \right) \frac{V_0}{L} \tag{53}$$

where

$$\xi \equiv \frac{\dfrac{\pi L}{2a}}{\sinh\left(\dfrac{\pi L}{2a} \right)} \tag{54}$$

and the second-order Taylor's expansion of the original function about $y=a$ is utilized. At the substrate interface (i.e., $y=a$) the magnitude of the electric field becomes

$$E(x,a) = |E_x(x,a)| = \frac{V_1}{L} + \xi \frac{V_0}{L} \cosh\left(\frac{\pi x}{2a}\right) \tag{55}$$

By definition of the plane $x=L_1$, $E(L_1,a) = E_c$. Combining this with (53) and solving for L_1 yields

$$L_1 = \left(\frac{2a}{\pi}\right) \ell n \left[\eta + \sqrt{\eta^2 - 1}\right] \tag{56}$$

where

$$\eta \equiv \left(\frac{2a}{\pi L}\right)\left(\frac{E_c L - V_1}{V_0}\right) \sinh\left(\frac{\pi L}{2a}\right) \tag{57}$$

Therefore, with a specified V_{ds} and V_1, L_1 can be explicitly determined.

The conduction current is determined as a function of potential by applying the Ψ solution to a specified location in the conducting channel (usually at the plane $y=a$). It follows that the source and drain conduction currents can be written in the form

$$I_s = AV_1 + B \tag{58}$$

and

$$I_d = A'V_1 + B' \tag{59}$$

where A, A', B, B' are nonlinear functions of V_1 and the device parameters. The V_1 term is the solution to the Poisson's equation evaluated at the drain side of the gate region. In the model it is assumed to be a variable and is calculated from current continuity requirements. Quasi-static conditions are assumed so that

$$\frac{\partial n}{\partial t} \to 0 \tag{60}$$

and neglecting the current through the gate contact the continuity equation yields the condition

$$I_s = I_d \tag{61}$$

Once a solution for V_1 is determined, all parameters of the model are calculated. The gate and drain displacement currents are related to the time derivatives of V_{gs} and V_{ds} by means of the capacitance matrix

$$\begin{bmatrix} i_g \\ i_d \end{bmatrix} = \begin{bmatrix} C_{11} & C_{12} \\ C_{21} & C_{22} \end{bmatrix} \cdot \frac{\partial}{\partial t} \begin{bmatrix} V_{gs} \\ V_{ds} \end{bmatrix} \tag{62}$$

where the capacitance matrix is, in general, non symmetric.

Gate Conduction. It has been observed that saturation in MESFETs depends upon a complex interaction between the dc bias, the RF operating voltage, and the gate-drain breakdown voltage. For example, Fig. 9 shows the dynamic v-i locus for one RF cycle at the drain terminal superimposed upon the dc I-V characteristics for a GaAs MESFET with a buried channel donor impurity profile. The device has a gate length of $L_g=0.5$ μm and a gate width of $W=0.8$ mm. For illustration the device is operated in a 50 Ω RF circuit and is biased for class A-B operation at a drain bias of $V_{ds}=7$ v. The three dynamic loci represent linear operation and compression levels of 1 and 3 db. Note that for linear operation the RF voltage swing is about 1 v and basically follows the dc load line. As the device is driven into saturation by 1 db the $v\text{-}i$ locus grows and the RF voltage swing obtains a magnitude of about 10 v. This swing is sufficient for the RF voltage to exceed the gate-drain breakdown voltage. Waveform clipping occurs, energy is transferred to harmonic frequencies, and saturation results. Note that the waveform clipping causes the average value of the waveform (i.e., the dc bias point) to shift. Since the drain is held at a fixed voltage by the bias source the current is shifted to increased drain current values to satisfy the average waveform requirements.

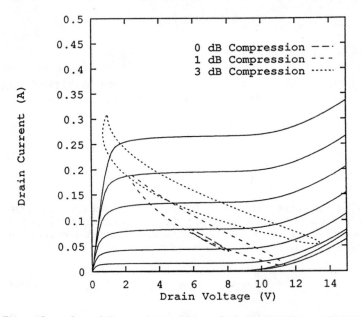

Figure 9   dc and Dynamic Load Lines for a Buried Channel MESFET

Increasing the drive to 3 db compression results in both forward and reverse conduction of the gate electrode. In this case waveform clipping occurs on both the forward and reverse RF voltage swing. Note the continued shift in the dc operating point. Also, note the shift in the dynamic load line (as represented by the slope of the dynamic $v\text{-}i$ locus) from the dc load line. The hysteresis in the dynamic loci when the device is operating in saturation is due to reactive energy that results from the impedance mismatch caused by the device impedance changing with RF drive.

In order to accurately calculate the large-signal RF operation of the MESFET it is necessary to include mechanisms for the forward conduction and reverse breakdown of the gate Schottky diode. These effects are generally modeled by shunt diodes connected between the gate-source and gate-drain terminals, respectively. This technique works well and is also used in the equivalent circuit models. Accurate formulations for these mechanisms are very important. Many of the non convergence problems encountered in harmonic balance simulators are due to difficulties caused by the method used to model the conduction characteristics of the shunt diodes.

A problem encountered in attempts to model the reverse breakdown of the gate Schottky diode relates to the physical mechanism responsible for the reverse conduction characteristics. Breakdown in MESFET's has never been well understood and most analyses assume standard avalanche ionization arguments are valid. Avalanche ionization, however, is not able to explain the detailed behavior of the measured breakdown data.

A breakdown model that is in good agreement with measured data can be formulated based upon a thermally assisted tunnel/avalanche argument [25]. The physical argument for the model follows. Under normal operating conditions breakdown occurs at the drain-side edge of the gate electrode. At this location the electric field is two dimensional with a large horizontal component. Under large drain voltage, but low gate bias (i.e., near Idss), the surface electric field can become sufficiently large that electrons tunnel from the gate metal onto the surface of the semiconductor in the gate-drain region. The tunneling mechanism is enhanced by elevated channel temperature at the gate electrode due to large dc power dissipation at this bias condition. An excess density of electrons accumulate on the surface of the semiconductor at the gate edge due to relaxation semiconductor effects and are free to flow to the drain contact, thereby providing a leakage current. The leakage current can become large and it is this current that is sometimes used to define breakdown using figures-of-merit such as *1 mA/mm*. Due to the energy band structure and increased scattering at the surface, conduction on the surface is generally reduced compared to that in the bulk. The effect can be modeled with increased electron effective mass. The surface resistance can be relatively high and the excess drain current is often indicated as 'soft' breakdown, although 'hard' breakdown and low surface resistance is sometimes also observed. The breakdown resistance is determined by the specific surface conduction mechanism.

As gate bias is increased towards pinch-off the channel current is reduced, resulting in a reduced channel temperature due to less power dissipation. The reduced temperature at the gate electrode produces an increased threshold voltage for the tunnel leakage and the surface current is reduced. The reduced gate leakage allows an increased drain bias to be applied to obtain a constant gate current (e.g., *1 mA/mm*), thereby providing an increased drain-source breakdown voltage as gate bias is increased towards pinch-off.

As the gate bias nears pinch-off the electric field under the gate electrode becomes more vertically oriented. The electric field can become sufficient for avalanche breakdown to occur. At this point, the breakdown can consist of a combination of tunneling and avalanche and the breakdown voltage may either increase or decrease with temperature. Both variations have been observed in experimental data. The onset of avalanche results in light emission from the edge of the gate electrode and the light intensity increases as the gate bias is increased to pinch-off. The light emission for this condition is due to electron-hole recombination.

Large-Signal Microwave MESFET Simulation

A block diagram for a comprehensive harmonic balance, large-signal MESFET integrated microwave simulator is shown in Fig. 10. The simulator allows the performance of a microwave integrated circuit to be determined as a function of material, process, and device design specifications. The simulator is useful for large-signal, non linear applications such as power amplifiers, oscillators, etc. The integrated simulator allows both MESFET design and embedding dc and RF circuits to be investigated. The physical MESFET model permits the dc and RF performance of a device or integrated circuit to be determined as a function of process and device design information and/or bias and RF operating conditions.

Verification of the Large-Signal Analytic MESFET Model

For verification purposes the analytic, large-signal MESFET model described in this chapter was used to simulate the dc and RF performance of a microwave power amplifier. A block diagram for the simulated circuit is shown in Fig. 11. The MESFET is embedded in a circuit that allows for arbitrary input and output impedances, and series and parallel feedback.

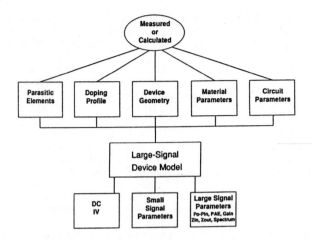

Figure 10   Block Diagram for an Integrated Microwave Simulator

A commercial GaAs MESFET designed for power applications was simulated. The MESFET has a gate length of $L_g=0.5 \, \mu m$ and a gate width of $W=1.125 \, mm$. The device is fabricated with an ion-implanted conducting channel as shown in Fig. 12. The channel impurity doping profile has a Gaussian shape due to the implant process. The free electron profile differs from the donor profile due to diffusion effects. That is, the free electrons will diffuse from regions of high concentration to low concentration. The process continues until the space charge field that is created is sufficient to prevent further diffusion. Under these conditions equilibrium is established. The differences in the donor impurity and free electron densities are important since the electron transport characteristics are a function of impurity density, but the current that flows is due to the free electrons. Accurate simulation requires that these effects be properly modeled.

The device was simulated in a microwave power amplifier circuit that was designed to operate class A at 10 GHz. The device was biased with a drain-source voltage of $V_{ds}=7 \, v$.

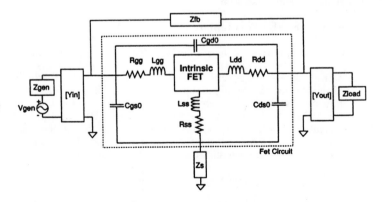

Figure 11   Block Diagram for the MESFET Power Amplifier

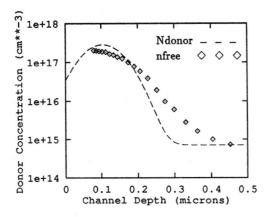

Figure 12   Donor Impurity and Free Electron Density Profiles for the MESFET

The RF performance for the class A amplifier at a frequency of 10 GHz is shown in Fig. 13. The RF output power, power-added efficiency, and gain are shown as a function of input RF power. The lines are the simulated data and the points are the measured data. The amplifier was operated over an input RF power drive range that was sufficient to cause saturation to occur. The amplifier produced a peak RF output power of about 27 dbm, a maximum power-added efficiency of about 34%, and a linear gain of 9.5 db. The dc gate and drain currents as a function of input RF drive power were also simulated and compared to the measured data. The results are shown in Figs. 14a and b. The gate current is observed to be zero at low RF input power drive, but is driven to negative values at an input power of about 10 dbm, and then to positive values at an input drive of about 19 dbm. This behavior indicates the main saturation mechanisms due to reverse and forward conduction of the gate electrode. As the gate is driven into reverse breakdown reverse conduction occurs and the dc gate current becomes negative. As input RF drive power is increased the gate is driven into forward conduction, and this process dominates at high RF power drive. The two mechanisms are in competition and each can occur every RF cycle. The saturation phenomena is also observed in the dc drain current versus input RF drive power as shown in Fig. 14b. The variation in the dc drain current reflects the competing gate conduction mechanisms.

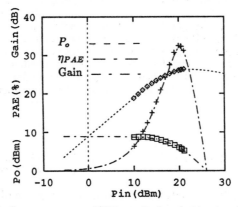

Figure 13   RF Performance versus RF Input Power for the Amplifier at 10 GHz

192

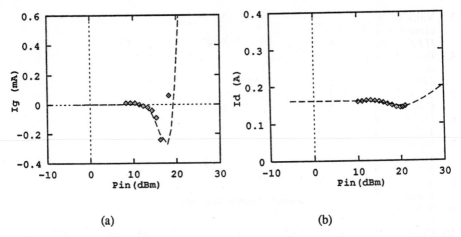

(a)   (b)

Figure 14   dc Rectified Current versus Input RF Power Drive for the Class A
             Amplifier at 10 GHz

             (a)   Gate Current

             (b)   Drain Current

SUMMARY

Large-signal device models are required for analysis of components such as power
amplifiers, oscillators, mixers, switches, etc. Large-signal conditions exist when the RF
or signal voltage is significant relative to the dc operating bias. Under these conditions
device operation is non linear and accurate models require descriptions of the
fundamental non-linearities. Appropriate models can be developed using either
equivalent circuit or physical modeling techniques. The former are generally empirically
based and element values are determined from measurement data by parameter
extraction. The models are widely used, are easy to implement in harmonic balance
microwave simulators, but demonstrate limited accuracy in certain applications. Physical
models, in contrast, are developed from the semiconductor device equations and contain,
therefore, an accurate description of the physical operation of the device. The models
tend to be complex and require significant computational resources to obtain solutions.
The models are, in general, difficult to apply in microwave circuit simulators. Physical
device models based upon analytic solution techniques provide a useful compromise
between model accuracy and simulation efficiency. The formulation of a suitable large-
signal analytic MESFET model has been presented. The model demonstrates good
accuracy compared with measured data for microwave power amplifiers.

REFERENCES

1. Bandler, J.W., Zhang, Q.J., and Chen. S.H., "Efficient large-signal FET parameter
    extraction using harmonics," *IEEE Trans. Microwave Theory Tech.*, 1989, **MTT-37**,
    pp. 2099-2108.
2. Bilbro, G.L., Steer, M.B., Trew, R.J., and Skaggs, "Extraction of the parameters of
    equivalent circuits of microwave transistors using tree-annealing," *IEEE Trans.
    Microwave Theory Tech.*, 1990, 33, pp. 1711-1718.

3. Vaitkus, R.L., "Uncertainty in the values of GaAs MESFET equivalent circut elements extracted from measured two-port scattering parameters," *Proc. IEEE/Cornell Conf. High Speed Semiconductors*, 1983, pp. 301-308.

4. Dambrine, G, Cappy, A., Heliodore, F., and Playez, E, "A new method for determining the FET small-signal equivalent circuit," *IEEE Trans. Microwave Theory Tech.*, 1988, **30**, pp. 1151-1159.

5. Trew, R.J., "Equivalent circuits for high frequency transistors," *Proc. IEEE/Cornell Conf. High Speed Semiconductors*, 1987, pp. 199-208.

6. Shockley, W., "A unipolar 'field-effect' transistor," *Proc. IRE*, 1952, **40**, pp. 1365-1376.

7. Shichman, H., and Hodges, D.A., "Modeling and simulation of insulated-gate field-effect transistor switching circuits," *IEEE J. Solid-State Circuits*, 1968, **SC-3**, pp. 285-289.

8. VanTuyl, R., Liechti, C.A., "Gallium-arsenide digital integrated circuits," *U.S. Air Force Tech. Rep. AFAL-TR-74-40*, 1974.

9. Taki, T., "Approximation of junction field-effect transistor characteristics by a hyperbolic function," *IEEE J. Solid-State Circuits*, 1978, **SC-13**, pp. 724-726.

10. Curtice, W.R., "A MESFET model for use in the design of GaAs integrated circuits," *IEEE Trans. Microwave Theory Tech.*, 1980, **MTT-28**, pp. 448-456.

11. Materka, A, and Kacprzak, T., "Computer calculation of large-signal GaAs FET amplifier characteristics," *IEEE Trans. Microwave Theory Tech.*, 1985, **MTT-33**, pp. 129-135.

12. Curtice, W.R., and Ettenberg, M., "A nonlinear GaAs FET model for use in the design of output circuits for power amplifiers," *IEEE Trans. Microwave Theory Tech.*, 1985, **MTT-33**, pp. 1383-1394.

13. Statz, H., Newman, P., Smith, I.W., Pucel, R.A., and Haus, H.A., "GaAs FET device and circuit simulation in SPICE," *IEEE Trans. Electron Dev.*, 1987, **ED-34**, pp. 160-169.

14. Shawki, T., Salmer, G., and El-Sayed, O., "MODFET 2-D hydrodynamic energy modeling: optimization of subquarter-micron-gate structures," *IEEE Trans. Electron Dev.*, 1990, **37**, pp. 21-30.

15. Snowden, C.M., and Loret, D., "Two-dimensional hot electron models for short gate length GaAs MESFETs," *IEEE Trans. Electron Dev.*, 1987, **ED-34**, pp. 212-223.

16. Awano, Y., Tomizawa, K., and Hashizumi, N., "Performance and principle of operation of GaAs ballistic FET," *IEEE IEDM Tech. Digest*, 1983, pp. 617-620.

17. Snowden, C.M., Howes, M.J., and Morgam, D.V., "Large-signal modeling of GaAs MESFET operation," *IEEE Trans. Electron Dev.*, 1983, **ED-30**, pp. 1817-1824.

18. Grubin, H.A., Kreskovsky, J.P., and Levy, R., "Modeling of large-signal device/circuit interactions," *Proc. IEEE/Cornell Conf. High Speed Semiconductors*, 1989, pp. 33-45.

19. Dacey, G.C., and Ross, I.M., "The field-effect transistor," *Bell Sys. Tech. J.*, 1955, **34**, pp. 1149-1189.

20. Yamaguchi, K, and Kodera, H., "Drain conductance of junction gate FETs in the hot electron range," *IEEE Trans. Electron Dev.*, 1976, **ED-23**, pp. 545-553.

21. Madjar, A., and Rosenbaum, F.J., "A large-signal model for the GaAs MESFET," *IEEE Trans. Microwave Theory Tech.*, 1981, **MTT-29**, pp. 781-788.

22. Kennedy, D.P., and O'Brien, R.R., "Computer aided two dimensional analysis of the junction field-effect transistor, *IBM J. Res. Devel.*, 1970, pp. 95-116.

23. Khatibzadeh, M.A., and Trew, R.J., "A large-signal, analytic model for the GaAs MESFET," *IEEE Trans. Microwave Theory Tech.*, 1988, **MTT-36**, pp. 231-238.

24. Khatibzadeh, M.A., Large-Signal Modeling of Gallium Arsenide Field-Effect Transistors, Ph.D. Thesis, 1987, North Carolina State University, Raleigh.

25. Trew, R.J., and Mishra, U.K., "Gate breakdown in MESFET's and HEMT's," *IEEE Electron Dev. Lett.*, 1991, **12**, pp. 524-526.

Noise Modelling

Alain Cappy
Université des Sciences et Technologies de Lille

INTRODUCTION

In physics and electrical engineering one often encounters fluctuating signals generated in electrical devices and circuits. These fluctuating signals are due to random events that modify the number and/or the velocity of carriers. So, at any point r of any device, the current density j fluctuates which induces a fluctuating voltage (or current) at the device electrodes. It should be noted that only these macroscopic current and voltage fluctuations can be measured. To illustrate this, figure (1) shows a DC biased device and AC signals are applied to some electrodes. Let us consider a small volume dV located at point r in the device. The fluctuations of the current density at point r induces voltage (or current density) fluctuations at any point r' and consequently at the external electrodes. So, the voltage fluctuation at electrode M, resulting from the current density fluctuation at r depends on two factors:

- the strength of the fluctuation at r (local or microscopic noise source)
- the device structure between r and M (impedance field).

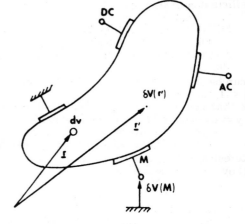

Fig (1) General view of
the noise modelling in
devices

So, in any noise modelling we have to know the microscopic noise processes involved and the device structure (material, doping...). The total noise voltage (or total noise current), due to all the microscopic noise sources located in the device, is obtained by integration over the whole device. This total noise voltage is referred to as the macroscopic noise source. When the macroscopic noise sources are known, the calculation of the noise performance (noise figure for an amplifier, noise to carrier ratio for an oscillator ...) can be carried out by simple circuit manipulations.

THE LANGUAGE OF NOISE

Random variables-Fourrier analysis - spectral intensity

In calculations about noise in electrical systems, one often must calculate the ensemble average of order m x^m of the random variable $x(t)$. x^m is calculated using the probability density function $f(x)$:

$$\overline{x^m} = \int x^m f(x)\,dx \qquad (1)$$

in noise study, the most important averages are $\overline{x}$ and $\overline{x^2}$. In addition, the noise processes encountered in physics are engodic. That means that ensemble average is equal to time average

$$\overline{x} = <x> = \lim_{T \to \infty} \int_0^T x(t)\,dt \qquad (2)$$

If we now consider two continuous random variables $x(t)$ and $y(t)$, one can define the averages in the same manner as for a single variable from the probability dP that x has a value between x and $x+dx$ and y has a value between y and $y + dy$. In practice, the most important average is $\overline{xy}$. If $\overline{xy} = 0$ the two variables are said uncorrelated and if $\overline{xy} \neq 0$ the two variables are correlated. The degree of correlation can be quantified by the correlation coefficient defined by :

$$C = \frac{\overline{xy}}{(\overline{x^2} \cdot \overline{y^2})^{1/2}} \qquad (3)$$

It can be shown that $|C| \leq 1$. If $|C| = 1$ the two variables are fully correlated. If $|C| < 1$, the two variables are partially correlated and we can split y into a part $a.x$ that is fully correlated with x and a part z that is uncorrelated with $x : y = ax + z$. a and z are given by $a = \overline{xy}/\overline{x^2}$ and $\overline{z^2} = \overline{y^2}(1 - |C|^2)$.

Since the frequency bandwidth of a device is not infinite, it is often interesting to introduce the spectral intensity of a continuous random variable $x(t)$ by :

$$\overline{x^2} = \int S_x(f)\,df \qquad (4)$$

From a physical point of view, $\overline{x^2}$ represents the power of the signal $x(t)$ for all frequencies while $S_x(f_0) \Delta f$ represents the power of $x(t)$ in a bandwidth Δf about f_0.

The noise sources

The current density is defined by $j = qnv$. In this expression, n and v represent average values that are obtained from electrokinetic and Poisson's equations. In fact, these quantities are fluctuating quantities. In practice, four different noise processes will be encountered :

(i) **Shot noise** : It occurs whenever a noise phenomenon can be considered as a series of independent events occuring at random. This noise is mainly encountered in vaccum tubes and junctions (PN, Schottky...). In semiconductor devices this noise is generally masked by diffusion noise. The spectral intensity of the fluctuating current I(t) of average value I is given by (Schottky's theorem)

$$S_i(f) = 2q \; \overline{I} \quad (5)$$

(ii) **Generation recombination noise** : Due to generation and recombination (band-to-band, deep level , shallow levels...), the number of carriers fluctuates. For a given G.R. process, the spectral intensity of the current fluctuations is given by :

$$S_I(f) = \frac{4a \; \overline{I^2}}{N} \; \frac{\tau}{1 + \omega^2 \tau^2} \quad (6)$$

where I is the average current, N the total number of carriers, a and τ are two typical constants of the G.R. process. In linear systems (not in oscillators), GR noise is only significant at low frequency, typically in the range 1 kHz-1 MHz.

(iii) **1/f or flicker noise** : Despite a large amount of work, the physical origin of 1/f noise is not very well known. This noise occurs in all physical or biological systems. In electronic devices it occurs in collision-free devices (vaccum tubes) as well as in collision dominated devices. The 1/f noise origin can be either the bulk or the device surface (An interesting review is given in Van der Ziel, 1988). The spectral intensity of flicker noise is given by :

$$S_I(f) = \frac{\overline{I^2}}{N} \; \frac{a_H}{f_\beta} \quad (9)$$

where a_h is the Hooge constant (Hooge 1976) and β is close to unity.

(iv) **Diffusion noise** : This noise is due to the random motion of carriers in conductors. At thermal equilibrium, this noise process is called thermal noise. In the high frequency range (f > 1 MHz), this noise process is predominant in most semiconductor devices. For a slice (section S(x), thickness Δx) of a current tube, the spectral intensity of the diffusion noise is expressed as :

$$S_I(f) = 4 q^2 \; S(x).n(x) \, D(x)/\Delta x \quad (8)$$

In this expression n(x) is the carrier density and D(x) the diffusivity. By definition, the diffusivity is proportional to the Fourrier transform of the velocity fluctuation autocorrelation function. At thermal equilibrium, the diffusivity to mobility ratio (DMR) obeys Einstein relation $D/\mu = kT/q$. So, equation (10) becomes :

$$S_I(f) = 4 kT S(x) \sigma(x)/\Delta x \quad (9)$$

where $\sigma(x)$ is the conductivity of the slice. Equation (11) is nothing but Nyquist law and it can be used to define the carrier noise temperature in the case of devices off equilibrium (J.P. Nougier, 1978). Generally the noise, as can be measured at a device terminals, is a superposition of all these noise processes. In practice, the frequency of operation allows us to consider the predominant processes : 1/f and G.R. noise at low frequency, diffusion and shot noise at higher frequency (the case of non linear systems, such oscillators, will be considered separately).

THE MACROSCOPIC NOISE SOURCE CALCULATION

In noise modelling, the main problem is to obtain the terminal noise in terms of macroscopic noise current or noise voltage sources. Two methods can be used : the Langevin method (LM) and the impedance field method (IFM). Although it has been shown that these two methods are strictly equivalent (K.K. Thornber, 1974) the latter is mostly used probably because the IFM is more suited for numerical analysis. For this reason, the IFM only will be described here.

The impedance field method

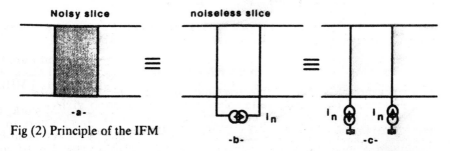

Fig (2) Principle of the IFM

For the sake of simplicity, a one dimension formulation will be first presented. Figure (2) shows a 1D noisy device. Let us assume that the slice $(x, x + \Delta x)$ is noisy whereas the remainder of the device is not. This noisy slice can be transformed into a noiseless slice bridged by a noise current source i_n (figure 2). In a second step, the noise current source is transformed in two current sources : i_n is introduced between $x + \Delta x$ and the ground while i_n is removed between x and ground. Assuming that $i_n = i_0 \exp (j\omega t)$, the voltage fluctuation $\delta V(L)$ at abscissa $x = L$ is expressed by :

$$\delta V^+(L) = Z (x + \Delta x, 0, \omega) i_0 \quad (10)$$

where $Z (x + \Delta x, 0, \omega)$ is the impedance between 0 and $x + \Delta x$ at angular frequency ω. When i_n is removed at x, the corresponding voltage fluctuation is :

$$\delta V(L) = - Z (x, 0, \omega) i_0 \quad (12)$$

As a result, the total voltage fluctuation $\delta V(L)$ resulting from the noise source located between x and $x + dx$ is given by

$$\delta V(L) = \delta V^+(L) + V^-(L) = \left[Z\left(x + \Delta x, 0, \omega \right) - Z\left(x, 0, \omega \right) \right] i_0 = \frac{\partial Z}{\partial X} (x, 0, \omega) i0.\Delta x \quad (12)$$

198

the partial derivative of Z with respect to x is called the impedance field. If we now consider i_n as a noise source with spectral density $Si(x)$, it comes from (12) :

$$S_v(x) = \left| \frac{\partial Z}{\partial X} \right|^2 . S_i(x) . (\Delta x)^2 \quad (13)$$

Assuming that two different local noise sources corresponding to two different slices are uncorrelated the spectral density of the total noise voltage S^{tot} can be calculated by integration over the whole length of the device.

$$Sv^{tot} = \int_0^L \left| \frac{\partial Z}{\partial x} \right|^2 . K(x).dx \quad (14)$$

where $K(x)$ is $S_i(x)/dx$. Equation (14), initially given by Shockley (Shockley 1966), shows that the macroscopic noise source $\overline{v^2} = S_v^{tot} \Delta f$ can be calculated by IFM when the local (microscopic) noise source $K(x)$ is known. To summerize, equation (14) shows that an important problem in the noise modelling in device is the calculation of the impedance field. Two main cases can be considered.

Impedance field using the local electric field

In that case, (Nougier 1985, 1987, 1991), the transport equations can be combined to give a non linear relation of the type (1D case)

$$I(x,t) = f(E, x, t) \quad (15)$$

where E represents the local electric field and I the current flow between the two electrodes. The steady state solution of this equation gives a relation between the electric field $E_0(x)$ and the DC current $I_0(x)$. The first order development of (15) is of the form

$$\delta I(x) = F \left| \delta E(x) \right| \quad (16)$$

where F is a linear operator. Eq(16) represents the variation of the current $\delta I(x)$ resulting from a variation of the electric field $\delta E(x)$. To solve (16) it is convenient to introduce the Green's function $G(x, x')$ of operator the F defined by

$$F \left(G(x, x') \right) = \delta(x - x') \quad (17)$$

Solving (17) we get :

$$\delta E(x) = \int_0^L \delta I(x') G(x, x') dx' \quad (18)$$

As a consequence, the voltage fluctuation $\delta V(L)$ is given by

$$\delta V(L) = - \int_0^L \delta E(x) dx = - \int_0^L \int_0^L \delta I(x') G(x, x') dx' \quad (19)$$

Since $\delta V(L)$ is related to the impedance field by

$$\delta V(L) = \int_0^L \frac{\partial Z}{\partial X} \, \delta I(x) \, dx \quad (20)$$

we have

$$\frac{\partial Z(x')}{\partial x'} = - \int_0^L G(x, x') \, dx \quad (21)$$

       This method was first introduced by Van Vliet (Van Vliet 1974) where the Green's function $G(x, x')$ is called the "transfer impedance". This method has been used to study the noise in space charge limited current diodes (Van Vliet 1974) as well as in FETs (Whiteside, 1986). This method is also well suited for numerical modelings (Nougier 1985). In that case, Eq (16) can be written as

$$\delta I_i = \sum_j a_{ij} \delta E_j \quad (22)$$

where the coefficients aij depend upon the operator itself, the discretization scheme, and the DC values. A numerical solution of (16) is :

$$\delta E_k = \sum_j b_{kj} \delta I_j \quad (23)$$

and the impedance field is given by :

$$\frac{\partial Z}{\partial X} \bigg|_j = - \sum_k b_{kj} \Delta X_k / \Delta X_j \quad (24)$$

Using this very general numerical method, all the AC and noise properties can be easily obtained in any 1D device.

Impedance field using the local potential

       In some cases, it is convenient to describe the device behavior by a relation of the form (Van Vliet 1979)

$$f(V, x) = I \quad (25)$$

After linearization, the current fluctuation $\delta I(x)$ is related to the voltage fluctuation by a linear operator :

$$F[\delta V(x)] = \delta I(x) \quad (26)$$

Let $g(x, x')$ be the Green's function of F, the solution of Eq. (26) is :

$$\delta V(x) = \int_0^L g(x, x') \, \delta I(x') \, dx' \quad (27)$$

A comparison between (20) and (27) readily gives

$$\frac{\partial Z(x')}{\partial X'} = g(L, x') \quad (28)$$

       To summerize, the calculation of the impedance field in 1D device is not difficult since it only needs the solution of a 1D linear equation. For that purpose, the use of Green's functions is often powerful in analytical problems while in numerical problems, the solution is obtained by simple matrix manipulations. In 1D devices (mainly diodes) the calculation of the macroscopic noise source does not set problems as long as the microscopic noise sources are known.

Two dimensional formulation of the IFM

For two dimensional (2D) devices such as FETs or HBTs the calculation of the noise properties by using two dimensional modelling is more difficult and needs a 2D formulation of the IFM. Let us consider a 2D device with all electrodes but the i-th one are shorted, while the i-th is open. A small current density $\delta J(\underline{r}, \omega)$ density is introduced in d^3r at $\underline{r}$. This produces, at any point $\underline{r}'$ of the device, a voltage fluctuation $\delta V(\underline{r}')$ given by

$$\delta v(\underline{r}', \omega) = \underline{z}(\underline{r}, \underline{r}', \omega) . \delta J(\underline{r}, \omega) d^3 r \quad (29)$$

once again $\underline{z} (\underline{r}, \underline{r}', \omega)$ is just a form of Green's function (Linear problem). Assuming that the noise sources are spatially uncorrelated, the power spectrum of the current density fluctuation is given by :

$$S_{kl}(\underline{r}, \underline{r}', \omega) = \overline{\delta J_l(\underline{r}) . \delta J_k(\underline{r}')} = K_{\delta J}(\underline{r}, \omega) \delta(\underline{r} - \underline{r}') \quad (30)$$

with (Van Vliet 1975)

$$K_{\delta J}(r, \omega) = 2q^2 \left[D(\omega) + D^\dagger(\omega) \right] n_0(r) \quad (31)$$

where the "dagger" stands for "transpose conjugate" tensor. From (29), the power and cross spectra of the macroscopic noise voltage sources are given by :

$$S_{eiei} = \int \underline{z}(\underline{r}_i, \underline{r}, \omega) K_{\delta j}(\underline{r}, \omega) \underline{z}^\dagger(\underline{r}_i, \underline{r}, \omega) d^3 r \quad (32)$$

$$S_{eiej} = \int \underline{z}(\underline{r}_i, \underline{r}, \omega) K_{\delta j}(\underline{r}, \omega) \underline{z}^\dagger(\underline{r}_j, \underline{r}, \omega) d^3 r \quad (33)$$

In (32) and (33) the vector Green's function $\underline{z}(\underline{r}_i, \underline{r}, \omega)$ is no else that the impedance field $\nabla_r Z(\underline{r}_i, \underline{r})$. An powerful method to compute the 2D impedance field proposed by Ghione (G. Ghione 89, 90) is based on the adjoint approach.

Noise modelling using the Monte-Carlo method

Monte-Carlo techniques are very useful, not only to calculate the properties of homogeneous semiconductors but also to calculate the macroscopic noise sources. As a matter of fact, a device modelling using the Monte Carlo method provides, at each time t, the instantaneous current i(t) including the diffusion noise. The Fourrier transform of the quadratic current fluctuations $\Delta i^2 = (i(t) - I)^2$ provides the spectral density of the current fluctuations. However, two problems have to be solved. Firstly, the observation time can be very long if the frequency of operation is small (10 GHz ~ 100 ps !). Secondly, the Monte Carlo technique introduces some numerical noise related to the time and space discretization scheme and to the number of particles introduced in the simulation. In order to give reliable results, the time and space meshes have to be small and the number of simulated particles has to be high which gives rise to problems related to the memory occupancy and computation time. For this reason, only few results have been published till now (Moglestue 1985) but progress can be expected in a near future.

THE DEVICE NOISE PERFORMANCE

The noise in one port

Fig (3) representation
of a noisy one port

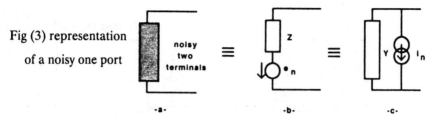

A noisy two terminal device can be represented by a noiseless device with admittance $Y_s(f)$ bridged by a noise current source i_n with spectral intensity $S_{in}(f)$ (Fig. 3c) or by a noiseless device with impedance $Z_s(f)$ in series with a voltage noise source e_n with spectral intensity $S_{en}(f)$ (Fig. 3b). For linear devices the current and voltage noise sources are related by :

$$\overline{e_n^2} = \left| Z_s(f) \right|^2 . \overline{i_n^2} \qquad (34)$$

At thermal equilibrium, Nyquist law holds and the noise current source is given by :

$$\overline{i_n^2} = 4\,kT\,\mathrm{Re}\left| Y_s(f) \right| \Delta f \qquad (35)$$

where Re stands for "Real part" and T is the device temperature. If the device is not at thermal equilibrium, it is possible to define a noise temperature by :

$$\overline{i_n^2} = 4\,k\,T_n\,\mathrm{Re}\left| Y_s(f) \right| \Delta f \qquad (36)$$

It can be readily shown that kT_n is the maximum noise power (per unit bandwidth) which can be delivered to an output matched circuit. As a consequence the noise temperature T_n can be easily measured. A full noise modelling of a one port device is performed in several steps.

(i)   the device is DC biased and the I-V characteristics is calculated ;
(ii)   for a bias point, linearization of the physical equations provides AC properties $Y_s(f)$ or $Z_s(f)$ or any equivalent representation
(iii) the microscopic noise sources are introduced, the impedance field is calculated and the macroscopic noise sources $\overline{e_n^2}(f)$ or $\overline{i_n^2}(f)$ are calculated by using IFM.
(iv)   the noise temperature is deduced from (ii) and (iii).

An important conclusion of this general representation is that "it is absolutely impossible to model the noise as long as one is not able to model first order (DC and AC) characteristics" (Nougier 1990). In addition, this representation is only valid for linear operation. For non linear systems (such as oscillators) the output noise power at frequency f can result from a noise source at frequency $f' \neq f$ due to the non linearities and the problem is much more complicated.

THE NOISE IN TWO PORT

The noise sources

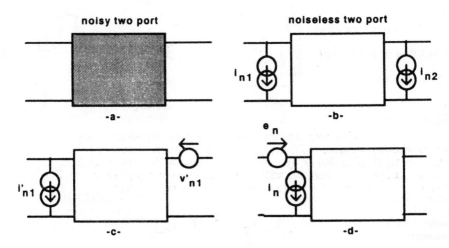

Fig (4) representation of a noisy two port

From a general point of view, a noisy two-port (fig. 4a) can be represented by a noiseless two port associated together with two <u>correlated</u> noise sources. Therefore, a full noise description of a two port needs <u>four</u> parameters : the spectral intensity of the two noise sources and the <u>complex</u> correlation coefficient. The noise sources can be either current or voltage noise sources (fig 4b, c, d) and simple transform formula can be found to pass from one representation to the other. However it should be noted that not only the spectral intensities change with changing representation but also the complex correlation coefficient. For the manipulation of the noise sources in two port, it is convenient to introduce the noise correlation matrix (Hillbrand 1976) by :

$$\left[C_{ii}\right] = \overline{\begin{bmatrix} i_{n1} \\ i_{n2} \end{bmatrix} \cdot \begin{bmatrix} i^{*}_{n1} & i^{*}_{n2} \end{bmatrix}} = \begin{bmatrix} \overline{i^{2}_{n1}} & \overline{i_{n1}i^{*}_{n2}} \\ \overline{i_{n2}i^{*}_{n1}} & \overline{i^{2}_{n2}} \end{bmatrix} \quad (37)$$

The concept of noise correlation matrix is interesting for several reasons

(i) for a two port at thermal equilibrium, the correlation matrix is given by the generalized Nyquist formula, valid for any linear two port (even non reciprocal)

$$\left[C_{ii}\right] = 4k\,T\,\Delta f \begin{bmatrix} Re(Y_{11}) & \frac{1}{2}(Y^{*}_{21} + Y_{12}) \\ \frac{1}{2}(Y_{21} + Y^{*}_{12}) & Re(Y_{22}) \end{bmatrix} \quad (38)$$

(ii) Let us consider two different representations of a linear two port, (i_{n1}, i_{n2}) and (i'_{n1}, v'_{n1}) for instance. These two representations are connected by a linear relation:

$$\begin{bmatrix} i_{n1} \\ i_{n2} \end{bmatrix} = \begin{bmatrix} T \end{bmatrix} \begin{bmatrix} i'_{n1} \\ v'_{n1} \end{bmatrix} \quad (39)$$

and therefore $[C_{ii}] = [T][C_{iv}][T]^\dagger$ (40)

(iii) the noise description of the association of several two port is simplified if the noise correlation matrix is used (Hillbrand 1976).

The noise figure

The noise figure of a two-port characterizes the degradation of the signal-to-noise ratio at a reference temperature $T_0 = 290$ K.

$$NF = \frac{(S/N)_{in}}{(S/N)_{out}} \Bigg|_{T_0 = 290 K} \qquad NF = \frac{N_a + N_{in} Ga}{N_{in} Ga} \Bigg|_{T_0 = 290 K} \quad (41)$$

where N_{in} is the available noise power at the two port input ($kT_0\Delta f$), Ga the available gain and Na the available noise power added by the two port. The noise temperature T_e of the two-port is given by $Na = kTeGa\Delta f$ and as a consequence

$$NF = 1 + \frac{T_e}{T_0} \text{ with } T_0 = 290° K \quad (42)$$

As a ratio between available powers, the noise figure is dependent on the generator (source) admittance. For a given generator admittance Y_{opt}, the noise figure is minimum and is noted NF_{min}. Since the full representation of the noise in a two port requires four parameters, the minimum noise figure NF_{min} is associated with three other parameters : the equivalent noise resistance R_n and the (complex) optimum admittance Y_{opt} (or optimum reflection coefficient Γ_{opt}).

The calculation of NF_{min}, R_n and Y_{opt} is straightforward if the macroscopic noises and their correlation are known. In a first step, the two port is described using a chain representation. This representation is characterized by two noise sources i_n and e_n located at the input of the noiseless two port (fig. 4d). Let $[C_{ie}]$ be the corresponding noise correlation matrix, we have :

$$[C_{ie}] = \begin{bmatrix} \overline{e_n^2} & \overline{i_n e_n^*} \\ \overline{i_n e_n^*} & \overline{i_n^2} \end{bmatrix} = 4kT_0 \begin{bmatrix} R_n & \frac{NF_{min} - 1}{2} - R_n Y_{opt} \\ \frac{NF_{min} - 1}{2} - R_n Y_{opt}^* & R_n |Y_{opt}|^2 \end{bmatrix} \quad (43)$$

An equivalent form for the four noise parameters is :

$$R_n = \overline{e_n^2}/4kT_0 \Delta f \quad (44)$$

$$Y_{opt} = \sqrt{G_{cor}^2 + \frac{G_n}{R_n}} - j B cor = g_{opt} + j B opt \quad (45)$$

$$NF_{min} = 1 + 2R_n \left[G_{opt} + G_{cor} \right] \quad (46)$$

with

$$Y_{cor} = G_{cor} + jB\,cor = \frac{\overline{i_n e_n^*}}{\overline{e_n^2}} \quad (47)$$

and

$$R_n G_n = \overline{i_n^2}\ \overline{e_n^2} - \left| \overline{e_n^* i_n} \right|^2 \quad (48)$$

For any generator admittance $Y_s = G_s + j\,B_s$, it is straightforward to show that the noise figure is given by:

$$NF = NF_{min} + \frac{R_n}{G_s} \left| Y_s - Y_{opt} \right|^2 \quad (49)$$

So, all the noise parameters can be readily deduced from the noise sources e_n and i_n. As in the case of one-port, the modelling of the noise performance of any two port needs the following steps:

(i) calculation of the DC characteristics
(ii) calculation of the AC performance
(iii) introduction of the microscropic noise processes, calculation of the impedance field and calculation of the macroscopic noise sources and correlation using IFM.
(iv) transformation of the noise sources into two sources e_n and i_n located at the input of the noiseless two-port (noise correlation matrix technique or direct method).
(v) calculation of the four noise parameters NF_{min}, R_n and Y_{opt}.

- **Application : the FET noise performance.**

- As it is well known, FETs are very well suited for low noise amplifiers in the centimeter and millimeter wave range. Several noise modellings of MESFETs and HEMTs have been proposed (Van der Ziel, 1962, Pucel, 1974, Cappy, 1985, 1988) and it has been shown that the noise sources exhibit the following behaviour : the gate noise current source increases as ω^2, the drain noise current source is frequency independant while the correlation coefficient between the gate and drain noise sources is almost purely imaginary. So, the noise performance of FETs can be described by three parameters P, R and C where C is the imaginary part of the correlation coefficient and the dimensionless parameters P and R are connected to the noise sources by the following expressions :

$$\overline{i_d^2} = 4kT\,g_m\,P\,\Delta f \quad (50)$$

$$\overline{i_g^2} = 4kT\,\frac{\omega^2\,Cgs^2}{g_m}\,R\Delta f \quad (51)$$

where g_m is the device transconductance and C_{gs} the gate-to-source capacitance. The variation of i_d^2, i_g^2 and C are shown as functions of the DC drain current in figure (5).

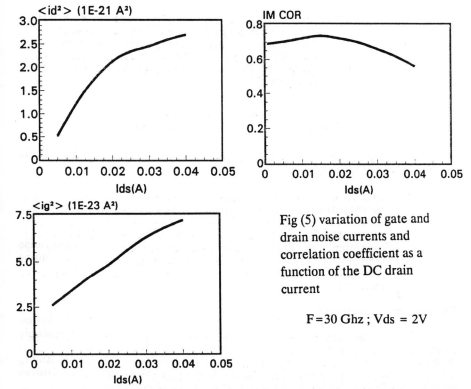

Fig (5) variation of gate and drain noise currents and correlation coefficient as a function of the DC drain current

$F = 30\ Ghz\ ;\ Vds = 2V$

From P, R and C, the determination of the noise figure and the other noise parameters is straightforward (Pucel 1974). Introducing K_1 and K_2 defined by

$$K_1 = P + R - 2C\sqrt{PR} \quad (52)$$

$$K_2 = PR(1 - C^2) \quad (53)$$

we have

$$NF_{min} = 1 + 2\sqrt{K_1} \cdot \frac{\omega C_{gs}}{g_m} \sqrt{g_m(R_s + R_g) + \frac{K_2}{K_1}} \quad (54)$$

$$R_n = R_s + R_g + \frac{P}{G_m} \quad (55)$$

$$G_n = \frac{\omega^2 C_{gs}^2}{g_m^2} \left[\frac{(R_s + R_g)g_m + K_2/K_1}{R_n} \right] \quad (56)$$

$$Y_{opt} = \left(\frac{G_n}{R_n} \right)^{1/2} + j\omega C_{gs} \cdot \left(\frac{C\sqrt{RP} - P}{g_m R_n} \right) \quad (57)$$

These simple expressions of the four noise parameters of FETs show several important features.

(i)   the gate noise influences the noise figure even at low frequency.

(ii) the noise figure (in linear scale, not in dB) keeps a linear variation versus frequency even if the gate noise is taken into account.

(iii) if Rs + Rg tends to zero, the noise figure, contrarely to the Fukui formula (Fukui, 1979) is no larger close to unity if the gate noise is taken into account. The noise figure of an intrinsic FET $(R_s + R_g = 0)$ is given by :

$$NF_{min} = 1 + 2 \ \frac{\omega C_{gs}}{g_m} \ \sqrt{PR(1 - C^2)} \quad (58)$$

This expression shows the great importance of the correlation on the intrinsic FET noise figure.

(iv) Due to the correlation between the drain and gate noise current sources, the gate noise is partially substracted from the drain noise. This fundamental effect is expressed in the noise parameters by the terms $P + R - 2C \sqrt{PR}$ and $PR(1-C^2)$. This reduction of the drain noise is the basic reason why the field effect transistor is a low noise device. Therefore, the gate noise and the correlation coefficient have to be include into a serious FET noise analysis.

The noise in oscillators

In the preceding paragraphs of this paper, all the device and structures investigated are considered to be linear. In non linear systems, noise study is very difficult and only few theories were proposed, especially in the case of non linear resistors and oscillators. In the case of non linear resistors, it was shown (Gupta 1982) that the spectral intensity of the voltage noise source is given by :

$$S_v(f) = 4kT \left[\frac{dV}{dI} + \frac{1}{2} \cdot I_0 \frac{d^2V}{dI^2} \right]_{I = I_0} \quad (59)$$

Equation (59) can be considered as a generalized Nyquist theorem and it can be used in diodes for instance.

The second important case concerns the calculation of the oscillator noise performance. The noise in the output RF power spectrum is characterized in terms of amplitude modulation (AM) and frequency modulation (FM) components. These arise from amplitude and phase fluctuations $\Delta I(t)$ and $\Delta \Phi(t)$ in the output current given by :

$$I(t) = \left| I_0 + \Delta I(t) \right| \cos \left(\omega_0 t + \Delta \Phi(t) \right) \quad (60)$$

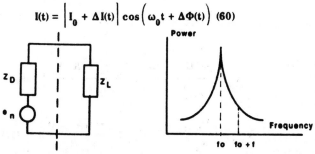

Fig (6) The noise in oscillators

the purpose of the noise analysis in oscillators is to calculate the noise power in a bandwidth Δf of a single side band at a frequency f from the carrier, divided by the power in the carrier (Fig. 6b). Since oscillators have a non linear behavior, the mechanism for noise generation is that low frequency device noise (frequence f) is mixed with the carrier signal (frequency f_0) via the non linearities of the FET to produce noise at frequencies $f_0 \pm f$. In this way, the low frequency noise generated by the active device is upconverted to microwave frequencies. According to Edson and Kurokawa (Edson, 1960, Kurokawa, 1969), a general view of an oscillator is shown in figure (6a). The active device and the associated network is treated as a negative resistance noisy one port. Introducing the spectral intensity of the noise source $S_e(f)$, the noise to carrier ratio (N/C) is given by (Kurokawa, 1969) :

- FM noise

$$\frac{N}{C}\bigg|_{SSB} (f_0 + f) = \frac{Se(f)}{8R_L P_0} \cdot \left(\frac{f_0}{Q_0 f}\right)^2 \qquad (61)$$

- AM noise

$$\frac{N}{L}\bigg|_{SSB} (f_0 + f) = \frac{Se(f)}{8R_L P_0} \cdot \frac{1}{\left(\frac{S}{2}\right)^2 + \left(\frac{Q_0 f}{f_0}\right)^2} \qquad (62)$$

where P_0 is the output power developed in R_L, Q_0 the external quality factor and S a dimensionless factor which characterizes the non linear behavior of R_D. Expressions (61) and (62) show that the choice of an active device for oscillator realization is mainly governed by the low frequency noise properties of the device.

CONCLUSION

In the case of one dimensional linear devices, the calculation of the noise performance can be carried out using the very general and very powerful impedance field method. This method can be used either in the case of analytical or numerical modelings for calculating the macroscopic noise sources (voltage or current) if the microscopic noise processes are known. Then, the macroscopic noise sources can be transformed to provide the parameters of practical interest such as noise figure, equivalent noise resistance, noise to carrier ratio... Since the noise properties of a device are first order properties, it should be emphasized that a noise modelling has to be based on reliable modeling for the DC and AC properties.

It seems also interesting to specify here some important problems concerning the noise modelling.

. Firstly, the microscopic noise sources are not always very well known. For instance, the diffusivity is not very well known even in bulk materials such as GaAs because only few measurements are available. Likewise, the microscopic 1/f noise sources are not very well known because a satisfactory interpretation of 1/f noise is still lacking.

. Secondly, the IFM is based on spatially uncorrelated noise sources. Obviously this assumption is not correct if the distance between the two considered points is of the order of magnitude of the mean free path. The effects of the microscopic spatial correlation are not presently well known although a mathematical derivation of the two points correlation function was proposed (Nougier 1983) for some spectific scatterings.

. Thirdly, as the device frequency of operation increases, the device characteristic length decreases. This decrease leads to noticeable quantum effects, non stationary electron transport and change of the nature of noise (shot noise versus diffusion noise).

Fourthly, accurate determination of the device behavior needs often 2D (or even 3D !) modelling. In such case the IFM is difficult to implement. The Monte Carlo method seems to be the best way for calculating the noise performance of such devices.

REFERENCES

Cappy A. et al (1985) Noise modeling in submicrometer gate two dimensional electron gas field effect transistor. IEEE Trans. Electron Devices 32, n° 12 : 2787-2795.

Cappy A. (1988) Noise modeling and measurement techniques. IEEE Trans. MTT 36, n° 1 : 1-10.

Cappy A., Heinrich W. (1989). The high frequency FET noise performance : A new approach IEEE Trans. Electron Devices 36, n° 2 : 403-410.

Edson W.A. (1960) Noise in oscillators. Proc. IRE, Vol. 48 : 1454-1466.

Fukui H. (1979) Optimal noise figure of microwave GaAs MESFET's. IEEE Trans. Electron. devices, Vol. 26, n° 7 : 1032-1037.

Ghione G. et al (1989). Proc. Int. Elec. Dev. Meeting : 505.

Ghione G. (1990). Proc. ESSDERC 90, Notthingham IOP Publ. Ltd : 225.

Hillbrand H., Russer P. (1976) An efficient method for computer Aided Noise Analysis of linear Amplifier Networks IEEE Trans. on Circuits and Systems, Vol. CAS 23, n° 4 : 235-239.

Hooge F.N. (1976) 1./f noise, physica 83B : 14-23.

Kurokawa K. (1968) Noise in synchronized oscillators. IEEE Trans. MTT, Vol. 16, n° 4 : 234-240.

Nougier J.P. et al (1985) Method for modeling the noise of submicron devices. Physica 129 B : 580-582.

Nougier J.P. et al (1983) Microscopic spatial correlations at thermal equilibrium in non polar semiconductors" Noise in physical systems and 1/f noise. North Holland, ed by M. Savelli, G. Lecoy : 15-19.

Nougier J.P. et al (1981) Mathematical formulation of the impedance field method. Application to the noise in the channel of field effect transistors in Proc of 6th Int. Conf. Noise phy. Systems : 42-46.

Nougier J.P. (1991) Noise in devices : Definition modelling III-V microelectronics. J.P. Nougier Editor. Elsevier Science publishers : 183-238.
Nougier J.P. (1978) in Noise in Physical systems. Springer Ser. Electrophys. Vol. 2, Ed by D. Wolf. : 71.

Pucel R.A. et al (1974) Signal and noise properties of GaAs field effect transistors. Advances Electron. Electron Physics. Vol. 38 : 135-265.

Shockley W. et al (1966) The impedance field method of noise calculation in active semiconductor devices. In quantum theory of atoms, Molecules and solid state. Academic Press, New York, 1966 : 537-563.

Thornber K.K. et al (1974), Structure of the langevin and impedance field methods of calculating noise in devices. Solid State Elect. Vol. 17 : 587-590.

Van Vliet K.M. et al (1975) Noise in single injection diodes. A survey of methods. J. Appl. Phys. Vol. 46, n° 4 : 1804-1813.

Van Vliet K.M. (1979) the transfer-impedance method for noise in field-effect transistors. Solid state Elec. vol. 22 : 233-236.

Van der Ziel A. (1988) Unified Presentation of 1/f Noise in Electronic Devices. Fundamental 1/f Noise sources. Proc. IEEE, vol. 76, n° 3 : 233-258.

Van der Ziel A (1962) Thermal noise in field effect Transistors. Proc. IRE 50 : 1808-1812.

Whiteside C. (1968) the DC, AC and noise properties of the GaAs/GaAlAs Modulation-doped Field effect transistor channel. IEEE Trans. on Electron. Dev. vol. 33 :

Monte Carlo Models and Simulations

Paulo Lugli
Universita' di Roma "Tor Vergata"

A critical review of the Monte Carlo (MC) simulation as applied to semiconductor device modeling is presented. The general principles of the method are analized and discussed in detail, together with the physical model implemented for the description of carrier transport in III-V semiconductors. Some special features that can be very useful in dealing with real devices are also examined. It is shown that the MC method is a mature technique for modeling, and can offer great advantages over more traditional approaches. Critical points are pointed out and analyzed. A variety of applications in the field of III-V device modeling is then outlined.

INTRODUCTION

Due to the continuous technological improvements in microelectronics, and to the constant push for miniaturization of devices, there is an increasing need for physical simulators, able to combine a realistic description of carrier transport to an intrinsic speed which allows their implementation in Computer Aided Design (CAD) tools. Indeed, computer programs are extremely important for technology development. CAD has become one of the keywords in microelectronics. The importance of such field can be greatly appreciated focussing on the steps required for the fabrication of integrated circuits (IC) There, the development of new technologies has been driven by an experimental approach. A useful alternative was offered by software tools, which can lead to a speed up of the development cycle and a reduction of the development costs. In fact, those calculations can be considered as simulated experiments, which can be much faster and less expensive than real experiments. Furthermore computer experiments allow a deep physical interpretion of the final results that leads to a better understanding of the problem at hand. The characteristic links between the different aspects of CAD can be summarized as follows. The output of the process simulation is fed directly into a device simulation program, which determines the electrical characteristics and the performance of the device. At this stage, the interplay between process and device simulation can suggest improvements on the processing steps deduced from the simulated device performance. The output of the device simulator is then inserted in a compact form into a circuit simulation program, which determines the characteristics of the overall circuit. MC simulators are finding wider and wider use as CAD tools, both at the level of device as well as process simulation, rapidly closing the gap with respect to more traditional, and widely used, Drift-Diffusion simulators, posing itself as the most valuable alternative. Indeed, the MC technique has moved a long distance since its first introduction as a tool for the study of charge transport in semiconductors. The outstanding improvements in MC device simulations and the vaste range of applications they have reached is due to the availability of very fast workstations and mainframes, which guarantee considerable speed at affordable costs, allowing the implementation of more and more sophisticated physical models, and to the optimization of the numerical algorithms. Furthermore, as the active region of devices shrinks and heterostructures come more heavily into play, non local and quantum effects become extremely important. The physics involved in charge transport of such structures then authomatically prevents the use of conventional simulators, and calls for more advanced modeling tools.

Although it will not be possible to exhaust the complexity of the MC device simulation in such a short review, the present paper is intended to give a critical overview of the MC algorithm for device modeling. The next session will deal with the general principles of the MC simulation, specialized to the case of semiconductor devices. A short description of a generic MC simulator will be given, together with special features that are needed for the simulation of very complicated systems. A complete overview of the Monte Carlo simulation of semiconductor devices can be found in Ref. 1, which also contains a detailed list of references. We will then present a simplified physical model which describes quite well the high field transport properties of bulk GaAs. An application section will follow, where we will concentrate on some recent results obtained for GaAs unipolar and bipolar devices, which illustrate quite effectively the strength and the capabilities of the MC simulation. It is impossible, in the limited space available, to account for the enormous amount of work originated in recent years in the field, and we have to refer the interested reader to the vaste literature existing on the subject, starting for example from the bibliography that can be found in Refs. [1-4].

THE MONTE CARLO TECHNIQUE

As the name suggests, the MC method is based on the selection of random numbers [5-7]. In its present form, the method is attributed to Fermi, Von Neumann, and Ulam, who developed it for the solution of problems related to neutron transport. In principle, the MC method can be considered as a very general mathematical tool for the solution of a great variety of problems.

Among the various applications of the method, the following are probably the most important:

* Integro-differential equations
* Matrix inversion
* Transport of nuclear particles
* Transport in semiconductors
* Modeling of semiconductor devices
* Process simulation

An important feature of the MC technique is that more precise results can be obtained by generating larger numbers of points. More generally, being based on random numbers, the results obtained with a MC procedure are never exact, but rigorous in a statistical sense: the exact result lies in given intervals with given probabilities. The uncertainty of the results is strictly related to the variance of the possible outcomes and it is smaller if the size of the sample (i.e. the amount of computations devoted to the solution of the problem) is larger. One basic element of the numerical procedure is the possibility to generate random numbers with given distributions starting from pseudorandom numbers uniformly distributed between 0 and 1. Modern computers provide sequences of numbers obtained with precise mathematical algorithms, starting from a given element (seed). For each seed, the sequence is perfectly predictable. However it satisfies a large number of statistical test of randomness. Those pseudorandom numbers offer two great advantages: they can be generated in a very fast way, and they are reproducible (which is essential for example in debugging a code).

The applications of MC methods can be divided into two major groups. One consists of direct reproduction on a computer of the microscopic dynamics of the physical process in a system which is already statistical in its nature. We use in this case the term "MC simulation". The second group consists of MC methods devised for the solution of well defined mathematical equations. In such cases the methods are used to solve the equations that describe the problem of interest. The majority of real cases are a mixture of the two extreme limits indicated above. The application to the study of semiconductor devices and processes is a good example [8]. In fact, transport processes are statistical in nature, but are also accurately described by well defined transport equations. A MC method applied to the solution of such equation may or may not correspond to the direct simulation of the physical system under examination. For instance, the MC solution of Boltzmann transport equation (BTE) not only gives the distribution function that verifies the equation, but also yields information that are lost in BTE itself (noise and fluctuations are the best example). On the other hand, the direct simulation is at times very inefficient, as for example in the analysis of situations that are rare in the actual physical system. One example that will be dealt with extensively later on is given by the impact ionization process. In such cases it is necessary to distort the simulation by applying some more sophisticated MC techniques that reduce the variance of the quantity of interest, giving up the advantages offered by the direct simulation.

The application of the MC that we will focus on, namely the simulation of semiconductor devices, is particularly important in light of the growth in the field of microelectronics achieved in recent years. Semiconductor devices are nowadays built with their active dimensions below one micron. The reduction in size leads to a higher integration level as more devices can be put into a single chip. Moving into the submicron scale, many new physical phenomena become important that require a sophisticated theoretical treatment. Furthermore, new possibilities for novel devices are offered by the capability to grow nanometer layered structures with extremely high quality by molecular beam epitaxy (MBE) and Metal organic chemical vapour phase epitaxy (MOCVD). There exists, therefore, a new challenge towards the understanding of the principles of operation of those novel devices. As we will see, the MC method offers great advantages in this direction. On the other side, the push towards smaller and more powerful devices (which immediately translate into higher levels of integration and enhanced performance of the single devices as well as of the overall circuit) has been substained by enormous advances in the area of fabrication and processing. A very precise control is nowaday possible on the device geometry and doping profile through techniques such as ion implantation, reactive ion etching, electron and X-ray lithography.

As pointed out earlier, the MC technique is a fairly new tool in the area of device modeling, traditionally dominated by simulators based on drift-diffusion or on balance-equation models (for an overview of the subject see Refs. 9-11). The application of MC techniques to the simulation of semiconductor devices started quite soon after the introduction of the method [12,13] but has received great attention only recently [14-16] since it requires very large amount of computations, made possible only by the most recent computers. With the recent advances in material growth, contact deposition and physical point of view. Incidentally, this has provided physical systems of extreme interest. At the same time, MC algorithms have gained in sophistication and are now able to handle phenomena and systems of great complexity. These are two fundamental steps since the necessary input for a MC simulation of semiconductor materials and devices is the physical system under investigation. Many semiconductor devices can be nowaday simulated with the MC method, which is becoming more and more a very useful modeling tool.

The most common (and also the most interesting) simulation of a semiconductor device is performed for many particles in parallel (Ensemble Monte Carlo) and coupled to Poisson's equation in order to obtain the self-consistent potential consistent with the charge distribution given directly by the Monte Carlo procedure. For systems of great complexity, a one particle Monte Carlo (OPMC) simulation can performed on a given fixed potential previously determined.

Since no a-priori assumptions are needed on the form of the real and k-space carrier distributions, a Monte Carlo simulator is the only reliable tool for the investigation of those physical phenomena that critically depend on the shape of the distribution, or on the details of its tail (such as electron injection over potential barriers). Furthermore, the Monte Carlo technique allows us to focus on particular physical mechanisms that might be of importance on the device performance (for example, intercarrier scattering, impact ionization, generation-recombination, etc.). The prices one has to pay are a very time-consuming algorithm, and the requirement of a complete knowledge of the physical system under investigation. Often many assumptions have to be made in order to reduce the complexity of the model describing a given device.

In recent years the Ensemble Monte Carlo (EMC) has been widely used to study the properties of semiconductor devices. Particular emphasis has been lately attributed to submicron structures, because of their performances in switching and high frequency operations [17]. Once the basic physics involved in the transport of such devices is known, EMC simulation provides a formidable tool to determine their limits and characteristics and can be very helpful in modeling. Together with the determination of the macroscopic properties of a device, EMC also gives a microscopic description of the local electric field, charge density,velocity distribution, etc. An excellent overview can be found in Ref. 18.

The Monte Carlo algorithm for device simulation

A flowchart of a generic EMC self consistent device simulation is show in Fig. 1. The basic steps are the following :

*i) Set up geometry and discretization scheme.*Two parameters that play an important role in the choice of the time step and the grid size are the plasma frequency and the Debye length [18].

ii) Charge assignement. The charge of each particle is assigned to a particular mesh point. Since it is not possible to simulate all the electrons present in a real device, each simulated particle represents a cloud of electrons for the purpose of estimating currents, charge and field distributions. For all other purposes, each individual particle carries its elementary charge e. The doping charge is also added to the mesh according to its distribution. Although the EMC simulation is inherently three dimensional, we usually deal with one or two dimensional grids. In such contest the assumption is the perfect homogeneity in the dimensions that are not considered explicitly.

iii) Potential solution. Poisson's equation is solved to determine the electrostatic potential at the mesh points. In connection to EMC simulations, a finite difference scheme is generally used. The solution can be obtained in several ways, the most efficient being the Fourier Analysis Cyclic Reduction (FACR) and the direct matrix inversion. The former method provides a very effective algorithm that allows the inclusion of special boundaries through the so called capacity method. The latter requires a matrix inversion at the beginning of the simulation. The new potential is calculated with a simple matrix multiplication at fixed times during the simulation. Such method is particularly efficient on computers with vector processing. The electrostatic field is then obtained from the potential with a finite-difference algorithm.

iii) Flight. Each particle,now treated as an individual electron, undergoes the standard MC sequence of scatterings and free flights, subject to the local field previously determined from the solution of Poisson's equation. The MC sequence is stopped at fixed times, when the field is adjusted following the steps described above. For OPMC techniques, the potential and field profiles are calculated at the beginning of the simulation, and only step iii) is performed. The differential scattering probability $S(\mathbf{k},\mathbf{k'})$ that an electron undergoes a transition from an initial state $\mathbf{k}$ to a final state $\mathbf{k'}$ is usually calculated within first order perturbation theory, using the so-called "Fermi Golden Rule".

214

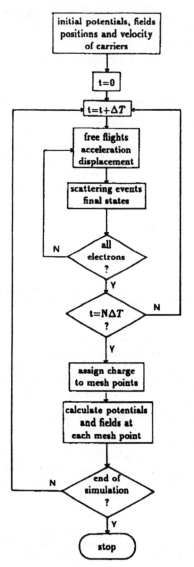

initial potentials, fields
positions and velocity
of carriers

t=0

t=t+ΔT

free flights
acceleration
displacement

scattering events
final states

all
electrons
?
N

t=NΔT
?
N

assign charge
to mesh points

calculate potentials
and fields at
each mesh point

end of
simulation
?
N

stop

Figure 1 Flowchart of a typical self consistent MC program for device simulation

The total scattering rate that determines the length of a free flight is obtained by summing $S(\mathbf{k},\mathbf{k}')$ over all final states and over all possible scattering mechanisms Several scattering agents can be included in a MC simulation, among which acoustic, optical (polar and non polar), intervalley, and interband phonons, ionized and neutral impurities, other carriers, surface roughness, alloy fluctuations. Furthermore, ionization, Auger, and generation-recombination, processes can be considered. During the free flight, carriers are accelerated by the electric field as classical particles. For this reason, the MC simulation described here is "semiclassical" in nature. The energy-wavevector dispersion $\varepsilon(\mathbf{k})$ is usually given by a non parabolic dispersion. It is possible to use also a full band structure description, taken for instance from a pseudo-potential calculation [3,16,19].

The description of the problem is completed by setting initial and boundary conditions. The initial conditions are not so important, since only the self-consistent steady-state result is usually retained. Boundary conditions are instead crucial, in particular in submicron devices, where contact properties drastically influence the whole behavior of the device. We will return to this point later.

The steady-state current is given directly by the net number of particles crossing one contact per unit time. An efficient technique for current calculation in MC two dimensional simulations have been proposed in Ref. 20. Such technique is based on an extension of Ramo's theorem. Once the current is calculated for a given bias configuration, it is possible, by performing several computer runs, to construct the I-V characteristics of the device. Important device parameters can also be extracted from the simulation. For example, the output resistance $R_{out} = \Delta V_D / \Delta I_D$ for constant gate bias V_G and the transconductance $g_m = \Delta I_D / \Delta V_G$ for constant drain bias V_D can be obtained from a series of runs, starting from normal operating conditions and varying the drain and gate bias respectively. Parasitic elements can also be calculated, referring to an equivalent circuit description of the device. For example, the source-to-gate C_{sg} and source-to-drain C_{gd} capacitances are obtained by applying a step change respectively to the gate (ΔV_G) and drain voltage (ΔV_D) as $C_{sg} = \Delta Q_G / \Delta V_G$ and $C_{gd} = \Delta Q_G / \Delta V_D$, where ΔQ_G is the total equivalent charge flowing from the gate in response to the step potential. From the time dependence of the charge flow it is also possible to estimate the parasitic source and drain resistance.

Traditionally, device simulators have been based on drift diffusion (DD) or on balance equation (BE) models. Both of them are fast and reliable as long as a local description of the physical phenomena in the device is possible, that is, when the carriers can be described by a distribution characteristic of the given field present in every region of the device. Such an assumption breaks down when the device dimensions are small (typically below one micron), and high fields are set up, leading to non-local phenomena. More specifically, when the field inside the device varies appreciably over lengths comparable with the electron mean free path, the electrons at a given position carry information about the field value at another position, and the transport process becomes non local. The inclusion of the energy balance equation allows to incorporate some of these effects, at the cost of a much heavier computation [10,15]. The Monte Carlo technique, which is inherently non local, lends itself very well to the simulation of non stationary transport in devices. The discrepancy between local models and the EMC have been clearly outlined by several authors [1,21] and will be briefly discussed below.

Examples of MC simulators present in the literature can be found in refs [22-66]. The prototype, and hystorically the first self consistent MC program, was applied to a MESFET structure by Hockney and coworkers at Reading University [12,18]. These algorithms have been so successful that they are adopted by most of the self consistent MC programs.

A large class of devices (MOSFETs, bipolar transistors, HEMTs, etc) are characterized by areas with high doping (and free electron) concentrations, low electric field and retarding barriers. The direct simulation of electronic motion in these regions can be terribly time consuming. In contrast, more traditional simulators, such as (DD) schemes, can be applied to such a situation in a reliable and straightforward way. A hybrid method (HYMC) has been proposed [67,68] that combines the two techniques, by relying on the fast DD simulators for low field areas, and on the direct MC simulation where steep gradients of the potential createthe condition for hot carriers. Although excellent in principle, the hybrid technique (also called regional MC) requires a very accurate handling of the boundary conditions at the interface between the various region which has not yet been obtained.

Special Features

In the following section, we focus on special aspects of the MC simulation that are not generally considered because of their difficulty, although they can be of great importance in the device performance.

Pauli Exclusion Principle

Electrons obey the Fermi-Dirac statistics and must satisfy the Pauli Exclusion Principle (PEP). This means that not all the states are available because only two electrons of opposite spin can reside in each state. This aspect is not very important in nondegenerate case and electrons are distributed in a large interval of states; in the degenerate case the problem becomes more conspicuous. In GaAs the electrons are degenerate for $T \leq 300K$ and for $n > 5 \ 10^{18} \ cm^{-3}$. This is the case for many devices of interest. Degeneracy is equivalent to a many body interaction which reduces the phase space available for the electron final state in an induced transition. If $p(\mathbf{k})$ and $p(\mathbf{k}')$ are the probabilities that the initial and final state are, respectively, occupied, the total rate of transition $P(\mathbf{k}, \mathbf{k}')$ between two different states is given by $P(\mathbf{k}, \mathbf{k}')=p(\mathbf{k}) S(\mathbf{k}, \mathbf{k}') [1-p(\mathbf{k}')]$. Normally a semiclassical Monte Carlo works with the approximation $p(\mathbf{k}')=0$ because all the states are considered available as final states. The inclusion of PEP is then essentially the inclusion of this term in the total scattering rate. In Ensemble Monte Carlo this is obtained very easily because the particle distribution is known step by step. The algorithm generating the distribution function is set up by multiplying the scattering probability by the correction factor $1-p(\mathbf{k}')$; the term $p(\mathbf{k}')$ is determined self consistently and a rejection technique is used after selecting the final state without the correcting Pauli factor [69,70].

Contact Simulation

The simulation of contacts is one of the most serious problems in MC device simulations, due in part to the limited knowledge of the physics of contacts. On the other side, contacts are of great importance in a number of semiconductor devices, whose applications range from high-speed logic to microvawes. As the dimensions of these devices reach the submicron limit, contacts become the limiting factor for the performance in the ballistic or quasi-ballistic mode of operation.

In general, semiconductor devices do not operate under charge neutrality conditions. The net charge inside the device is directly related through Gauss' law to the flux of the electric field on the boundaries, and consequently to the potential inside the device. Therefore, charge neutrality (that is conservation of the number of particles) cannot be enforced in the simulation. Rather, an appropriate handling of the boundaries is required to simulate a number of electrons that varies in time self consistently with the potential distribution. In a device such as a field effect transistor the most significant boundaries are at the source, drain and gate electrodes. Source and drain contacts are usually treated as ohmic contact by absorbing all the electrons that hit the electrodes and by injecting a number of electrons which maintains a neutral region in the adjacence of the electrodes [1]. The Schottky barrier at the gate is treated as a perfectly absorbing electrode with a potential equal to the applied potential minus the barrier height. Although commonly assumed, the above conditions have never really been tested. One attempt to deal with the problem of contacts in a simulation of a 1-D metal -n-n$^+$ structure has been presented in Ref. [71]. This is an interesting system in that the device is never charge neutral, except under flat-band condition. This is due to the presence of a depletion or an accumulation region near the interface. Since the value of the electric field at the two boundaries is related through Gauss' law to the net charge inside the device, it is necessary to allow the number of the electrons simulated to vary during the simulation, as a constant number of the electrons would give incorrect results.

Tunneling across a thin barrier can also been included [71]. The probability that an electron with energy ε at a distance x from the barrier (located at W) will tunnel through is given, within WKB theory, by

$$T(\varepsilon) = \exp \{ -\frac{2}{h} \int_{x}^{W} dx\, 2\, m^* \, [eV(x) - \varepsilon]^{1/2} \tag{1}$$

where m^* is the effective mass, V the potential seen by the electron, and h is Plank's constant. As a MC electron reaches the barrier, the tunneling probability is calculated from Eq. (1) using a parabolic least square interpolation of the potential $V(x)$ obtained from the solution of Poisson's equation. A random number is then used to decide wheter the electron will tunnel or not. It is important to notice that no assumptions on the electron distribution function near the contact, nor on the shape of the potential barrier are needed. Quantum mechanical reflection and field emission from the metal into the semiconductor couls be easily included along the same lines.

Carrier-carrier scattering

Many devices are characterized by very high electron concentrations. In such situation one might have to worry about the possible influence of the interaction among the conducting electrons. A good example is provided by the Tunneling Hot Electron Transfer Amplifier (THETA) [72] or by the Heterojunction Bipolar Transistor (HBT) [73] that will be examined later. The former one belongs to a series of new devices generically called ``hot electron transistors", based on the idea of improving the device performance by injecting fast electrons into thin base regions. In a standard device, such as a MESFET or HEMT, electrons are injected into the channel with a thermal energy distribution and a small initial velocity. In order to reduce the transit time through the channel or base, it is adventageous to increase their initial velocity. Electrons are then injected into the base with energies some hundreds meV 's greater than the thermal energy. The active part of the device consist of a GaAs-AlGaAs-GaAs quantum well, with a very thin (few hundred A) and highly doped GaAs base. Electrons are injected into the base by tunneling through the potential barrier between emitter and base, and they are collected as they overcome the barrier between base and collector. There, if the hot electrons could maintain their high speed through the active region of the device, then very good performance (fast switching , high cut-off frequencies) should be expected. On the contrary, it has been suggested, see for example [73], that the interaction of the injected electrons with the base electrons might be a strong source of degradation of device performance. As well as in metals, plasma phenomena can be of great importance also in polar semiconductors [74,75]. Electrons injected into highly doped regions can be scattered by the collective excitations of the electron gas, as well as through normal binary collisions with the other electrons. A similar situation is found in HBTs, the main difference being the p-type doping of the base, which causes the injected electrons to scatter with a degenerate hole gas in the base.

Two main contributions to the carrier-carrier scattering can be identified:

- the individual carrier-carrier interaction via a screened Coulomb potential which accounts for two-body short-range interaction;

- the electron-plasmon interaction, which accounts for the collective long-range behaviour of the electron gas.

In semiconductors, the plasmon energy at a reasonable electron density can be of the same order of magnitude as the characteristic phonon energies. In a device simulation, the scattering rates for electron-electron and electron-plasmon processes can be tabulated at the beginning of the simulation. Carrier-carrier scatterings are then treated as any other mechanisms in the MC algorithm. A full selfconsistent simulation of the THETA device, that includes the e-e scattering as described above, has been performed by Antonelli and Lugli [34].

Noise and fluctuations

Noise in a semiconductor device is related to the fluctuations of the output current around its mean value and is due to the stochastic nature of the microscopic processes that govern carrier transport in the device. The MC simulation is particularly well suited for the study of noise, since it naturally incorporates all the microscopic noise sources.

The most important figure of merit in a transistor is the minimum noise figure NF, given by the signal-to-noise power ratio at the gate $P_1/\delta P_1$ divided by the corresponding ratio at the drain $P_2/\delta P_2$, or [18,27]:

$$NF = \frac{P_1/\delta P_1}{P_2/\delta P_2} = \frac{\delta P_2}{\delta P_1} \frac{P_1}{P_2} \tag{2}$$

which can be expressed in terms of the Y parameters as

$$NF = 1 + \left(\frac{Y_{11}}{Y_{22}}\right)^2 \left(\frac{\delta I_2}{\delta I_1}\right)^2 \tag{3}$$

The second factor on the right end side of Eq. (3) is the ratio of the mean-square current fluctuations at the drain to those at the gate, which is multiplied to the ratio of the Y parameters. Such quantities are calculated directly during the MC simulation

Optimization procedures

An original, efficient algorithm has been implemented to calculate the appropriate duration of the free flights (depending on the actual carrier status). The method which is based on a space dependent definition of the scattering rate [57], leads to a drastic reduction in the number of self scatterings thus allowing large saving in computation time (more than one order of magnitude compared with the conventional approaches).

In areas where the electron population is very small, it is possible to extend a technique originally proposed by Phillips and Price [76], that allows to obtain a good statistics in rarely visited energy regions. Two situations are of particular interest. If a device presents regions with a high doping density N^+ connected to regions with low densities N^-, the carrier concentration will reflect (except at the interface between the different regions) the doping distribution. Therefore, most of the simulated carriers (roughly in the ratio N^+/N^-) will populate the highly doped, low field regions, requiring an extensive amount of computation for the simulation of a quasi-thermal distribution. This is the case for example of the MOSFET or a MESFET with ion implanted source and drain contacts.

A similar case is found in k-space, when we are interested in the population of high energy states, which are rarely touched by the carriers, but might cause very important physical phenomena (a typical example is the carrier injection into SiO_2 for the channel of a MOSFET or impact ionization processes). The latter situation is the one examined by Phillips and Price. The population of the high energy states can be enhanced by generating a fixed number N of carrier histories every time one of the simulated particles enters the rarely populated region. Each one of the N generated particle will have the same initial condition (equal to the state of the "parent" particle at the moment of the multiplication) and a weight $1/N$ for the calculation of the average quantities. The multiplication algorithm can be repeated several times at higher energies, originating an "avalanche" of carriers that fill up the the tail of the distribution function at higher and higher energies [56,57]. A similar multiplication technique has been also used in real space in [36, 56,77].

A further implementation of the multiplication idea has been presented in [78] and is illustrated in Tab. 1. Such approach can be seen as a weighted MC algorithm, where .each particle is assigned a statistical weight which varies with its position in the device and its energy. With such approach, it is possible to account for regions with very different doping levels (as in bipolar transistors) and to obtain a reliable statistics of rare processes, keeping at the same time a constant number of particles.

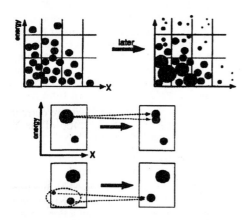

AT THE BEGINNING OF THE SIMULATION

- *set up a discretization in energy and real space*
- *assign a prescribed number of psuedoparticles to each cell*

DURING THE SIMULATION

- *duplicate a psuedoparticle if the cell is not sufficiently populated*
- *condense two particles in cells overly populated*
- *conserve mean position, velocity and energy during duplication or condensation by introducing variable statistical weights of the pseudoparticles*

Tab. 1 Multiplication algorithm for the weighted Monte Carlo algorithm

Comparison with traditional simulators

As pointed out earlier, the Monte Carlo technique is a fairly new tool in the area of semiconductor device modeling. Traditionally, two alternative methods are used: the drift-diffusion (DD) model, and the hydrodynamic (HD) model. A detailed analysis of the two methods is out of the purpose of this article (a very good discussion can be found in Ref. 9-11. It is nevertheless interesting to look at basic assumptions underlining the DD and HD models, and to compare their results with those obtained with a MC procedure. All methods are based on Boltzmann transport equation. By taking the first moments of the Boltzmann equation, we obtain (after few assumptions) three coupled equations:

$$\frac{\partial n}{\partial t} + \nabla\cdot(n\mathbf{v}) = 0 \qquad (4)$$

$$\frac{\partial \mathbf{p}}{\partial t} + (\mathbf{v}\cdot\nabla)\,\mathbf{p} + e\mathbf{E} + \frac{1}{n}\nabla(nK_BT_e) = -\frac{\mathbf{p}}{\tau_p} \qquad (5)$$

$$\frac{\partial \varepsilon}{\partial t} + \mathbf{v}\cdot\nabla\varepsilon + e\mathbf{E}\cdot\mathbf{v} + \frac{1}{n}\nabla\cdot(K_BT_en\mathbf{v}) - \frac{1}{n}\nabla\cdot(K\nabla T_e) = \frac{\varepsilon-\varepsilon_0}{\tau_\varepsilon} \qquad (6)$$

where n , **p**, and ε represent respectively the carrier density, average momentum and average energy. T_e denotes the carrier temperature, **E** the electric field, and K the thermal conductivity. Equation (4) is the usual current continuity equation, and expresses charge conservation (here no generation and recombination processes are considered).Equations (5) and (6) express respectively momentum conservation (with **p** =m **v**) and energy conservation within the relaxation time approximation.   The DD approach assumes that carriers are always in equilibrium with the lattice temperature (T_e=T), which leads to the usual equation, for steady-state conditions,

$$\mathbf{v} = -\mu\mathbf{E} - \frac{1}{n}D\nabla n \qquad (7)$$

where the carrier mobility and diffusivity can be field-dependent quantities related through a generalized Einstein relation. The HD approach uses all three equations (4-6), and relies on reasonable models for the relaxation times. Clearly, the HD scheme is far superior to the DD one, since it can account (when all terms of Eq. (6) are considered) for carrier heating and non-homogeneous distributions of the carrier temperature. The MC procedure stands on an ever higher level, since it provides (even in non-homogeneous, non-stationary conditions) an exact solution of the Boltzmann equation. It correctly describes non local effects (in space **r** or time t) where an electron contributes to the current in **r** at t coming from a position **r'** at t ' where the distribution function $f(\mathbf{r'},t')$ was different from $f(\mathbf{r},t)$. Unfortunately, the complexity and cost of each approach is inversely proportional to the refinement of the physical model it is based on. Therefore, the use of one approach or another depends on the specific device under investigation. In conclusion, the Monte Carlo simulation of semiconductor devices has made considerable progress over the last few years, and is today a very valuable tool in the area of devices modeling. Furthermore, the discussion of the last section indicates that MC is the best technique to study situations where non-stationary effects are important (as for example in submicron devices). It is safe to anticipate that, as the tendency toward the miniaturization of devices will continue in the future, MC simulators will progressively extend their applicability. The success of MC in device modeling will ultimately depend on the compromise between two tendencies, one to use very sophisticated physical models (which lead to very costly, but extremely accurate simulations), the other to rely on simplified models, sacrificing a bit of accuracy for a reduced complexity and cost of the algorithms. Along this line, it will be extremely useful to be able to combine different methods, in order to fully exploit the potential of each approach.

THE PHYSICAL MODEL FOR GaAs

The transport model developed for GaAs is based on a three-valley description of the conduction band and on three valence bands. Two separate simulations are performed in general for electrons and holes, respectively. Non parabolic isotropic dispersions are used, with non parabolicity factors treated as fitting parameters and adjusted as to reproduce a variety of experimental results, including those provided by time resolved spectroscopy.

The carrier interaction with polar optical, acoustic, equivalent and nonequivalent intervalley, intraband and interband phonons, and ionized impurities is considered. In the presence of heavily doped regions, carrier-carrier scattering (including plasmon contributions) are taken into account. A list of all relevant parameters is presented in Table 2. As mentioned previously, a full band structure description can be used for both conduction and valence band [16,19,54]. In such case the discretized $\varepsilon(\mathbf{k})$ relationship coming from a pseudopotential calculation is used for the electron dynamics. The scattering matrix elements can also be directly calculated using the electronic states coming out of the band calculation [16]. The MC simulation based on a full band structure description requires considerable memory and CPU time. It might be therefore preferable to use simplified models, such as the one presented here.

A very important effect in devices is impact ionization, a process is responsible for the breakdown of the device. A lot of effort is currently been devoted to the study of such effect [78-80] We have adopted the model proposed by Kane [81], where the energy dependence of the ionization process is given by $P(E)= (P/\tau) (E-E_{th})^a$, E being the total carrier energy. Kane model gives a softer threshold for the ionization with respect to the more standard Keldish formalism (Fig. 2) [82]. The free parameters have been determined through the fit of the bulk ionization coefficient and are given in Table 2. Band-to-band tunneling and its dependence on electric field has also been accounted for on the basis of Kane's theory [83].

Electron parameters (300K)	Γ (000)	L (111)	X (100)
Acoustic deformation potential (eV)	7.0	7.0	7.0
Effective mass (m^*/m_0)	0.063	0.222	0.58
α (nonparabolicity) (eV^{-1})	0.610	0.244	0.061
Band-edge energy (eV)	1.439	1.739	1.961
Intervalley deformation potential (eV/cm)			
from Γ (000)	0.0	$7.0 \cdot 10^8$	$1.0 \cdot 10^9$
from L (111)	$7.0 \cdot 10^8$	$1.0 \cdot 10^9$	$1.0 \cdot 10^9$
from X (100)	$1.0 \cdot 10^9$	$1.0 \cdot 10^9$	$1.0 \cdot 10^9$
Intervalley phonon energy (meV)			
from Γ (000)	0.0	27.8	29.3
from L (111)	27.8	27.8	29.3
from X (100)	29.3	29.3	29.3
Number of equivalent valleys	1	4	3

Hole parameters (300K)	H.H.	L.H.	S.O.
Acoustic deformation potential (eV)	7.0	7.0	7.0
Effective mass (m^*/m_0)	0.45	0.085	0.154
Band-edge energy (eV)	0.0	0.0	0.35
α (nonparabolicity) (eV^{-1})	0.55	4.0	0.3
Interband deformation potential (eV/cm)			
from H.H.	$9.0 \cdot 10^8$	$7.0 \cdot 10^8$	$5.0 \cdot 10^8$
form L.H.	$7.0 \cdot 10^8$	$9.0 \cdot 10^8$	$5.0 \cdot 10^8$
from S.O.	$5.0 \cdot 10^8$	$5.0 \cdot 10^8$	$9.0 \cdot 10^8$
Interband phonon energy (meV)			
from H.H.	27.8	27.8	29.9
from L.H.	27.8	27.8	29.3
from S.O.	29.3	29.3	29.3

Parameter	Value
Density (g/cm^3)	5.36
Sound velocity (cm/s)	$5.22 \cdot 10^5$
ε_∞	10.92
ε_0	12.90
Optical phonon energy (meV)	35.36
Electron ionization coefficients	
P/τ (s^{-1})	$6.22 \cdot 10^{12}$
a	3.2
E_{th} (eV)	1.439
Hole ionization coefficients	
P/τ (s^{-1})	$3.11 \cdot 10^{11}$
a	6.35
E_{th} (eV)	1.439

Table 2

Parameters used in the MC simulation for electrons (upper part), and holes (middle part). Other parameters, including impact ionization ones, are reported in the lower part.

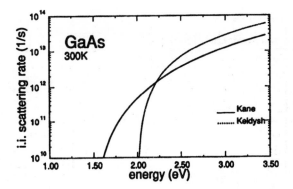

Figure 2 Comparison of the room temperature scattering rates for electron impact ionization calculated from Kane (solid line) and Keldish (dotted line) models.

High Field Transport in bulk GaAs

We present now the results of the MC simulation of electron and hole transport in homogeneous high electric fields based on the model previously described. All results are presented for 300 K and 500 K. The higher temperature is significant since devices often operate at a temperature much higher than the lattice one, due to Joule heating. The drift velocity versus electric field characteristics (Fig. 3) show the well known negative differential mobility region for electrons, with peak field around 4 kV/cm. The overall velocity is decreased for high temperature operation The hole velocity also shows a slight decrease, after reaching a maximum around 7 kV/cm, in correspondence to a sizable transfer from the heavy-hole to the light-hole and split-off bands. Our results are in good agreement with other MC calculations that use full band structure models and with available experimental results.

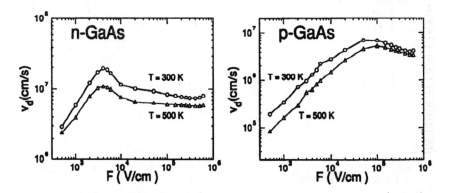

Figure 3 Drift velocity versus electric field curve for electrons and holes at 300 K (solid lines) and 500 K (dashed lines)

Carrier heating at the highest electric fields is clearly displayed in Fig. 4, with average energies reaching few tenths of eV for fields exceeding 100 kV/cm. Holes tend to be hotter than electrons above 50 kV/cm.

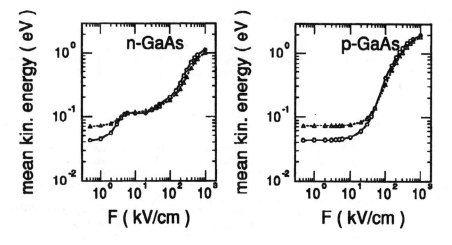

Figure 4 Electron and hole mean energy as a function of the electric field at 300 K (solid lines) and 500 K (dashed lines)

The field dependence of the carrier energy leads to the ionization rates shown in Fig. 5. We notice that, in agreement with room temperature experimental results, the ionization rates of electrons and holes are quite similar over the whole field range. At the higher temperature, the ionization rate decreases because of the decrease of the carrier mean energy due to the more effective scattering mechanisms.

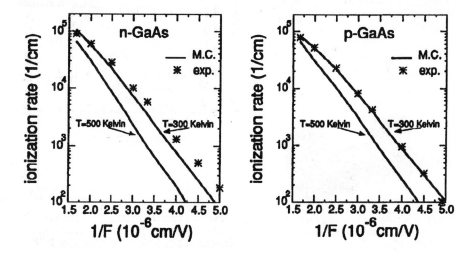

Figure 5 Ionization rates calculated from the Monte Carlo simulation (solid lines) at 300 K and 500 K, compared with the room temperature experimental values (symbols).

MONTE CARLO SIMULATION OF GaAs DEVICES

In the present section we illustrate some results obtained with the model previously introduced applied to self consistent simulations of GaAs MESFETs, pin diodes, and AlGaAs/GaAs HBTs. A more detailed discussion of the results is found in Refs. [73,78,84].

Selfaligned FETs

Selfaligned FETs (SAGFET) are one of the most promising candidates for GaAs power applications. As illustrated in Fig. 6 the active part of the device is characterized by four different regions, obtained via four different implants. The doping profile is also shown in the figure. The proper operation of the device is limited by the occurence of breakdown at high source-to-drain voltages, caused by avalanche multiplication. The onset of such regime can be monitored through the variation of the gate current at increasing drain bias. Figure 7 shows the drain and gate current characteristics for two different devices, which differ only for the gate length, respectively 1530 and 530 nm. In both cases, above 6 V the calculated gate current increases, due to the collection at the gate electrode of holes which are generated by impact ionization in the LDD region. The reduction of gate length leads to a more rapid rise of the gate current. This comes from the higher electric field of the shorter device, which in turn leads to higher average energies of the channel electrons and therefore to the enhancement of the ionization processes. It should be pointed out that the excellent agreement with the measured characteristics (solid lines in Fig. 6) and the comparison with the results of a Drift-Diffusion simulator have clearly indicated the occurence of a considerable velocity overshoot in the shorter devices, which affects directly the output current of the transistor and enhances its high frequency performance.

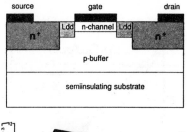

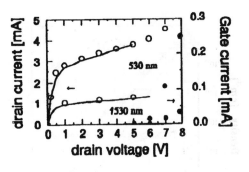

Figure 6 Cross section and doping profile of the simulated self-aligned implanted FET.

Figure 7 Drain (left scale, open circles) and gate (right scale, closed circles) current characteristics for two different gate lengths.

Heterojunction Bipolar Transistors

HBTs are receiving considerable attention becuase of their high speed performance and high current handling capability. The device shown in Fig. 8 has been simulated, and the results compared with measured data. By varying the base-collector voltage (in the common-base configuration), the electric field profile changes as depicted in Fig. 9. Very high values are reached in the collector region, with the maximum occuring at the base-collector interface. Electrons, injected from the emitter, cross the base where they strongly interact with the dense hole plasma. As they enter the collector, they are ballistically accelerated by the junction field, reaching velocities as high as 6×10^7 cm/s. The spatial extent of the velocity overshoot is limited to about 100 Å, as the electrons are rapidly scattered into the satelite valleys where they move at saturated velocity.

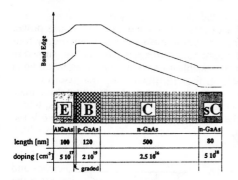

Figure 8 Energy band diagram and cross section of the simulated HBT.

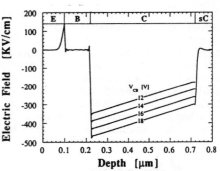

Figure 9 Electric field profile of the simulated HBT at various V_{BC}.

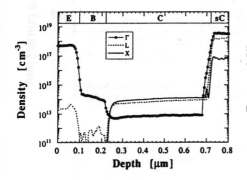

Figure 10 Electron concentration in the different valleys for $V_{BC}=16$ V

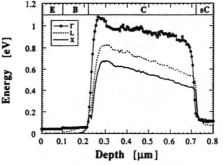

Figure 11 Electron average energy in the different valleys for $V_{BC}=16$ V.

Figure 10 indeed shows that for a collector voltage of 16 V almost all electrons in the collector populate, in equal number, the L and X valleys. There, they are strongly heated by the collector field, obtaining the high values of average energy illustrated in Fig. 11.

226

Correspondingly, the calculated (M-1) factor, which measures the relative increase of collector current due to multiplication phenomena, is around unity at this voltage, in excellent agreement with the measured value [73]. As V_{BC} is further increased, the holes created by primary ionization processes are in turn able to ionize, marking the onset of breakdown, which is predicted around 18 V.

PIN diodes

Probably the thoughest test of the proposed physical model and numerical algorithm is the simulation of the breakdown regime in PIN diodes. There, both electrons and holes contribute to the device current and to the multiplication mechanisms. Furthermore, the current level can change by several orders of magnitudes as the reverse bias increases. By varying the width of the intrinsic region, it is possible to switch from a soft to a hard breakdown. This is illustrated by the symbols in Fig. 12, illustrating the experimental results obtained for a 20 nm and a 50 nm intrinsic regions respectively. The MC simulation (solid lines) agrees well with the measured characteristics, and allows the identification of tunneling as the mechanism responsible for the smoother current increase, previous to the actual breakdown caused by impact ionization.

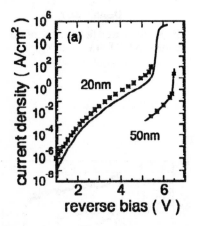 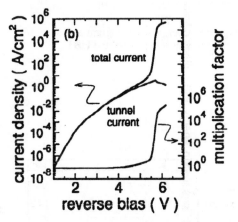

Figure 12 Comparison of the simulated and measured I-V characteristics for two diodes of different intrinsic region.

Figure 13 Temperature dependence of the simulated I-V characteristics for the PIN 20 nm structure.

The asterisks in Fig. 12 show the measured current density versus voltage at room temperature near breakdown for the 20 nm and 50 nm pin diode, respectively. As can be deduced from Fig. 12, the measured reverse current in these diodes exhibit a soft breakdown suggesting a significant tunneling contribution. The results for the 200 nm structure are not included in the figure; they show a hard breakdown at -10 V, in contrast to the short i-zone diodes. In Fig. 12, the calculated current density is indicated by continuous lines and is generally seen to be in good agreement with the experimental data. In the tunneling regime for low bias, the simulation underestimate somewhat the current for the 20 nm structure. This may be partly caused by band gap narrowing in the doped regions which is not accounted for in the tunneling employed in the simulation. For the 200 nm structure, the theory predicts the breakdown voltage to be -10 V, in excellent agreement with the experimental value.

227

Fig. 13 shows the separate calculated contribution of tunneling and ionization to the total
reverse current in the 20 *nm* structure. Up to a voltage of -4 V, the total current is mainly
due to particle generation via band-to-band tunneling. For higher reverse bias, carrier
generation in the intrinsic region is enhanced due to impact ionization processes.
Conventionally, such current increase is measured by the avalanche multiplication factor.
Fig. 14 elucidates the physical mechanisms leading to the observed behavior. The curves
refer to two different voltages applied to the 20 *nm* structure (full line: -4.5 V; dashed line:
-5.5 V). It can be seen that the electric field in the near intrinsic region exceeds 1 MV/cm,
as breakdown is approached (Fig. 14a). Consequently, the electron and hole average
energies reach values of several hundred meV's with energy maxima near the p-i and i-n
junction regions, respectively (Fig. 14b). Associated to such high energy values, the
highest impact ionization rates for electrons and holes also occur near the junction region
(Fig. 14c).This leads to a strong *dark space* effect, that is a displacement of the maximum
ionization rate from the peak of the electric field

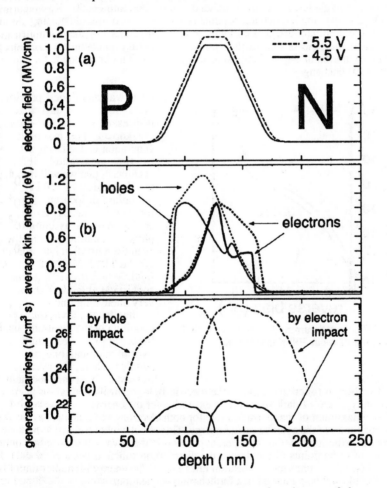

Fig. 14. Electric field, average kinetic energy, and generated carrier rate of electrons and
holes as a function of position in the 20 *nm* pin-diode.

In Fig. 14b, the full lines show the carrier energies for a bias of -4.5 V. In the middle of the intrinsic zone, the dominant mechanism is tunneling, whereas band-to-band impact ionization dominates near the boundaries. This is the origin of the doubly-peak shape of these curves. For the hgher bias, however, the impact ionization dominates already in the whole space-charge region and leads to a smoother energy profile (dashed lines).

At higher voltages the device breaks down. From a microscopic point of view, this occurs when ,on the average, a pair of carriers created by an impact ionization event is able to create another one pair by the same process. Such multiplication in the number of carriers causes the space-charge region to be flooded with newly generated carriers. As a consequence, the electric field and the tunneling current are reduced as the generated space charge approaches the contact doping. A related effect, visible in Fig. 14b around 6V, is the saturation of the total current, which has been observed also experimentally.

The 20 *nm* pin diode has been also simulated at a temperature of 500 K, which is more likely to be the operatng internal temperature of the device due to Joule heating. As shown in Fig. 15, the dominant effect is an increase with temperature of the calculated current density. Such increase is due to the enhanced tunneling contribution caused by the narrowing of the band gap at increasing temperatures. The breakdown for the 20 *nm* structure is left unchanged.

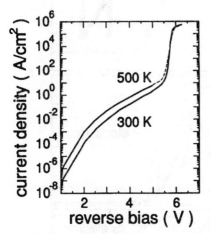

Figure 15 Calculated current density vs voltage for the 20 *nm* pin diode at 300 K and 500 K

As it was mentioned previously, a quite pronounced dark space effect is present in the pin diode near breakdown. The maximun carrier generation is displaced with respect to the maximum field. Such occurence, that we will find again in the discussion of HBTs, leads to the failure of local models (such as for instance the DD model), which are based on the assumption that all physical quantities, and therefore even the ionization rate, depend *only* on the local value of the electric field. When this is not true, more sophisticated models are needed. A very interesting finding of our study is that in the extreme operation condition of the pin diode examined here, approaches such as the hydrodynamical one are also not suitable. There, the average energy (related to the second moment of the carrier distribution function), and not the electric field, controls ionization phenomena. Unfortunately, a sort of dark space effect exist also for the energy, as clear from Fig. 14. Indeed, the maximum of the ionization rate for both electrons and holes does not coincide with the maximun of the average carrier energy. This effect could be clarified by looking at the carrier distribution function in different position in the device. It is possible, in fact, to verify that in some points of the device the distribution function has a pronounced high energy tail (which implies high ionization rate) even if the energy is smaller than in other regions. Details will be published in a forthcoming publication.

CONCLUSIONS

We have illustrated the basic principles of the Monte Carlo technique, and outlined a series of applications of the Monte Carlo method to device simulation. The technique provides a very useful tool for the understanding of the physical phenomena involved in carrier transport in semiconductor devices and is becoming a fundamental aspect in the area of CAD. This is particularly true in connection to submicron structures, where the non local features of the electron transport have to be accounted for. A simplified model for electron and hole high-field transport in GaAs have been presented. The good agreement with available experimental results confirms the validity of such model. The possibility to simulate by Monte Carlo methods near-breakdown and breakdown regimes has been also demonstrated.

ACKNOWLEDGMENTS

The contributions of A. Di Carlo, D. Liebig, M. Saraniti, P. Vogl, and G. Zandler to part of this work are gratefully acknowlededged.

REFERENCES

1. C. Jacoboni, and P. Lugli, *The Monte Carlo Method for Semiconductor Device Simulation*, Springer Verlag, Wien (1989).
2. C. Jacoboni, and L. Reggiani, Rev. Mod. Phys. **55** , 645 (1983).
3. M.V. Fischetti and S.E. Laux, Phys. Rev. **B38**, 9721 (1988).
4. K. Hess, J.P. Lebourton, and U. Ravaioli, Eds., *Computational Electronics,* Kluver Academic, Norwell, MA (1991).
5. J. M. Hammersley and D.C. Handscomb, *Monte Carlo Methods,* Methuen, London (1964).
6. Yu. A. Shreider, Ed., *The Monte Carlo Method*, Pergamon, Oxford 1966.
7. Jerome Spanier and Ely M. Gelbard, *Monte Carlo Principles and NeutronTransport Problems,* Addison-Wesley, Reading (1969).
8. B.R. Penumalli, in *Process and Device Modeling*, Ed. W.L. Engl, p. 1, North Holland, Amsterdam (1986).
9. S. Selberherr, *Analysis and Simulation of Semiconductor Devices*, Springer Verlag, Wien (1984).
10. G. Baccarani, M. Rudan, R. Guerrieri, and P. Ciampolini, in *Process and Device Modeling*, Ed. W.L. Engl, p. 107, North Holland, Amsterdam (1986).
11. W. Haensch, *The Drift Diffusion Equation and its Applications in MOSFET Modeling*, Springer Verlag, Wien (1991).
12. R.W.Hockney, R.A.Warriner, and M.Reiser, Electron. Letters **10**, 484 (1974).
13. G.Baccarani, C.Jacoboni, and A.M.Mazzone, Solid State Electr. **20** , 5 (1977).
14. P.Lugli and C.Jacoboni, in ESSDERC 87, Proc. 17th EuropeanSolid State Device Research Conference, edited by P.U.Calzolari andG.Soncini, p.97 (1987) and references therein.
15. A Yoshii, and M. Tomizawa, in *Process and Device Modeling*, Ed. W.L. Engl, p. 195, North Holland, Amsterdam (1986).
16. M. V. Fischetti and S.E.Laux, IEEE Trans. Electr. Dev. **ED38**, 650 (1991).
17. R. Castagne, in *High Speed Electronics*, Eds. B. Kallback and H. Beneking, Springer Verlag, p.2 (1986).
18. R.W. Hockney, and J.W. Eastwood, *Computer Simulation using Particles*, Mc Graw-Hill, N. York (1981).
19. K. Brennan and K. Hess, Phys. Rev. **B29**, 5581 (1984).
20. V. Gruzinskis, S. Kersulis, and A. Reklaitis, Semicond. Sci. Technol. **6**, 602 (1991).
21. W. R. Curtice, IEEE Trans. Electr. Dev. **ED 29** , 1942 (1982).

22. J.Zimmermann and E.Constant, Solid-State Electron. **23** , 915 (1980).
23. A. Yoshii, M. Tomizawa, and K. Yokoyama, IEEE Trans. Electron Dev. **ED30** , 1376 (1983).
24. Y. Awano, K. Tomizawa, and N. Hashizume, IEEE Trans. Electron Dev. **ED 31** , 448 (1984).
25. C. Moglestue, COMPEL **1** , 7 (1982).
26. M. Tomizawa, A Yoshii, and K. Yokoyama, IEEE Electron Dev. Lett. **EDL 6** , 332 (1985).
27. C. Moglestue, IEEE Trans. Computer-Aided Design, **CAD-4** , 536(1985)
28. P.Hesto, J.F.Pone, and R.Castagne, Appl.Phys.Lett. **40** , 405 (1982).
29. Y. Park, T. Tang, and D.H. Navon, IEEE Trans. Electr. Dev. **ED 30** , 1110 (1983).
30. C. Moglestue, IEEE Trans. Computer-Aided Design, **CAD-5** , 326 (1986).
31. Y. Park, D. H. Navon, and T. Tang, IEEE Trans. Electron Dev. **ED-31**, 1724 (1984).
32. M.A. Littlejohn, R.J. Trew, J.R. Hauser, and J.M. Golio, J. Vac. Sci. Technolog. **B1** , 449 (1983).
33. T. Wang, K. Hess, and G.J. Iafrate, J. Appl. Phys. **59** , 2125 (1986).
34. F. Antonelli, and P. Lugli, in Proc. of XVII European Solid State Device Research Conference ESSDERC87, Eds. P. Calzolari and G. Soncini, p. 177, Tecnoprint, Bologna (1987).
35. K. Brennan, IEEE Trans. Electron. Dev. **ED-32** , 2197 (1985).
36. M.V. Fischetti, IEEE Trans. Electr. Dev. **ED38**, 634 (1991).
37. K. Yokoyama, M. Tomizawa, A. Yoshii and T. Sudo, IEEE Trans. El.Dev. **ED-32** , 2008 (1985).
38. Sanghera, A. Chryssafis and C Moglestue, IEE Proc. **127** , 203 (1980).
39. C.K. Williams, T.H. Glisson, J.R. Hauser, M.A. Littlejohn and M.F. Abusaid, Sol.St.Electron. **28** , 1105 (1985).
40. R. Fauquembergue,M. Pernisek and E. Constant, El.Lett. **18** , 670 (1982).
41. I.C. Kizillyaly and K. Hess, Jap. J. Appl.Phys. **26** , 9 (1987).
42. A.P. Long, P.H. Beton and M.J. Kelly, J. Appl. Phys. **62** , 1842 (1989)
43. T. Wang and K Hess, J. Appl. Phys. **57** , 5336 (1985).
44. R. Fauquembergue, J.L. Thobel, P. Descheerder, M. Pernisek and P. Wolf, Solid-State Electron. **31** , 595 (1988).
45. M. Tomizawa, A. Yoshii and K. Yokoyama, IEEE El.Dev.Lett. **EDL 5** , 362 (1984).
46. M. Tomizawa and N. Hashizuma, in Proc. NASECODE IV Conf., p. 98, Ed. J.J. Miller, Boole, Dublin (1985).
47. U. Ravaioli and D.K. Ferry,Superlattices and Microst. **2** , 75 (1986).
48. U. Ravaioli and D.K. Ferry,Superlattices and Microst. **2** ,377 (1986).
49. U. Ravaioli and D.K. Ferry,in *High Speed Electronics*, Eds.B. Kallback and H. Beneking, p.136, Springer, Berlin (1986).
50. U. Ravaioli, and D.K. Ferry, IEEE Trans. Electr. Dev. **ED 33** , 677 (1986).
51. M. Mouis, J.F. Pone, P. Hesto and R. Castagne, in Proc. IEEE/Cornell Univ. Conf. on Advanced Concepts on High Speed Semiconductor Devices and circuits", Ithaca, N.Y. (1985).
52. R. Castagne, Physica **134B** , 55 (1985).
53. M. Mouis, P. Dolfus, B. Mougel, J.F. Pone and R. Castagne, in *High Speed Electronics*, Eds. B. Kallback and H. Beneking, p. 35, Springer, Berlin (1986).
54. H. Shichijo and K. Hess, Phys. Rev. **B23**, 4197 (1981).
55. K. Throngnumchai, K. Asada, and T. Sugano, IEEE Trans. Electron Dev. **ED-33** , 1005 (1986).
56. B. Ricco', E. Sangiorgi, F. Venturi, and P. Lugli, in Proc. Int. Electron Dev. Meeting (IEDM), Los Angeles, Ca, p. 559 (1986).
57. E. Sangiorgi, B. Ricco' and F. Venturi, IEEE Trans., Computer-Aided Design, **CAD-7** , 259 (1988).

58. F. Venturi, R.K. Smith, E. Sangiorgi, M.R. Pinto and B. Ricco', in *Simulation of Semiconductor Devices and Processes*, vol. 3, Eds. G. Baccarani and M. Rudan, Tecnoprint, Bologna (1988).
59. F. Venturi, E. Sangiorgi, and B. Ricco', IEEE Trans. Electron. Dev. **ED-38**, 1895 (1991)
60. M. Tomizawa, K. Yokoyama and A. Yoshii, IEEE Trans. Computer-Aided Des., **CAD-7**, 254 (1988).
61. Y-J. Chan, and D. Pavlidis, IEEE Trans. Electron. Dev. **ED-38**, 1999 (1991).
62. I. C. Kizilyalli, M. Artaki, and A. Chandra, IEEE Trans. Electron. Dev. **ED-38**, 197 (1991)
63. G.U. Jensen, B. Lund, T.A. Fjeldly, and M. Shur, IEEE Trans. Electron. Dev. **ED-38**, 840 (1991).
64. K.W.Kim, H. Tian, and M.A. Littlejohn, IEEE Trans. Electron. Dev. **ED-38**, 1737 (1991)
65. J. Hu, D. Pavlidis, and K. Tomizawa, IEEE Trans. Electron. Dev. **ED-39**, 1273 (1992).
66. H. Nakayama, M. Tomizawa, and T. Ishibashi, IEEE Trans. Electron. Dev. **ED-39**, 1558 (1992).
67. C. Hwang, D. H. Navon, and T. Tang, IEEE Trans. Electron Dev. **ED-34** , 154 (1987).
68. S. Bandyopadhyay, M.E. Brown, C.M. Maziar, S. Datta, and M.S. Lundstrom, IEEE Trans. Electron Dev. **ED-34** , 392 (1987).
69. S. Bosi, and C. Jacoboni, J. Phys. C **9** , 315 (1976).
70. P. Lugli and D.K.Ferry, IEEE Trans. Electron Dev. **ED-32** , 2431 (1985).
71. U. Ravaioli, P. Lugli, M. Osman, and D. K. Ferry, IEEE Trans. Electron. Dev. **ED 32** , 2097 (1985).
72. M. Heilblum, Solid State Electr. **24** , 343 (1981).
73. A. Di Carlo, P. Lugli, P. Pavan, E. Zanoni, and R. Malik, Microelectronic Eng. **19**, 135 (1992)
74. P.Lugli and D.K.Ferry, Physica B **129** , 532 (1985) .
75. P.Lugli and D.K.Ferry, IEEE El.Dev.Lett. **EDL6** , 25 (1985).
76. A. Phillips, and P.J. Price, Appl. Phys. Lett. **30** , 528 (1977).
77. Y. Yamada, S. Ikeda, N. Shimojok, in ESSDERC 87, Proc. 17th European Solid State Device Research Conference, edited by P.U.Calzolari and G.Soncini, p. 111 (1987)
78. D. Liebig, P. Lugli, P. Vogl, M. Claassen, and W. Harth, Microelectronic Eng. **19**, 127 (1992)
79. R. Thoma, H.J. Peifer, W.L. Engl, W. Quade, R. Brunetti, and C. Jacoboni, J. Appl. Phys. **69**, 2300 (1991)
80. J. Bude, K. Hess, and G.J. Iafrate, Phys. Rev. **B45**, 10958 (1992)
81. E.O. Kane, Phys. Rev. **159**, 624 (1967).
82. N. Sano, and A. Yoshii, Phys. Rev. **B45**, 4171 (1992).
83. E.O. Kane, J. Appl. Phys. **32**, 83 (1961).
84. P. Lugli, Microelectronic Eng. **19**, 275(1992)

Quasi-Two-Dimensional Models for MESFETs and HEMTs

Alain Cappy
Université des Sciences et Technologies de Lille

INTRODUCTION

For the physical modelling of MESFETs and HEMTs, the choice of a modelling technique is strongly guided by the aimed objectives. Roughly speaking, two dimensional (2D) models are mainly devoted to physical analysis while more simple models (analytical or one-dimensional for instance) are rather devoted to electrical engineering, CAD, IC design... The aim of the quasi two dimensional approach (Q2D) is to be a modelling taking into consideration the most important phenomena occuring in the device while staying simple and fast enough to be used in CAD. In addition one objective of Q2D approach is to provide not only the I-V characteristics but also the AC and noise performance as well as the non linear behavior of the device. The main particularities of the Q2D approach can be summerized as follows.

- Introduction of "exact" transistor topology (recess structure, position of the gate in the recess...) ;

- Introduction of "exact" layer structure including the cap layer, the different layers in a HEMT, the donor distribution in an implanted MESFET...

- Introduction of the non stationary electron dynamics for the modelling of submicrometer gate length devices ;

- Introduction "as far as possible" of the two dimensional effects occuring at the drain edge of the gate ;

- Short calculation time allowing a good interactivity.

In the Q2D approach, the modelling is performed in three steps.

- first, the analysis of the active layer is carried out to provide the charge control law of the active layer (C-V relationship) ;

- second, this charge control law is combined with the electron transport, current continuity and Gauss law to give the DC characteristics ;

- third, the AC, noise and non linear analysis are carried out after introduction of the time derivative in the physical equations and/or circuit concepts.

THE CHARGE CONTROL

The aim of the charge control analysis in MESFETs and HEMTs is to calculate the sheet carrier density in each layer constituting the active layer as a function of the gate voltage. A one dimensional calculation is carried out in the direction perpendicular to the gate. No longitudinal electric field (i.e. $V_{ds} = 0$ or infinite structure) is assumed. The case of MESFETs and HEMTs will be successively considered.

MESFET

In the case of epitaxial MESFETs with constant doping N_d in a layer of thickness A, the first idea is to use a simple analytical model for the charge control. For this purpose, Shockley's model (Shockley 1952) is well suited. Using this model, the sheet carrier density is given as a function of the gate voltage by :

$$N(V_g) = AN_d\left(1 - \sqrt{\frac{2\varepsilon(V_b - V_g)}{q \, Nd \, A^2}}\right) \quad (1)$$

The gate capacitance per unit area is simply proportional to the derivative of $N(V_g)$ with respect to V_g

$$C(V_g) = \frac{\varepsilon}{A}\left(\sqrt{\frac{2\varepsilon(V_b - V_g)}{q \, Nd}}\right)^{-1} \quad (2)$$

In the case of non uniform doping, the same approach can be used if an equivalent doping density Nd and an equivalent active layer thickness A are defined from the actual doping profile Nd(y) by (Shur and Eastman, 1980) :

$$\overline{A} \, \overline{N_d} = \int_0^\infty N_d(y) \, dy \quad (3)$$

$$q \, \frac{\overline{N_d} \, \overline{A^2}}{2\varepsilon} = \frac{q}{\varepsilon} \int_0^\infty N_d(y) \, y \, dy \quad (4)$$

The main advantage of this method is its simplicity while its main drawback is a bad description of the capacitance-voltage characteristics near pinch-off. As a consequence such a model can introduce errors and inacurracy for low drain current operation (low noise condition). To overcome this difficulty a full numerical modeling is well suited. For any doping profile $N_d(y)$, the N-V and C-V characteristics can be deduced from the resolution of Poisson's equation and Fermi statistics

$$\frac{d^2 Ec}{dx^2} = \frac{q2}{\varepsilon}\left(N_d^+(y) - n(y)\right) \quad (5)$$

$$n(y) = N_c F_{1/2}\left(\frac{E_F - E_C}{kT}\right) \quad (6)$$

where $F_{1/2}$ denotes Fermi integral. Starting from boundary conditions taken in the substrate far from the schottky barrier, the resolution of Eq(5) and (6) by using Runge-Kutta method for instance, provides the variations of the sheet carrier density and gate capacitance as a function of the gate potential. This calculation has to be made only once for each layer structure and the time needed for this calculation does not increase significantly the calculation time needed for the whole (DC, AC, Noise) performance calculation.

HEMT

The charge control problem in HEMTs is much more complicated due to the higher complexity of the active layer structure. Basically a HEMT active layer is made of two materials ; a highly doped high gap material associated together with a non intentionally doped small gap material. Even in the case of multiheterojunction layers, we have in fact two electron populations : a high mobility electron population corresponding to the electrons located in the small gap material and a low mobility electron population corresponding to the electrons located in the large gap material. Hence, the problem of charge control in HEMTs is to calculate the sheet carrier density in each material as a function of the gate potential. Unfortunately this problem is complicated by the presence of the so called DX centers in large gap materials such as AlGaAs. These DX centers are deep levels associated to the L valley of the material and are responsible of different parasitic effects such as : collapse of I-V characteristics, persistent photoconductivity, pinch-off voltage shift (Drummond, 1984) (Morgan, 1986). The introduction of DX centers in the charge control law of HEMTs is essential, especially if the low temperature behavior study is planned. To account for DX centers effects in the charge control law, two different Poisson's equations have to be written. For DC or low frequency operation, the electrostatic potential V(y) is a function of the ionized impurity density as well as the free carrier density while in the case of high frequency operation the ionized impurities do not contribute to the variations of the electrostatic potential V(y) :

$$\frac{d^2 V(y)}{dy^2} = \frac{q}{\varepsilon}\left(N_d^+ (y) - n(y)\right) \text{ for DC} \qquad (7)$$

$$\frac{d^2 V(y)}{dy^2} = -\frac{q}{\varepsilon} n(y) \text{ for HF} \qquad (8)$$

Hence, the calculation of the HF gate capacitance for a HEMT has to be performed using Eq.(8).

A second problem encountered in HEMT layers is related to quantum effects. As it is well known, the narrow potential well present at the heterojunction interface induces a quantization of the energy levels. A rigorous treatment of this problem can be made by solving self consistently Schrödinger and Poisson's equations (Vinter, 1982). However this calculation is rather time consuming especially if both the DC and AC behaviors have to be calculated.

A second possibility is to perform a classical calculation neglecting all quantum effects. In this case Eq(6), (7) and (8) are solved taking into consideration the conduction band discontinuity at the heterojunction. A comparison between quantum and classical calculations was presented by Yoshida (1986).

A third possibility was proposed by Delagebeaudeuf (Delagebeaudeuf, 1982). In this method, the quantum well is treated with quantum effects by using conventional approximations (triangular well, square well, W.K.B. method) while the high gap material is treated classically.

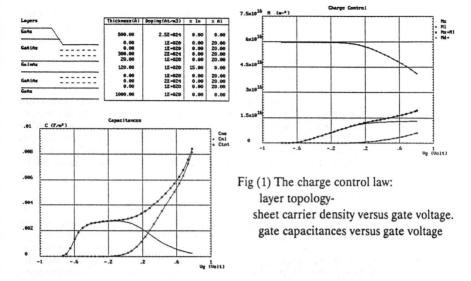

Fig (1) The charge control law:
layer topology-
sheet carrier density versus gate voltage.
gate capacitances versus gate voltage

Continuity of electric field and electrostatic potential is used to connect the quantum world (small gap material) to the classical world (large gap material). This method is more accurate than the classical calculation and faster than the full quantum calculation. As a consequence, it is very well suited for the device modelling. Figures (1b), and (1c) show the charge control low in terms of sheet carrier density and gate capacitance in the case of a pseudomorphic layer having the topology shown in figure (1a). The calculation assumes a linear dependence of the subband energy versus the sheet carrier density Ns (Alamkan 1990, Happy 1993). Figure (1b) and (1c) are important for several reasons. First, it gives the basic insight into the device behavior and it defines precisely the pinch-off voltage. Second, the total capacitance, as shown in fig. (1c), can be directly compared with the experimental capacitance in order to check the validity of the technological parameters of the active layer introduced in the simulation. At this step it should be emphasized that for a comparison between modelling results and experimental ones it is useless to calculate DC or AC performance as long as the theoretical and experimental C-V characteristics are not in good agreement. Obviously, this assertion is true for any modelling technique.

In the development of a charge control model, the physical and mathematical problems are rather simple and the main problems are concerned with material data. In the following references, reliable data concerning band structures, conduction band offsets and other useful parameters can be found : (Adachi, 1982, Alamkan 1990).

THE ELECTRON TRANSPORT

A device modelling is basically based on the self consistent solution of a current continuity equation together with Poisson's equation. Hence an electron transport model is needed. In the case of submicrometer gate length devices, the non stationary electron dynamics is known to be important and has to be taken into account. For that purpose, the relaxation time approximation (RTA) is well suited because it is simple and accurate enough for gate length down to 0.15-0.2 micron. Assuming a one dimensional transport and a single electron gas we have :

$$m^*(\varepsilon)\,v\,\frac{dv}{dx} = qE - \frac{m^*(\varepsilon)\,v}{\tau_m(\varepsilon)} \qquad (9)$$

$$\frac{d\varepsilon}{dx} = q\,E - \frac{\varepsilon - \varepsilon_0}{v.\tau_\varepsilon(\varepsilon)} \qquad (10)$$

the effective mass m* and the momentum and energy relaxation times are supposed to be functions of the average energy ε and these parameters are deduced from steady state Monte Carlo calculation (Shur 1976, Carnez 1980). Once again, these parameters are not always very well known especially in the case of complex structures such as pseudomorphic AlGaAs/InGaAs/GaAs layers. As a first approximation and due to the lack of reliable data, bulk values can be used (Thobel, 1990).

BASIS OF THE Q2D APPROACH

In order to clearly explain the different approximations and assumptions made in the Q2D approach, a general view of a full two dimensional approach is interesting. For that purpose figure 2 shows the charge concentration total electron energy, electrostatic potential and current density in the case of a conventional AlGaAs/GaAs HEMT (Shawki, 1988, 1990). These figures have been obtained using a two dimensional modelling based on the hydrodynamic approach. The following features can be emphasized :

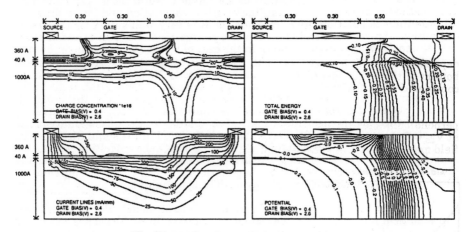

Fig (2) results of two-dimensional modelling

(i) the equipotential lines are nearly parallel to the gate in the depleted region under the gate ($E_y \gg E_x$).

(ii) the equipotential lines are nearly perpendicular to the gate in the GaAs layer ($E_x \gg E_y$) except near the heterojunction.

(iii) the current density vector is mainly parallel to the gate (x direction) and an injection of carriers in the GaAs buffer (delocalisation) occurs at the drain end of the gate).

These features constitutes the basis of the Q2D approach. To describe the assumptions made in Q2D modellings let us consider a slice of the device under the gate. For the sake of generality, a pseudomorphic AlGaAs/InGaAs/GaAs HEMT is considered (figure (3)).

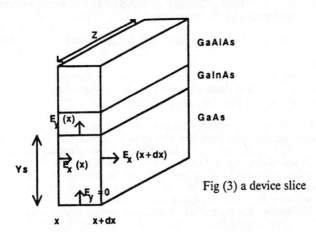

Fig (3) a device slice

* First, it is assumed that the electric field component E_y is only dependent on the difference between the gate potential V_g and the potential at the bottom of the buffer layer $V(x)$. For the calculation of Ey, two cases have to be considered according to the value of $V_g - V(x)$.

If $V_g - V(x) > V_p$, the active channel is not pinched-off and the perpendicular electric field Ey is simply given by the charge control law obtained at $V_{ds} = 0$. In that case, Ey is zero in the GaAs buffer. The band structure and the field map are shown in figure (4) and (5).

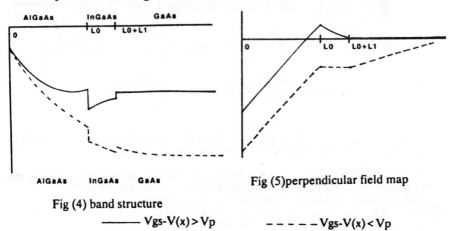

Fig (4) band structure

Fig (5) perpendicular field map

——— $V_{gs} - V(x) > V_p$ – – – – – $V_{gs} - V(x) < V_p$

If $V_g - V(x) < V_p$ the channel is completely pinched-off and the current is only due to the injected carriers in the buffer. In this case, we assume that the field map undergoes a translation of the value $E_y (L_0 + L_1)$. In the GaAs buffer Ey increases from zero to $E_y (L_0 + L_1)$. The band structure and the field map of this case are shown in figure (4) and (5).

This assumption concerning E_y allows us to define the whole function $E_y(y)$ from the knowlegde of the difference $V_g - V(x)$. In addition this behavior constitutes a good approximation of 2D modellings. In the case of other FET structures, like conventional HEMT or MESFET, the same method can be used with only some changes in the notations.

* Second, it is assumed that the longitudinal electric field E_x is independent of y.

* Third, in the x direction the diffusion current is supposed to be small as compared with the drift current (in the y direction, the diffusion of carriers is taken into account in the charge control model).

* Fourth, in the source-to-gate and gate-to-drain regions, far from the gate, the same analysis is used but the gate potential V_g is replaced by the surface potential V_s corresponding to the Fermi level pinning by the surface states. On both sides of the gate, near the gate, the potential at the surface varies from V_g to V_s (edge effects). It is rather difficult to specify the variation of the surface potential in these regions. We believe that any approximation (linear, quadratic, cubic) can be used without any dramatic changes in the device performance. Obviously, the potential variations in the regions located on both sides of the gate are really two-dimensional and the different approximations used of the Q2D approach are rather poor in these regions. The poor modelling of these regions constitutes one limitation of Q2D approach especially in the case of recess structures when the recess width is close to the gate length.

THE DC CHARACTERISTICS

For calculating the DC characteristics, the following set of equations has to be solved

$$I_{ds} = qZ \left(N_1(x) v_1(x) + N_2(x) v_2(x) \right) \quad (11)$$

$$\Delta x \, Ey \, (0, x) + \left(L_0 + L_1 + L_2 \right) \left[E_x(x) - E_x(x - \Delta x) \right] = \frac{q \Delta x}{\varepsilon} \left[N_d^+(x) - N_1(x) - N_2(x) \right] \quad (12)$$

$$m^*(\varepsilon) v_2(x) \frac{dv_2(x)}{dx} = q \, E_x(x) - m^*(\varepsilon) \frac{v_2(x)}{\tau_m(\varepsilon)} \quad (13)$$

$$\frac{d\varepsilon}{dx} = q \, E_x(x) - \frac{\varepsilon - \varepsilon_0}{v_2(x) \tau_\varepsilon(\varepsilon)} \quad (14)$$

Eq(11) is the current continuity equation. Z is the device width N_1 the sheet carrier density in the large gap material, v_1 the electron velocity in the large gap material (low mobility) while N_2 and v_2 are the sheet carrier density and electron velocity in the high mobility material. As it is clearly shown in Eq (11) the velocity v_1 or v_2 corresponds to the average velocity of all the carriers of a layer (sheet carrier density N_1 or N_2).

Eq (12) is simply Gauss law expressed in a slice.

Eq(13) and (14) are the electron transport equations. In principle, these equations have to be solved in each layer (subscript 1 and 2). However in normal operation, the large gap material is fully depleted and solving Eq(9) and (10) for this material is unecessary. In addition, it can be noted that the relaxation times τ_m and τ_ε are not very well known in AlGaAs, while the exchange of particles between layers 1 and 2 is poorly described in Q2D modelling. For these reasons it seems sufficient to describe the electron transport in the large gap-low mobility material with a simple v(E) relationship. In the case of MESFET the problem is more simple and equation (11) to (14) can still be used with $N_1 = 0$ and $v_1 = 0$.

Introducing a finite difference scheme, it is possible to write equations (11) - (14) as follows :

$$I_{ds} = q\,Z\left[N_{1i}\,v_{1i} + N_{2i}\,v_{2i}\right] \quad (15)$$

$$\Delta x\,E_{yi} + \left(L_0 + L_1 + L_2\right)\left[E_{xi} - E_{xi-1}\right] = \frac{q\Delta x}{\varepsilon}\left[N_{di}^+ - N_{1i} - N_{2i}\right] \quad (16)$$

$$v_{2i} = \left(\frac{q\tau_{mi}}{m_i^\bullet}\,E_{x,i} + \frac{\tau_{mi}\,v_{2,i-1}^2}{\Delta x}\right)\Big/\left(1 + \frac{\tau_{mi}\,v_{2,i-1}}{\Delta x}\right) \quad (17)$$

$$\varepsilon_i = \varepsilon_{i-1} + \Delta x\left(q\,E_{x,i-1} - \frac{\varepsilon_{i-1} - \varepsilon_0}{v_{2,i-1}\cdot\tau_{\varepsilon i-1}}\right) \quad (18)$$

Combining (15), (16) and (17) it comes :

$$\alpha_{i-1}\,E_{xi}^2 + \beta_{i-1}\,E_{xi} + \gamma_{i-1} = 0 \quad (19)$$

where α_{i-1}, β_{i-1} and γ_{i-1} are only functions of the physical quantities at step i-1. For calculating the DC performance, the following method is used.

(i) the gate voltage vgs is fixed as well as the drain current I_{ds}. The calculation begins at the source (x = 0). As boundary condition we have :

$$I_{ds} = q\,Z\left(N_{10}\,\mu_1\,(E_{x0}) + N_{20}\,\mu_2\,(E_{x0})\right)E_{x0} \quad (20)$$

(ii) from source to drain, the following sequence is carried out. First, Eq(18) is solved. This provides the average energy as well as the energy dependent parameters τ_{mi}, $\tau_{\varepsilon i}$ and m^*_i. Second, the gate-to-channel voltage $V_g\text{-}V_i$ is calculated from V_{i-1} and $E_{x,\,i-1}$ while $E_{y,i}$ is deduced from the charge control law. Third, Eq(19) is solved to give $E_{x,i}$. Fourth, the electron velocity and sheet carrier density in each layer are calculated using (16) and (17).

(iii) the drain-to-source voltage (channel potential at the drain end) is determined as well as the total stored charge (flux of the electric field at the gate electrode).

For each DC point, the Q2D approach yields the variations of the physical quantities in the channel (electrical field, average velocity, average

240

energy) as functions of the distance. The I-V characteristics of a quarter micron gate pseudomorphic HEMT are shown in figure (6).

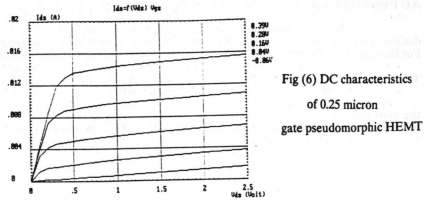

Fig (6) DC characteristics

of 0.25 micron

gate pseudomorphic HEMT

Another interesting feature of Q2D approach is its ability to provide data concerning the breakdown in the case of power devices (Snowden, 1989). Both gate and drain breakdown can be considered. For channel breakdown it is usually assumed that breakdown occurs if the multiplication factor M

$$M = 1 + \int_{S}^{D} \alpha\, dx \quad (20)$$

is larger than 1. The ionization rate α is assumed to be function of the local average energy

$$\alpha = C \exp\left[-\left(\frac{E_c}{E_S(\epsilon)}\right)^\beta\right] \quad (21)$$

E_S is the stationary electric field corresponding to the average energy ϵ and C, E_c and β are coefficients provided by the literature (Wroblewski, 1983).

The gate Breakdown occurs if

where

$$V_{ds} \geq V_{gd}^0 - \left| V_{gs} \right| \quad (22)$$

$$V_{gd} = \frac{2\epsilon\, E_c^2}{C\sigma} \quad (23)$$

and where σ is the sheet density of ionized donors including surface states (Frensley, 1981).

The heat dissipation is also an important effect which has to be introduced in the modelling of the DC characteristics of power MESFETs and HEMTs. Assuming that the lattice temperature increases proportionaly to the power dissipated in the device and to the thermal resistance, a good fit between theroetical and experimental I-V characteristics can be obtained (Snowden, 1992). One objective of Q2D approach is to yield results in short computation times. One I-V point can be typically obtained in 2-4 seconds, according to the number of slices, using a 386 or 486 based machine. This time is reduced by a factor of 10 if a work station is used. As a consequence Q2D approach can be very interactive.

AC PERFORMANCE

Bearing in mind that Q2D approach is intended to be used for the design of microwave circuits, the calculation of the AC performance is essential. For that purpose, several methods can be used.

The quasi static approach

In this method, the main small signal elements g_m, g_d, C_{gs} and C_{gd} of the conventional small signal equivalent circuit are deduced by the difference between DC points (variational method).

$$C_{gd} = \frac{\Delta Q}{\Delta V_{ds}} \bigg|_{V_{gs}} \quad (24) \qquad C_{gs} + C_{gd} = \frac{\Delta Q}{\Delta V_{gs}} \bigg|_{V_{ds}} \quad (25)$$

$$g_m = \frac{\Delta I_{ds}}{\Delta V_{gs}} \bigg|_{V_{ds}} \quad (26) \qquad g_d = \frac{\Delta I_{ds}}{\Delta V_{ds}} \bigg|_{V_{ds}} \quad (27)$$

where ΔQ and ΔI_{ds} represent the total charge and drain current variation around a DC point. This method has several drawbacks. First, it is limited to the low frequency range by nature. Second, only the main small signal elements are provided. Third, the calculation of the gate capacitance and consequently the gate current is obtained from a calculation assuming constant drain current and hence zero gate current. So such a method is basically questionable, especially in the calculation of the high frequency performance (cutoff frequencies).

Time dependent modelling

This method is very general and it can be used either for large signal or small signal analysis (Snowden, 1989, Halkias, 1989). In this method, the current continuity equation (11) is replaced by the time dependent continuity equation

$$\mathbf{V} \cdot \left(\underline{\mathbf{J}} + \varepsilon \frac{\partial \mathbf{E}}{\partial t} \right) = 0 \qquad (28)$$

It should be noted that for usual frequencies of operation, the time derivative in the energy and momentum conservation equation can be neglected ($\tau_\varepsilon \approx 1$ pS and $\tau_m \approx 0.1$ pS). To perform time dependent modelling, the gate and drain terminals are connected to RF loads. For instance (Snowden, 1989) the gate is driven with a sinusoïdal voltage superimposed on the DC bias and the drain is held open circuit for RF signals. After some iterations the A and C parameters of the ABCD two-port matrix can be evaluated after Fourier transform of the waveforms. The gate is then RF short circuited while the source is driven with a sinusoïdal waveform. Fourier analysis of the drain voltage and gate current allows the parameters B and D to be extracted (Snowden, 1989). Another possibility is to fix the drain load, and to carry out an iterative process based on a prediction/correction scheme for the gate voltage and source current waveforms. When the convergence of the procedure is achieved, the drain voltage and drain current are known as function of time (Halkias, 1989). Hence, such a method can be used to investigate the large signal device behaviour (compression, intermodulation, basic mechanism for the generation of upper harmonic components...). The time dependent modelling is very powerful but it needs an iteration procedure which can increase dramatically the computing time.

The local small signal equivalent circuit

This concept has been introduced recently for Q2D approach (Cappy, 1989). The basic idea of this method is to described a slice of the device by a local small signal circuit. For that purpose all the physical quantities are written :

$$X = X_0 + \delta_n \exp (j\omega t) \quad (29)$$

where X0 is the time independent part of X and δ_n the time dependent part of X. Introducing (29) in the time dependent set of physical equations (Eq (12), (13), (14), (28)) yields the wave equation. The wave equation is not solved from source-to-drain but is solved in one slice to give two relations of type :

$$i_g(x) = Y_{11}(x)\left(\delta v_g - \delta v(x)\right) \quad (30)$$

$$i_d(x) = Y_{21}(x)\left(\delta v_g - \delta v(x)\right) + Y_{22}(x) \, \Delta x \, \delta E_x(x) \quad (31)$$

The parameters $Y_{ij}(x)$ can be considered as the admittance parameters of a slice. They are functions of the steady state values of the physical quantities and of the frequency. It should be emphasized that these parameters are directly determined from a first order development of the time dependent physical equations which is very fast from the calculation time point of view. For moderate frequencies, the parameters $Y_{ij}(x)$ reduce to :

$$Y_{11}(x) = j \Delta C_g(x) \quad (32)$$

$$Y_{21}(x) = g_m(x) \quad (33)$$

$$Y_{22}(x) = \frac{1}{\Delta R_d(x)} \quad (34)$$

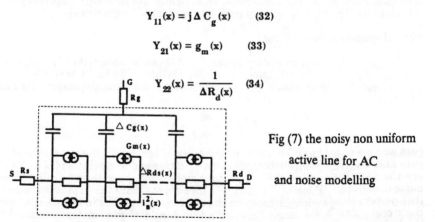

Fig (7) the noisy non uniform active line for AC and noise modelling

The physical signification of Eq (32), (33) and (34) is immediate and hence the whole device is described by a non uniform ΔC_g - gm - ΔR_d active line (figure (7)). The AC performance (admittance matrix or any equivalent representation) calculation is reduced to the calculation of the electric parameters of a ligne of a number of cascaded two-ports. This calculation has to be made for each frequency of interest but is very fast. It should be noted that the output parameters of the calculation are not the elements of a small signal equivalent circuit but the Y-parameters (or any other representation of a linear two-port) as functions of the frequency. Hence, it is possible to determine the small signal elements that give the best fit of the frequency dependence of the Y parameters.

This method provides "de facto" all the small signal elements (primary g_m, g_d, C_{gs}, C_{gd}, secondary R_i, τ, C_{ds}, C_{dc}) and also their frequency dependence in a short calculation time and small computing effort. Typically, the calculation of the AC device performance needs less than 1 second per DC point and per frequency point on a 386/486 based machine.

MESFET AND HEMT NOISE PERFORMANCE

As previously shown, Q2D approach is a well suited tool for calculating the DC and AC performance of micrometer or submicrometer gate length MESFET and HEMTs. However, one of the main features of this approach is to provide the device noise performance in a large frequency range. For this purpose (Carnez, 1981 ; Cappy, 1985 ; Cappy, 1989) a microscopic noise source that represents the carrier velocity fluctuations in a slice is introduced in the local small signal equivalent circuit previously described (Figure 7). The spectral intensity of this noise source is given by :

$$Si(x) = 4\,q^2\,Z\left(N_1(x)\,D_1(x) + N_2(x)\,D_2(x)\right)/\Delta x \quad (35)$$

where $N_1(x)$ and $N_2(x)$ are the sheet carrier densities in layers 1 and 2 (HEMT case) while $D_1(x)$ and $D_2(x)$ are the diffusivities. In order to be consistent with the relaxation time approximation the diffusivities are supposed to be dependent on the average electron energy : $D(x) = D\,[\varepsilon(x)]$. Data concerning the diffusivities can be found in the literature (Fischetti, 1991). Assuming that the noise sources located in two different slices are uncorrelated, it is possible to calculate the effects of the noise source located at x, on the gate and drain electrodes. For gate and drain in open circuit we have :

$$S_{gg}(x) = \left|\frac{\partial Z_g(x,\omega)}{\partial x}\right|^2 . Si(x)\,\Delta x^2 \quad (36)$$

$$S_{dd}(x) = \left|\frac{\partial Z_d(x,\omega)}{\partial x}\right|^2 . Si(x)\,\Delta x^2 \quad (37)$$

$$S_{gd}(x) = \left(\frac{\partial Z_g(x,\omega)}{\partial x}\right)\left(\frac{\partial Z_d(x,\omega)}{\partial x}\right) Si(x)\,\Delta x^2 \quad (38)$$

In these expressions S_{gg} and S_{dd} are the spectral intensities of the gate and drain noise voltage fluctuations, S_{gd} is the cross correlation between the gate and drain voltage fluctuations while the derivative of $Z_g(x,\omega)$ and $Z_d(x,\omega)$ are the so called "impedance field" (see the chapter of this book devoted to the noise modelling). The impedance Z_g and Z_d are calculated directly from the active line parameters (Cappy, 1989). The whole noise voltage sources $\overline{v_g^2}$ and $\overline{v_d^2}$ and their cross correlation $\overline{v_g\,v_d^*}$ are then calculated by the quadratic summation of the influence of all the noise sources located in the device channel. From the noise sources, the cross correlation and the small signal equivalent circuit, it is easy to calculate the noise performance in terms of minimum noise figure NF_{min}, equivalent noise resistance R_n and optimum source admittance Y_{opt} by simple circuit manipulation (Rothe, 1956). The noise performance calculation using Q2D modelling is interesting for several reasons.

First, it is a physical modeling without any fitting parameter (although the diffusivity is not very well known in complex structures such as pseudomorphic layer on GaAs or InP). Hence, it is possible to specify the physics of the noise generation in the channel and to study the fundamental limitations of the device performance in terms of noise.

Second, the noise calculation is performed in a short computation time. Hence, it is easy to change any intrinsic (device and layer topology) or extrinsic (access resistance, pad capacitance) parameter and to see the influence of this change on the noise performance. For this reason, the noise performance calculation using Q2D technique is very well suited for the CAD of microwave circuits.

LIMITATIONS OF Q2D MODELS

The limitations of Q2D modelling for MESFETs and HEMTs can be divided in two groups : the limitations due to the use of 1D relaxation time approximation for the electron transport and the limitations due to the simplified resolution of Poisson's equation.

(i) the electron transport

The problems connected to the use of RTA are twofold. First the material parameters (m^*, τ_m, τ_ε) are only well known in bulk materials (GaAs, InP, ...). In the case of complex active layer structures like AlGaAs/InGaAs/GaAs or AlInAs/InGaAs/InP, the material parameters are not very well known. In addition, the subband energy values in the quantum well are functions of the gate voltage and consequently the material parameters are functions of the gate voltage. Fortunately, it has been shown that the bulk value represents a reasonable approximation (Thobel, 1990). Assuming that the material parameters are perfectly known, the second problem concerns the validity of RTA. In the case of very short gate length devices ($L_g < 0.2$ or 0.15 µm) it is not obvious that 1D RTA is still accurate enough because for such structures the transit time under the gate and the energy relaxation time are of the same order of magnitude.

(ii) two dimensional effects

In Q2D modelling, Poisson's equation is solved using several simplifications. In the case of high aspect ratio (gate lengh to epilayer thickness) devices, the simplifications are in good agreement with full 2D modellings for the electric fields under the gate. Hence, the main difficulties concerns the modelling of the electric fields at the drain end of the gate. As a matter of fact, the electric field variations in this region are very difficult to describe using simple formulation. As a consequence, the fit of the experimental variation of the output conductance versus gate voltage can be difficult in the case of quarter micron gate HEMTs (Happy, 1993) while the experimental and theoretical values are in good agreement in the case of halfmicron or micron gate MESFETs (Snowden, 1989).

REFERENCES

Adachi S. (1982), Material parameters of $In_{1-x}Ga_xAs_yP_{1-y}$ and related binaries" J. Appl. Phys. 53(12) : 8775-8792.

Alamkan J. et al (1990) Modelling of Pseudomorphic AlGaAs/InGaAs/GaAs layers using self consistent approach. European Transactions on telecommunications and Related Technologies, Vol. 1 (4) : 59-62.

Cappy A. et al (1985) Noise modeling in Submicrometer-gate two Dimensional Electron gas Field Effect Transistors. IEEE Trans. Elec. Dev. Vol. ED-32 (12) : 2787-2796.

Cappy A., Heinrich W. (1989) High freuqency FET noise performance : A new approach IEEE Trans. Elec. Dev. Vol. 36(2) : 403-408.

Carnez B. et al (1980) "Modeling of a submicrometer gate field effect transistor including effects of non stationary electron dynamics". J. Appl. Phys. vol. 51(1) : 784-790.

Delagebeaudeuf D., Linh N.T. (1982) "Metal (n) AlGaAs/GaAs two dimensional electron gas FET" IEEE Trans. Elec. Dev. vol. ED 29 (6) : 955-960.

Drummond et al (1983) "Bias dependence and light sensitivity of AlGaAs/GaAs MODFET at 77° K" IEEE Trans. Elec. Dev. Vol. ED-30 : 1806-1811.

Frensley W.R. (1981) "Power-limiting breakdown effects in GaAs MESFET's" IEEE Trans. Elec. Dev. Vol. ED-28 : 962-970.

Happy H. et al (1993) "HELENA, a friendly software for calculating the DC AC and noise performance of HEMTs" to be published in Int. J. of Microwave and millimeter wave CAE.

Morgan T.N. (1986) "Theory of the DX centers in AlGaAs and GaAs crystals" Phys. Rev. B, Vol. 34 : 2664-2669.

Rothe H. and Dahlke W. (1956) "Theory of noisy fourpoles" Proc. IRE, Vol. 44 : 811-817.

Schawki T. et al (1988) "2D simulation of degenerate hot electron transport in MODFET including DX-center trapping" Proc. 3rd Int. Conf. Simulation Semiconductor Devices Processes (SISDEP 88).

Shawki et al (1990) "MODFET 2-D Hydrodynamic Energy Modeling Optimization of subquarter-Micron-Gate structure" IEEE Trans. Elec. Dev. Vol. 37(1) : 21-30.

Shockey W. (1952) "A Unipolar Field Effect Transistor" Proc. IRE, 40 : 1365.

Shur M. Eastman L. (1980) "I-V characteristics of GaAs MESFET with non uniform doping profile" IEEE Trans. Elec. Dev. Vol. ED-27 : 455-461.

Shur M. (1976) "Influence of non uniform field distribution on frequency limit of GaAs field effect transistors" Elect. Lett. Vol. 12, n° 23 : 615.

Snowden C.M. and Pantoja R.R. (1989) "Quasi-two-Dimensional MESFET simulations for CAD" IEEE Trans. Elect. Devices, Vol. 36 (9) : 1564-1574.

Thobel J.L. et al (1990) "the electron transport properties of strained $In_xGa_{1-x}As$". Appl. Phys. Lett. Vol. 56(4) : 346-348.

Vinter B. (1984) "Subbands and charge control in a two-dimensional electron gas field-effect transistor". Appl. Phys. Lett. 44 (3) : 307-309.

Wroblewski R. et al (1983) : Theoretical analysis of the DC avalanche breakdown in GaAs MESFET" IEEE Trans. Elect. Devices Vol. ED-30 (2) : 154-159.

Yoshida J. (1986) "Classical versus mechanical calculation of the electron Distribution at the n-AlGaAs/GaAs heterointerface" IEEE Trans. Elec. Devices Vol. 33(1) : 154-156.

Application of Modelling to Microwave CAD

D. Michael Brookbanks
GEC-Marconi Materials Technology Ltd.

INTRODUCTION

In this chapter we will attempt to link the device models discussed elsewhere to the engineering world of practical microwave circuit CAD. To this end we will look at the problem from the CAD point of view and show where the device models are required. The discussion will mostly deal with the MESFET as this is the device most often used in Monolithic Microwave Integrated Circuits (MMICs) at present but the principles apply to any device.

We can begin by asking ourselves the question 'What do we mean by CAD for microwave systems'. There is obviously no easy answer to this question as it can depend strongly on our system requirements. Whereas there are CAD packages which may be able to analyse the performance of a complete subsystem the results will only be as good as the weakest link in the chain. In this chapter the intention is to give some of the background to the development and use of microwave CAD packages, so that the building blocks that will be necessary for the complete analysis of the subsystem can be evaluated. It is thus the microwave engineer and subsystem designer's intention to use whatever variety of CAD packages that may be available to provide a complete description of the microwave devices, circuit and subsystem, from these well defined elements. From this we can see that, unless one has access to a completely integrated CAD package, there must be well defined interfaces between the different components if we are to make any significant advances in our subsystem design. We shall return to this point later in the chapter.

The pace of the development of microwave CAD can be seen from the following example: in April 1972 Hewlett Packard issued the now classic application note AN154 - S-parameter design, in which they showed how transistor amplifiers could be designed for both gain and noise figure. It should be noted that this was before the introduction of even the simplest scientific calculator and so analysis would have very often taken place by the use of a Smith Chart aided by a slide rule. On this basis it would have been very difficult to design broadband components or to achieve any reasonable degree of complexity in circuit design. However within 5 years microwave designers had access to mainframe programs such as 'Compact' in which much more complex circuit descriptions could be achieved by the combination of series and shunt interconnections of circuit elements. To the engineer this was a considerable improvement, even if interaction with the program was on the basis of a single line at a time on a teletype. In 1980 the release of 'Super-Compact' gave the engineer access to a more flexible circuit description and access to a VDU display for file editing. Both 'Compact' and 'Super-Compact' are trademarks of Compact Software Incorporated. By far the greatest change in microwave CAD was in fact brought about by the introduction in 1981 of the IBM PC personal computer. Here at last was a computer that had sufficient capability to run fairly complex CAD programs with good graphics performance, and which was within the price range that could be afforded by individual microwave groups and design teams. As we shall see later in this chapter the latest versions of the PC can be comparable to mainframe computers in the speed at which they execute programs. Thus within a 15 year period we have moved from Smith Chart and slide rule design procedures to very sophisticated CAD programs running on personal computers, advanced multiuser workstations and large mainframe computers. The choice of which of these to use then becomes a combination of personal preference, the number of users that must be supported and the overall cost of the installation.

WHAT DO WE MEAN BY MICROWAVE CAD?

The easy answer to this question is any program or software package that assists us in the analysis or design of microwave circuits and subsystems. In this definition I include software packages that are involved in the layout of microwave circuits as there is very little point in having sophisticated design and analysis tools if the results of this analysis can not be translated to a physical layout for manufacture. Microwave CAD packages can be broken down into the following broad areas:

- Linear circuit simulation
- Non-linear circuit simulation
  (i) Time domain
  (ii) Frequency domain
- Filter and network synthesis
- Design database
  (i) Measurement
  (ii) Data extraction and manipulation
- Layout
  (i) Polygon based
  (ii) Line based
- Custom software
  (i) Internal programs
  (ii) External programs (e.g. University software)

All of these areas of CAD software are used at GEC-Marconi in the design of MMICs and associated subsystems. It is also true that the range of available software does not always form an integrated package and it has been necessary to develop in house software so that data can be transferred from one software package to another, and in many cases from one computer to another. This latter point arises as our CAD programs are available on the following mix of computers.

- VAX mainframe cluster (DEC)
- PC
- Hewlett Packard
- MicroVax workstation (DEC)

Although it is possible to purchase CAD packages that operate on one computer system (generally PC or a workstation) we shall see later in this chapter that this can present severe limitations during the design cycle. However for a small company or an individual this more integrated environment can be very useful. However the advent of specialist workstations does mean that for the new entrant into Microwave CAD, or for existing users upgrading, much greater levels of integration are available.

Before discussing individual microwave CAD packages it is pertinent to ask the question 'What do we require from our CAD package?' The following points form a rather obvious list, but we show later that some of the points will need clarification and that it is always necessary to appreciate the limitations of the CAD package that is being used.

- Accurate
- Fast
- User friendly
- Compatible with other CAD packages

Accuracy is an obvious requirement but this implies much more than simply numerical precision, which can be achieved by good programming techniques, as many microwave CAD packages have to use closed form expressions or look-up tables in order to achieve the computational speed required. Accuracy is then dependent upon the quality of these algorithms and formulae.

An exhaustive list of available microwave CAD software is beyond the scope of this chapter, but the following represents many of the programs that are in use at the present time.

(i)   Devices

Gates	- 1-D
PISCES	- Stanford Univ
CADIS	- Q-2-D, University of Leeds
BIPOLE	- Stanford University
SEDAN	- Stanford University
LUSTRE	- Cornell University
BAMBI	- University of Vienna
CADDET	- 2-D
MINIMOS	- 2-D, University of Vienna
HFIELDS	- 2-D
CURRY	- 2-D, Philips
WATMOS	- 3-D, University of Waterloo
FIELDAY	- 2-D, 3-D, IBM
GEMINI	- Stanford University
SPICE	- Stanford University (equivalent circuit)

(ii)   Linear circuit simulators

Super-Compact	- Compact Software Inc.
Touchstone	- EEsof Inc.
LINMIC+	- Jansen Microwave
Microwave Design System	- Hewlett Packard
Octopus	- ArguMens

(iii)   Non-linear circuit simulators

(a)   Time domain
          Spice
          Astec   -   SIA Computing Services
          Microwave Spice   -   EEsof
          Circec   -   Thom 6
          Anamic   -   University of Kent

(b)   Frequency domain
          Microwave Harmonica   -   Compact Software
          Libra   -   EEsof
          Microwave Design System   -   Hewlett Packard

(iv)   Data measurement

          MicroCat   -   Cascade Microtech
          Netcom   -   Compact Software
          Anacat   -   EEsof
          Automated noise figure   -   ATN
               measurement

(v)   Parameter extraction

          FET Fitter   -   Cascade Microtech
          Scout   -   Compact Software
          Xtract   -   EEsof
          Gascode   -   Gascode Ltd.
          Soptim   -   University of Kent
          MESFETDC   -   University of Kent

(vi)   Layout

          AutoCAD   -   Autodesk Inc.
          GaS Station   -   Barnard Microsystems Ltd.
          AutoArt II   -   Compact Software
          Academy   -   EEsof

(vii)   Filter and Network Synthesis

          E-syn   -   EEsof
          Complex Match   -   Compact Software

Note that Spice is a generic software that is available from many sources and it is very often customised by individual companies to suit their own requirements. Microwave Spice and Anamic have been developed so as to appeal to the microwave CAD user by providing the appropriate input and output formats and internal elements.

HOW DOES A COMPUTER ANALYSE MICROWAVE CIRCUITS?

In order to analyse any microwave circuit the user must provide for the computer a description of how the circuit is connected together from a set of basic building blocks. These building blocks can generally be any interconnection of the internal elements that the CAD program recognises, external measurements (in matrix form) and internally generated sub-circuits. The building blocks are connected together in a nodal fashion with each interconnection point (or node) given a unique number. Normally node assignments are cleared when a subcircuit is defined, the node numbers may then be reused without conflict. As an example we shall consider the development of the circuit description of an equivalent circuit model for a gallium arsenide MESFET and compare the S-parameters of this model to a set of measured data. This equivalent circuit model is shown in Figure 1. In order to show impartiality we shall develop this model in both Super-Compact and Touchstone formats. It is worth noting at this point that the computer does not understand our system of units or preferred values; as microwave engineers we tend to default to the following units, GHz, pF, nH, ps whereas the computer needs to be told what these units mean. Super-Compact uses the fundamental units of Hz, F, H etc. and so all values must be referenced to these units, whereas Touchstone will default to the 'microwave units' unless told otherwise in the circuit description. As a user of both CAD programs this can be confusing when you enter, for example, a shunt capacitance of 0.05 in a Super-Compact file and thus wonder why the input looks like a short circuit at 10 GHz! In the latest workstation based variants of the principal simulators, schematic entry can be used which bypasses the need for the user to set up a description based on a nodal connection. This chore is now handled by the schematic capture part of the program which generates the netlist and passes this to the simulator for analysis. The long term aim is of course to bypass the need for a netlist entirely.

Returning to our original question of 'how does the computer analyse microwave circuits?' the answer is by some form of matrix analysis by which ever method the individual software company prefers. However we can obtain a feel for what is going on if we choose the work with admittance matrices, as these tend to have fewer problems with singularities than impedance matrices. There is however one exception to this, and this is within Super-Compact; when the network can be simply described as a cascade of simple two ports Super-Compact has a construction known as a LAD (Ladder) network which uses a simple ABCD chain matrix. As this then only involves the multiplication of simple 2x2 matrices circuit analysis can be very fast. If one examines the chain matrix properties then it can be shown that this form of analysis precludes the use of shunt feed back in the network. The LAD network can be useful however in forming some of the subnetworks that are needed in complex circuits. In a general circuit the connection of a large number of elements can lead to a large admittance matrix, and computational speed decreases as the matrix size increases. However in many cases a large number of the elements of this matrix are zero and the computer in calculating the properties of the network and reducing the resultant matrix to a 2, 3 or 4 port network is multiplying by zero or adding zero. Under these conditions, when there are a large number of zeros in the matrix, the matrix is said to be sparse. For example, during the analysis of a branch line coupler a 20x20 complex matrix is generated in which about two thirds of the 400 elements are zero. Touchstone recognises this fact and gives the user the option of using conventional S-parameter matrix reduction or sparse Y matrix reduction. Unfortunately it is up to the user to choose which route to take, without knowing at the outset which will be the most suitable. In this case only experience will help but EEsof do give some guidelines in one of their

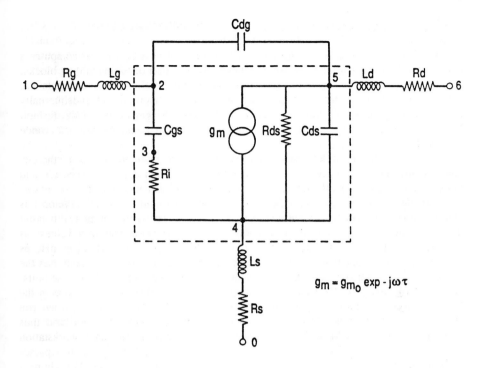

Figure 1. The equivalent circuit model for a GaAs MESFET

supporting documents, where they show the relative performance of sparse Y and S-parameter reduction for a range of circuit topologies. For the simplest circuits S-parameter reduction is the optimum choice but sparse Y reduction becomes advantageous as the circuit complexity increases. EEsof state that the sparse Y approach is the optimum method when one has a flat network where there are no sub-circuits. This situation may arise in tolerance analysis when we do not want to force components to track in different parts of a circuit. For our FET model example the use of sparse Y techniques will reduce the analysis and optimisation time by a factor of 1.3.

The sparse matrix reduction achieves its improvement in performance by only storing the row and column indices and the values of the non-zero elements. The trade off is then between mathematical operations involving zero and the time taken to find the appropriate element from the reduced matrix. Whichever method of matrix analysis is chosen the computer then stores the resultant circuit matrix in an internal format, so that conversion to the appropriate output format only takes place when the user chooses to either plot or print.

Statistical analysis

Tolerance analysis of microwave circuits is essential if designs are not to be put forward to manufacture that may not be tolerant to small changes in some of the component values. An initial estimate of potential problems can be obtained from a sensitivity analysis by adjusting a single parameter and observing the changed response. In Touchstone and Super-Compact this can be obtained by using the Tune and Tweak facilities. However both Super-Compact and Touchstone offer a more sophisticated means of assessing the influence of component variations by allowing Monte-Carlo analysis of the circuit in which component values are allowed to vary randomly between preset limits. In this case the circuit description must contain either a STAT (Super-Compact) or TOL (Touchstone) block within which the statistical requirements are specified. This is shown in Figure 2 where we have chosen to vary the intrinsic device gate-source capacitance of our FET model by +/- 10% about the nominal value. This could be used to simulate the effects of gate length variation. Tolerance analysis can also be applied to externally provided data, as well as the circuit variables, by placing the perturbations that must be applied to the S-parameters in the STAT or TOL block. In this case we have chosen to plot S_{11} and S_{21} as these are the most influenced by this variation.

In a complete circuit analysis the MMIC designer will need to use both the equation and statistical blocks if he is to undertake a thorough analysis. This arises as in an MMIC process not all parameter changes are statistically independent. As a typical example, consider the overlay capacitor. The capacitance of this component is almost entirely determined by the geometrical shape of the capacitor plates and the dielectric constant/unit area of the interlayer dielectric. Process variations will result in some random changes in this capacitance but this will affect all of the capacitors equally and so the circuit analysis must not let these capacitors vary in a totally random manner but must ensure that the capacitance value tracks across the circuit. This is then the advantage of linking all the capacitors through an equation area of the circuit description file. A similar philosophy could apply to resistors etc. The value of the Monte-Carlo analysis then comes from the ability to treat the variations in capacitance/unit area and resistance as statistically independent and thus perform a tolerance analysis.

Noise analysis

So far we have only discussed the linear analysis of circuits. However in many microwave systems the overall noise performance is critical and we may wish to optimise the circuit for low noise operation or to examine the trade-offs between gain and noise figure. For many years the noise analysis of microwave circuits has been more restricted than circuit analysis; in that a more restricted circuit topology was allowed for noise analysis than was allowed for circuit analysis. The reasons for this will be shown later, but all the major simulators now have unrestricted noise analysis, for linear circuits, as standard. It is then possible to perform a noise analysis on the same topology as the circuit analysis.

Figure 3 shows the circuit topology that was used for the restricted noise analysis, in both Super-Compact and Touchstone, as this provides an ideal method of introducing the concept of noise analysis.

Super-Compact uses a specific block called NOI, whereas Touchstone ascribes specific names to the sub-circuits shown in Figure 3. The results of the analysis are of course very similar apart from the fact that Touchstone can analyse a cascade of 4 stages as opposed to 2 in Super-Compact. If there are more than 2 stages in the NOI block, noise analysis is performed on the first two stages and electrical analysis on the complete block. The major limitation of this restricted analysis is that feedback across stages is not permitted.

In order to understand the method of noise analysis it is necessary to digress into the theory of noisy two ports: in any two port there will be sources of noise, which may simply be resistors in a passive network or the internal noise that is generated within an active device. These various noise sources may be totally independent of each other or related through some physical coupling mechanisms. For example the noise in the channel of a MESFET produces noise at both the gate and drain, which are obviously not independent. In order to proceed this noisy two port is transformed into

```
* Intrinsic Elements
* Vary gate-source capacitance by +/- 10% about nominal value
Cgs: 0.25pF # UD  10% #
Ri:  5
Gm:   0.045
Tau: 2.5E-6uS
Rds: 250
Cds: 0.06pF
* Parasitic Elements
Rg:  2
  .
  .
  .
  .
STAT
  FETMODEL
  * Set performance windows
  F 10GHZ  MS11  0.75  0.80  PS11  -100 -80
  F 10GHZ  MS21  2.15  2.25  PS21   103  107
END
```

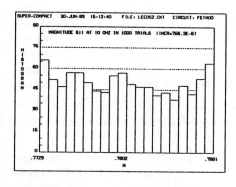
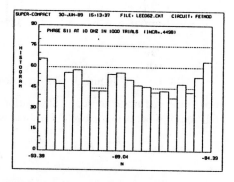
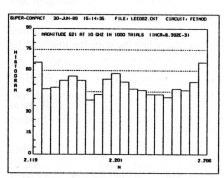
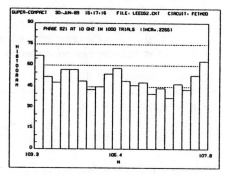

Figure 2. Tolerance analysis of a FET model (C_{gs} varied) - circuit syntax,
behaviour of S_{11} and S_{21}

a noise free two port with two noise sources at the input. One of these noise sources is
a voltage source and the other a current source, and the two sources are expected to
show some degree of correlation. This transformation s shown in Figure 4.
The overall noise properties of the two port can then be described by reference to these

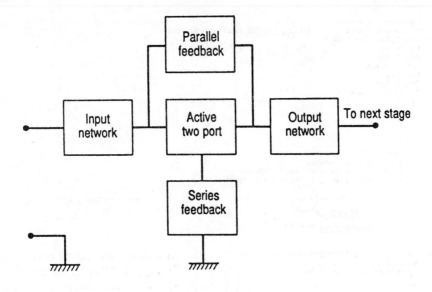

Figure 3. The structure used for noise analysis

two correlated noise sources. Straightforward circuit analysis leads to the following relationships between the two noise sources and the overall noise figure of the two port.

$$F = 1 + \frac{G_n}{G_s} + \frac{R_n}{G_s} \mid Y_s + Y_{cor} \mid^2$$

where Y_s is the source admittance

R_n, G_n are the Johnson noise equivalent resistance and conductance of the two noise sources

Y_{cor} is the correlation admittance that describes the correlation of the two noise sources.

The noise properties of the two port can then be fully described from a knowledge of the parameters G_n, R_n, and Y_{cor}. The expression for the noise figure is usually put into the more familiar format of

$$F = F_{\min} + \frac{R_n}{R_s} \left| \frac{(R_{opt} - R_s)^2 + (X_{opt} - X_s)^2}{R_{opt}^2 + X_{opt}^2} \right|$$

showing that the four parameters $F_{\min}$, R_{opt}, X_{opt} and R_n are all that are needed to fully describe the noise properties of the network. Our CAD problem has thus become one of finding how these four noise parameters, which apply to the central two port of Figure 3, are modified by the addition of the input/output networks and the feedback networks. Noise figure data is normally supplied as the parameters $F_{\min}$ (in dB units), magnitude and angle of Z_{opt} and R_n normalised to 50 ohms. In Touchstone this data follows the S-parameter data in the data file and in Super-Compact as an entry in the DATA block.

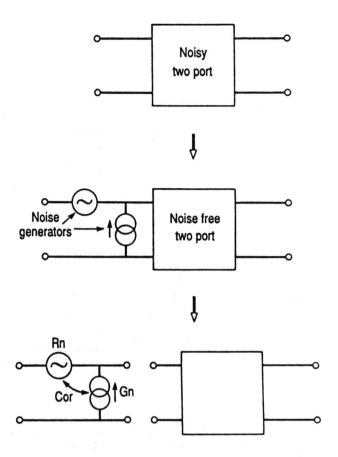

Figure 4. The conversion of a noisy 2-port into a noise free 2-port
and correlated external noise generators

The relationships between these circuit parameters and the fundamental noise parameters is as follows:

$$F_{\min} = 1 + 2R_n \, (G_{cor} + G_{opt})$$

$$G_{opt} = \left[G_{cor}^2 + \frac{G_n}{R_n} \right]^{1/2} - jB_{cor}$$

giving us the problem of determining the transforming equations for R_n, G_n and Y_{cor}. It has been shown that the transformation between the original noise voltage and current and the new noise voltage, and current, is given by the following matrix relationship

$$\begin{bmatrix} v' \\ i' \end{bmatrix} = \begin{bmatrix} N_{11} & N_{12} \\ N_{21} & N_{22} \end{bmatrix} \begin{bmatrix} v \\ i \end{bmatrix}$$

The expressions for the transformation matrix elements [N] have been given in the literature in terms of the S-parameters of the noise free two port and the external networks, and appear rather complicated and daunting. However, if we revert to our use of the ABCD matrix formulation these transformation parameters take on a much simpler and elegant form. The networks for which these transformation matrices have been derived are shown in Figure 5.

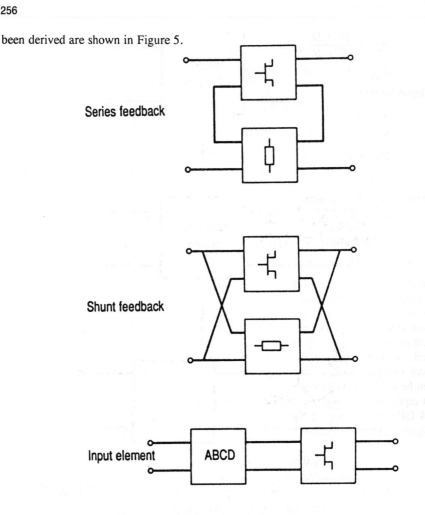

Series feedback

Shunt feedback

Input element

ABCD

Figure 5. Noise transformation topologies

The transformation matrix elements for the three cases are listed below, where the prime denotes the external network.

(a) Series Feedback

$$N_{11} = 1$$

$$N_{12} = \frac{A' - A}{C' + C} \qquad A' = 1$$

$$N_{21} = 0 \qquad C' = \frac{1}{Z}$$

$$N_{22} = \frac{C'}{C' + C}$$

(b) Shunt Feedback

$$N_{11} = \frac{B'}{B' + B}$$

$$N_{12} = 0 \qquad B' = 1$$

$$N_{21} = \frac{D'-D}{B'+B} \qquad D'= \frac{1}{Y}$$

$$N_{22} = 1$$

(c) Input Network

$$N_{11} = A'$$

$$N_{12} = B'$$

$$N_{21} = C'$$

$$N_{22} = D'$$

The transformed noise parameters are then given by:

$$R'_n = R_n \mid N_{11} + N_{12} \, Y_{cor} \mid^2 + G_n \mid N_{12} \mid^2$$

$$G'_n R'_n = G_n R_n \mid N_{11}N_{22} - N_{12}N_{21} \mid^2$$

$$Y_{cor} 'R_n '= R_n \, (N_{21} + N_{22}Y_{cor})(N_{11}^* + N_{12}^* Y_{cor}) + G_n N_{22} N_{12}^*$$

permitting the evaluation of the noise properties of the transformed network. It is possible to obtain from these equations the fact that any lossless network on the input of a two port does not change the minimum noise figure but only changes the value of the optimum source impedance. The transformed network can now be treated as a simple cascaded network of noisy two ports and the overall noise performance of the circuit found. In order to optimise the noise figure it is necessary to provide the optimum source impedance, Z_{opt}; and generation and optimisation of this input network can often be enhanced if an equivalent circuit for Z_{opt} is available. Figure 6 shows a one port equivalent network for a GEC-Marconi MESFET which can be used within the 2-14 GHz range. One of the advantages of this network approach is that the circuit designer can see the form of the network that needs to be synthesised.

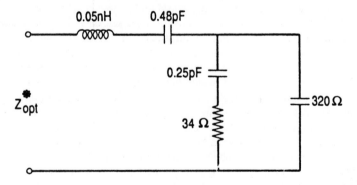

Figure 6. An equivalent circuit model for Z_{opt}

The more general unrestricted noise analysis that is now available uses a different approach to calculating the noise properties. Here the method is to use noise correlation matrices which can give a more exact formulation and are capable of dealing with feedback across stages. The disadvantage of this approach is probably in the speed of analysis and thus if the circuit designer can satisfactorily use the existing circuit structure then it would be advantageous to do so. However one major advantage of this approach is that the losses associated with transmission lines, resistors etc. are automatically taken into account in the noise analysis. It is also possible to ascribe different temperatures to different elements in the analysis.

Recently automated noise measurement systems have become available for both packaged devices and RF on wafer (RFOW) users. These rely on the principle of placing a series of known impedances at a reference plane at the input to the device and then measuring the noise figure of the device or circuit. From this information on noise figure at known source impedances we can solve the equations given earlier and obtain the four basic noise parameters that are required for the full noise description of the network. We thus have the information we require for complete noise analysis of our circuits.

Linear simulation - speed and accuracy

In many cases these are the principal questions that are asked. The answer of course depends upon many factors but it can generally be assumed that by using double precision on a mainframe and a dedicated maths co-processor on a PC numerical precision will not be a problem. Super-Compact will, for example, place a large (100 Mohm) resistor in a Z-matrix to avoid singularities during inversion. At least the user is warned when this happens so that the results can be treated accordingly. Execution speed can be enhanced by making the circuit as simple as possible and by limiting the number of frequencies at which analysis and optimisation takes place. In the first case, if there are parts of the circuit which are not going to change during optimisation consider an initial analysis which treats these sub circuits as individual n-ports which are then stored to a disc file or databank. These circuits can then be introduced into the main circuit as S-parameter data blocks and will only need to have been calculated once, as a complex nodal interconnection. In this case our benefit is doubled as we do not have to calculate the nodal connections but as importantly, we did not have to put this network through a syntax checker to see if we had made any errors in the interconnection or names of the keywords. Obviously reducing the number of frequencies at which analysis takes place will enhance speed and in particular this will be invaluable during optimisation where the circuit must be analysed once at each trial. Here one should remember that interpolation within a data set will be less accurate than if the analysis frequencies coincide with the measured data points. Also avoid if at all possible the use of extrapolation as there you have no control over the data that is then used as this depends on the method by which the extrapolation was undertaken. Occasionally this problem can be alleviated by the user inserting a data line at a suitable frequency into the data set so as to effectively force interpolation. I have used this with success in FET S-parameter files where it is not too difficult to provide reasonable values at very low frequencies so as to prevent for example the estimate of S_{11} exceeding unity. Returning to the problem of optimisation; if the circuit has a reasonably smooth behaviour with frequency it may not be necessary to use many frequencies in order to obtain adequate convergence and minimisation. Even if the final response requires calculation at a large number of frequencies initial optimisation to reduce the error value may only need a relatively few frequency points.

The question of speed is often asked, and it can be very difficult to give direct answers. A PC is generally dedicated to one task and so the user has all of the available CPU time; with a mainframe computer the user is competing with others for access to the CPU and memory. Thus on an interactive basis a PC may give a faster turn around than a large mainframe. The mainframe does however have the advantage of batch operation so that many calculations can be run overnight and the results made available the next day. As a guide I have compared the time taken for a 5 iteration gradient optimisation of a FET model, similar to the example earlier, using Super-

Compact on a range of PCs and DEC VAXs. The times used are the elapsed times for the PC and the CPU time printout that Super-Compact gives on the mainframe version. The times have been normalised to the time taken by the VAX 11/780, and the speed is the simple inverse of these relative times.

Computer	Processor	Clock Speed	Speed
IBM PC/AT	80286	6 MHz	0.18
Compaq 386/25	80386	25 MHz	1.2
DEC VAX 11/750	-	-	0.45
" " 11/780	-	-	1.0
" " 8700	-	-	6.1

These speeds are only to be taken as a snapshot at the present time and may change as the program versions change. Also it should not be forgotten that the PC has only a limited memory and thus the PC versions of the software can not necessarily handle as large a circuit description as the mainframe versions. The recent advances in PC operating environments with DOS extenders, Windows 3 and OS/2 overcome the inherent memory limitations and with the 33 MHz 386 and 486 processors performance comparable to workstations is available.

Customised simulators and user defined models

All of the principal simulators support the use of user defined elements to assist in circuit simulation. In the cases of Libra and Harmonica these models are compiled by the user and included as part of the executable code of the simulator, whereas in Hewlett Packard's Microwave Design System the operation is more similar to a function call with passed parameters to an uncompiled sub program. The advantages of the compiled code option are execution speed and the protection, and control, of proprietary models.

As an example of the benefit of a user defined element, consider the use of such an element for the description of an FET. If we describe the element by the use of equations then we have a section of the netlist as shown below.

```
W:  ?50 100 175?   *Finger width in microns
ID: ?0.1 .5 1?     *Fraction of Idss (range 0.1 - 1)
NV: 2              *Number of vias to ground
*
L_G:  0.3810
L_D:  0.2360
L_SI: 40.4
L_SV: 3
R_D:  90.000
R_G:  0.0213
R_S:  178.50
RI_0: 4.6700
C_DG: 0.0054
C_GS: 0.0600
C_DS: 0.0184
GM_0: 0.0000
TAU:  2.83E-3NS
*           ----000OOO000000----
```

```
*
*RDS
 A1:789.452433
 A2:2053.301
 A3:1934.121426
 A4:758.2776
 A5:373.19731
*GM
 B1:-52.3293
 B2:143.306285
 B3:-155.0835
 B4:98.441735
 B5:10.064519
*CDG
 C1:3.850481E-4
 C2:8.739112E-4
 C3:7.702526E-4
 C4:4.480064E-4
 C5:4.067016E-4
*CGS
 D1:8.02888E-3
 D2:0.019385
 D3:0.016134
 D4:7.397165E-3
 D5:5.195582E-4
*          ----000000OOO0000----
*
* Circuit model of FET using equations
* Polynomial coefficients taken from previous labels
*
* A new circuit block is needed for each independent FET
* in the circuit
*
BLK
 IND 1 2 L=(L_G*W*1E-12)
 RES 2 3 R=(R_G*W)
 CAP 3 5 C=((C_DG+(W*((C1*ID**4)-(C2*ID**3)+(C3*ID**2)-(C4*
+ ID)+C5)))*1E-12)
 CAP 3 8 C=((C_GS+(W*((D2*ID**3)-(D1*ID**4)-(D3*ID**2)+(D4*
+ ID)+D5)))*1E-12)
 RES 8 9 R=((RI_0-2.833*ID)*75/W)
 VCG 3 5 8 9 G=(GM_0+(1E-3*((B1*ID**4)+(B2*ID**3)+(B3*
+ ID**2)+(B4*ID)+B5)*W/75)) R1=1E6 R2=(((A1*ID**4)-(A2*ID**3)+(A3*
+ ID**2)-(A4*ID)+A5)*W/75)T=TAU
 CAP 5 9 C=((C_DS+(((E1*ID**3)-(E2*ID**2)+(E3*ID)+E4)*W))*1E-12)
 RES 5 6 R=(R_D/W)
 IND 6 7 L=(L_D*W*1E-12)
 RES 9 10 R=(R_S/W)
 IND 10 0 L=(((L_SI/(3*NV-2))+(L_SV/NV))*1E-12)
FETMOD:2POR 1 7
END
```

However if we wish to use, for example, four FETs in a circuit with possibly different bias or geometries then we would have to reproduce this section of the netlist four times. This then becomes impractical when we realise that we will also need similar expressions to describe the inductors, capacitors etc. in the circuit. Once included as a user defined element this FET can be called just as if it were one of the simulator's internal elements with, for example, the following syntax.

F_FET 1 2 NF=4 NV=2 W=75 ID=.5 NOI=1

In this case the allowed variables are the number of FET fingers (NF), the number of grounding via holes (NV), the unit gate width (W), the drain current, as a fraction of Idss, (ID) and a flag that determines whether noise analysis is performed or not.

This much simpler form then has the benefit of reduced execution time, a simpler circuit file description and less chance of error in the netlist. Also, as we shall see later, these elements can be linked to layout descriptions.

DESIGN DATABASE

In all of the foregoing references have been made to n-port S-parameter files which can be used as part of a database for design. As we have indicated these design databases can contain either measured data or sub circuit simulations. The main interest in this section is in the area of the measurement of database files. As MMIC designs become more complex progress will only be made if microwave engineers can adopt some of the techniques that are employed by their compatriots in the Si design industry. By this I mean that we shall have to consider making many circuits from existing, characterised, data blocks. The files in the design database will then consist of standard components, which in the case of the GEC-Marconi MMIC activity consists of overlay capacitors, spiral inductors, interdigital capacitors and FETs. These files will have well defined reference planes for inclusion in the circuit file. The measurements will be made with a vector network analyser such as the Hewlett Packard 8510B or the Wiltron 360 and are normally two port measurements. These vector network analysers are controlled by computers and can make measurements in test fixtures or direct on wafer. This controlling computer can have purpose written software, as we do at GEC-Marconi, or one of the commercially available programs, such as ANACAT or Netcom. The disadvantage of these commercial packages is that they only output data in the format that the vendor requires for his own CAD program (Touchstone from ANACAT and Super-Compact from Netcom), which is also PC format. Depending upon the software package the output data can be manipulated so as to give average values and standard deviation information. This standard deviation information can then be used in the STAT or TOL blocks of the circuit file for tolerance analysis. In the case of GEC-Marconi this data measurement and analysis takes place using Hewlett Packard computers and the resultant ASCII database files can be put in a format suitable for Touchstone, Super-Compact or LINMIC+ for linear analysis. The files can also be put into the appropriate format for non-linear analysis and parameter extraction; which will be discussed in a subsequent section.

NON LINEAR CIRCUITS

Compared to linear circuits the CAD of non linear microwave circuits has only recently become viable. There are many reasons for this but it is true that the added complexity of the circuit design has put off all but the most stout-hearted microwave engineer. This is now changing and CAD programs are becoming available that are reasonably easy to use by the experienced microwave engineer. The microwave subsystems designer's requirement falls into two categories when he considers non linear effects. These being the extension of small signal analysis to large signal analysis, such as the gain compression of amplifiers or the design of Class A power amplifiers, and the analysis of circuits in which device non-linearities play a dominant role, such as mixers, oscillators and Class B/C power amplifiers. As we shall see later it is only in the more recent CAD software that the requirements of the microwave subsystem designer have been considered. Non linear CAD programs fall into two distinct categories, with each of the categories having its champions. Basically these categories relate to the method of analysis and lead to the classification of programs as 'time domain' or 'frequency domain'. The method of analysis differs in each case and each is suited to a particular class of problems. Ideally the microwave engineer should have access to both classes of program in a format that is compatible with existing linear analysis programs.

Time Domain Programs

In these programs calculations take place, as one might expect, entirely in the time domain. These programs are typified by SPICE which was developed for the Silicon IC activity many years ago. These programs are ideally suited to the analysis of circuits in which there are strong non linear effects, or where the time development of the circuit is required. Such a situation is the start-up conditions in an oscillator. The one major disadvantage of all the time domain related programs is that frequency dependent parameters cannot be easily incorporated. Thus transmission lines tend to be non dispersive and lossless, or have losses that do not scale with frequency in the way in which we would expect from the transmission line analysis routines of the linear programs. The output format of these time domain programs tends to be currents and voltages, at the available circuit nodes, as a function of time. It is only recently that versions of the programs have become available that treat S-parameters as an optional output format. Similarly, power transfer characteristics have only recently become available. All non linear programs require the non linear parameters to be characterised in terms of a response to an applied voltage or current and it is the development of adequate models for these non linear elements that has restricted the use of those programs by the microwave subsystem designer.

At the present time Microwave SPICE from EEsof is the program that has the most user friendly interface in that it shares many common syntax blocks with Touchstone. Properly written circuit files can in fact be run on Microwave SPICE for non linear analysis and Touchstone for linear analysis. Anamic, Astec 3 and Circec all support user defined models but only Anamic has the microwave engineers requirements of S-parameters, microstrip lines etc. as built in features. The other programs require the user to develop his own post processing modules in order to obtain the S-parameter information. All of these time domain programs contain small signal AC analysis in which the non linear equations are linearised at the operating point and a determination of the frequency response made. The OS/2 and workstation variants of Microwave SPICE will allow a user defined non-linear device model but do not support the linear user defined elements that are available in Touchstone. This makes the development of complex circuit descriptions difficult.

The major advantage of the time domain programs is that strongly non linear circuits, such as oscillators, can be analysed without any preconceived views to the frequency of operation, and without having to assume that the circuit is in a steady state.

Frequency Domain Programs

These programs, of which Libra from EEsof, Harmonica from Compact Software and MDS from Hewlett Packard are the commercially available versions, use the harmonic-balance method of non linear analysis. This method is iterative and is based upon the crucial assumption that for a given sinusoidal excitation there exists a steady state solution that can be approximated to sufficient accuracy by a finite trigonometric series. Immediately we can see that if the analysis is to be kept within reasonable bounds then the distortions of the waveforms must not be too great. These methods are then applicable to rather more weakly non linear systems than the time domain programs. In addition the user must specify the frequency of interest for the harmonic analysis. Whereas this may be acceptable in an amplifier, or mixer, there is a potential problem when oscillators are being considered. As we are dealing with a Fourier series solution the method is obviously only applicable to the steady state of the system. In fact the harmonic-balance method does perform a time domain analysis, in that the non linear parts of the circuit are analysed in the time domain whereas the remainder of the circuit is analysed in the frequency domain. The method of solution is to assume a set of node voltages at the fundamental and harmonic frequencies and then to calculate the currents at the nodes using linear analysis for the linear elements and time domain analysis for the non linear elements. These time domain currents for the non linear elements are then converted to the frequency domain. The nodal voltages are then iterated until Kirchoff's current law is satisfied at all the nodes. We then have the steady state solution to the circuit response. From this discussion we can see that in the linear part of the network we can use all the available elements of our linear programs, and so we are not restricted to non dispersive media as we were in the case of the time domain programs. Touchstone and Super-Compact can therefore be thought of as subsets of their respective harmonic balance programs, Libra and Harmonica. Thus the harmonic-balance method has the ability to use all of the available linear circuit elements but is unable to obtain information on spurious oscillation, start up transients or strongly non linear circuits.

Since the non linear analysis takes place in the time domain the same comments apply here as to the use of models for the non linear elements as we gave for the time domain analysis. Both Libra and Harmonica are available on all of the standard computing platforms, although of course the speed will be much slower than when we undertake simple linear analyses. However when the harmonic content is low, as in an amplifier nearing saturation, very few iterations will be needed for convergence. This would not be the case for the time domain program where a full steady state solution would have to be determined before Fourier analysis showed the anticipated low level of harmonics. In a harmonic balance analysis oscillator simulation differs from the time domain analysis as the oscillation cannot build up from noise (in this case numerical noise). We need to know the frequency if we are to solve using harmonic balance and so we need to have a way of finding this out. Having chosen a frequency the circuit is allowed to converge to the steady state; which if there is no fundamental frequency component indicates that oscillation is not allowed at the given set of circuit conditions. Iteration then allows us to determine the envelope of frequency and circuit conditions. One approach to estimating this initial starting condition is to assume that the active element (FET or BIPOLAR transistor) is operating in a large signal mode, which gives us our first estimate of the rf amplitude. In order to estimate the frequency one can assume that there will be a relatively small frequency shift between the small signal starting conditions and the final oscillation frequency. This can be estimated using simple linear analysis programs by introducing a break in the oscillator feedback path at a suitable point and plotting the conjugate of the device impedance and the circuit impedance. If oscillations are going to start then one must have some net negative resistance and equal reactances. This follows simply from the conjugate match conditions for oscillation. Figure 7 shows an example of this approach as applied to an MMIC VCO.

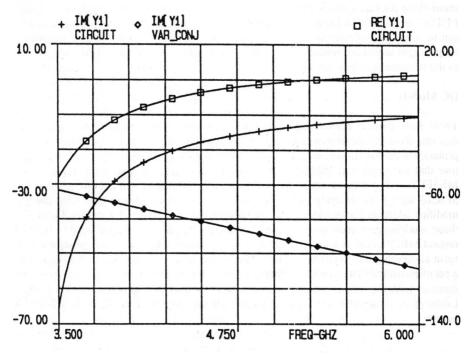

Figure 7. The small signal starting conditions in a VCO

The Figure shows the susceptance at the source of the FET in this circuit along with the conjugate of the susceptance of the tuning varactor. Oscillation will start at the frequency where these curves cross. Note also that there is negative conductance at this frequency and so oscillation can begin. In fact in the absence of a non linear CAD package this method was used successfully to design an MMIC VCO and to investigate the available tuning bandwidth. These frequencies can then be used as the starting frequencies for our large signal analysis. If the user requires information on the starting conditions or the phase transients during the movement from start up to steady state then he must resort to the time domain method of analysis.

Device Models and Parameter Extraction

From the discussion given above it is obvious that if we are to analyse a non linear circuit then we need to be able to parameterise the non linear elements as a function of applied current or voltage. It is thus necessary to derive the equations that govern the non linear behaviour of the device of interest. These equations and models fall into two categories, those that describe the dc conditions in the device and those that describe the ac conditions. The former is of critical interest to the user of a time domain program as he will need this if he is to converge to a satisfactory dc operating point. These two models are fitted to measurements of the dc I-V characteristics of the device for the dc model and to an equivalent circuit model for the ac parameters. As one might expect EEsof and Compact Software offer suitable packages for use with their respective CAD programs - these being Xtract and Scout respectively. These programs are quite complex and rely upon controlling parameter analysers and vector network analysers for the dc and rf measurements. For the user who may not wish to use these commercial packages from EEsof and Compact Software there are various useful stand-alone packages which can be used. Of particular interest are the programs MESFETDC and SOPTIM from the University of Kent which, although allied to Anamic, can be used quite effectively in model generation. We shall concentrate our discussion on packages which are related to MESFETs as these are the devices of principal interest to the microwave subsystem designer.

DC Models

There have been a considerable number of equations published in the literature to describe the I-V characteristics of MESFETs with varying degrees of success, and it is probably true that no one equation can be considered to be universal. It is certainly true that ion implanted MESFETs tend to need a different model to epitaxial grown MESFETs. The models favoured by EEsof are the models of Curtice (Microwave SPICE) and more recently Statz (Libra), whereas Compact Software are using a modified Materka-Kacprzak model. Users of these programs are thus constrained to these models until such time as fully user definable models are permitted. In this respect MESFETDC is a useful program for the generation of dc models as its modular form allows the user to define his own model equations, which can then be optimised to a set of measured data. Those same models can then be used in Anamic, or as we have done at GEC-Marconi, coded into the defined models in ASTEC3, Harmonica and Libra. One model that we are currently using for ion implanted MESFETs is given below.

$$I_{ds}(V_{gs}, V_{ds}) = I_{dss} V^{Exp} \tanh(\alpha V_d) +$$
$$(1 + V_g/V_{t1})[G_{dst} - (G_{dst} - G_{dso})V^{Exp}]V_d$$

Where:

$$V_g = V_{gs} - I_{ds}R_s$$
$$V_d = V_{ds} - I_{ds}(R_s + R_d)$$
$$V = 1 + V_g/V_t$$
$$V_{t1} = V_t(1 + \beta V_d)$$

The TANH function is a feature which is present in many models as it provides a sharp knee to the I-V characteristic as required. The terms G_{dst} and G_{dso} provide the correct behaviour of the output conductance. One of the modifications that we have introduced here is to make the threshold voltage V_t drain voltage dependent, which improves the fit to the I-V characteristics near to pinch off. Figure 8 shows the excellent results that can be achieved using MESFETDC.

Having found an acceptable model for the dc characteristics it is then necessary to

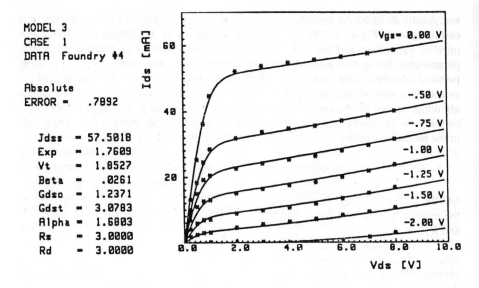

Figure 8. The DC current voltage characteristics of a GaAs MESFET
fitted to a non linear model - using MESFETDC

consider the ac characteristics. This will be discussed in the next section, where we introduce the general question of parameter extraction as this has its applications to linear as well as non linear circuits.

Equivalent Circuit Parameter Extraction

The approach that is taken is to generate an equivalent circuit model for the device, such as that shown in Figure 1, and to determine the model elements as a function of drain-source and gate-source voltage. Having obtained these model element values a parameterised set of equations can be developed so as to describe the element values as a function of the applied voltages. We see then that the extraction of the model element values is common to the linear analysis module, where we may wish to scale device size, or the requirements of our non linear program. For a long time the method that was employed by the majority of people was to use Super-Compact or Touchstone to optimise the equivalent circuit to the measured S-parameters as described previously in this chapter. This was in fact a very time consuming process, although the batch mode of mainframe Super-compact was invaluable to us in this respect. The main reason for this was that the optimisers treated all the element values as independent variables, whereas this is not the case. A more efficient way of optimising the model is to develop a software package that recognises that there is some dependency amongst the element values and to constrain the optimisation accordingly. This method has to make certain assumptions and so the models tend to be only valid for a MESFET in the saturated region. This however is a considerable step forward and we have found this approach to be advantageous at GEC-Marconi in analysing and modelling our ion implanted MESFETs. This again was an example of in-house software but similar model extraction capability is available in Fet Fitter from Cascade Microtech and GasCode from GasCode Ltd. Our approach has been to develop the equations that analytically describe the terminal characteristics of the MESFET whose equivalent circuit is shown in Figure 1. We thus generate a set of constraining equations between the intrinsic parameters and the parasitic elements which can be fixed to the measured S-parameters. Using this method the only variables that have to be optimised are the parasitic elements, and thus we can reduce the time required for optimisation. In essence this method does not waste time looking for combinations of parameters which are not permitted by the analytic equations. As an example of the constraints that apply consider the externally measured transconductance G_m^{ext} ; this relates to the intrinsic G_m in the following manner :

$$G_m = G_m^{ext} \left| 1 - G_m^{ext} \left[R_s + \frac{(R_d + R_s)}{G_m^{ext} R_{ds}^{ext}} \right] \right|^{-1}$$

and the product $G_m^{ext} R_{ds}^{ext}$ is given by $G_m R_{ds}$. Thus our estimate of G_m only requires the externally measured parameters G_m^{ext} and R_{ds}^{ext} and the two parasitic resistances R_s and R_d.

Thus when we have derived the equivalent circuit model element values a suitable set of equations can be derived to describe the voltage dependence of the important elements for use in our non linear program. SOPTIM, from the University of Kent, is designed to enable simultaneous optimisation of sets of S-parameters at various bias conditions to the non linear model elements. This would normally be associated with the models used in Anamic but these could be a user defined model. This is where one of the problems arises with much of the existing non linear FET models in that they assume a depletion capacitance behaviour for the gate-source capacitance, which may well be a reasonable approximation for a long ($> 1\mu m$) gate MESFET using epitaxial material but which breaks down when one is dealing with a short ($< 0.5\mu m$) gate length ion implanted device. Similarly a depletion capacitance approximation for C_{dg} , the drain gate feedback capacitance, is also erroneous. The overall importance of this discussion is to stress that for non linear analysis the results are only as good as the quality of the model that is used. This is particularly true for estimates of the harmonic distortion in circuits where the derivations of the non linear model equations become important.

INTEGRATED ENVIRONMENTS

We have not yet reached our goal of a fully integrated CAD and Layout environment. However the availability of Serenade from Compact Software, Academy from EEsof and MDS from Hewlett Packard is moving us in this direction. These software suites, which are available on the latest 32 bit workstations, allow both schematic or netlist entry and will generate a layout from the schematic. Also it is possible to include 'hooks' between the electrical simulator and the layout package so that customised versions, such as we have discussed with the user defined models, can put the appropriate layout cell in the correct place in the overall layout. At the present time, these packages can only produce an initial layout leaving final adjustments to the user.

THE FUTURE?

To date layout takes place after circuit design, but further developments in schematic capture and bi-directional data transfer offer the exciting possibility of circuit design from a layout orientated point of view. In order to achieve this we will require the integration of CAD packages such as Super-Compact, Touchstone, Harmonica, Libra, AutoArt II, MiCAD II, GaS Station, LINMIC+ and EMSim etc. At the present time this appears rather futuristic but if we look at the developments that have taken place in microwave CAD and computing power in the last 10 years then there could well be exciting times ahead for the microwave subsystem designer. We also look forward to the inclusion of physical models for the active devices so that we have a truly interactive design and optimisation capability.

BIBLIOGRAPHY

Snowden, C.M., 'Computer-aided design of MMICs based on physical device models' IEE Proc. Vol. 13, Pt.H, pp.419-427 1986

Besser et al 'Computer-aided design for the 1980s', IEEE MTT-S Int. Symposium Digest, 1981, pp.51-53.

Sobhy, M.I., Jastrzebski, A.K., Pengelly, R.S., Jenkins, J.A. and Swift, J.B. 'The design of microwave monolithic voltage controlled oscillators' Procs. of 15th European Microwave Conf., Paris, 1985, pp.925-930.

Baden Fuller, A.J. 'Computer optimisation of circuits applied to the modelling of microwave IC passive components' IEE Proc. Vol. 133, Pt.H., No. 5, October 1986.

Jansen, R.H. 'A novel CAD tool and concept compatible with the requirements of multilayer GaAs MMIC technology' IEEE MTT-S Symposium Digest, 1985, pp.711-714.

Suckling, C.W. 'Procedures examined for successful GaAs MMIC design' MSN and CT March 1986, pp.79-87.

Pucel, R.A. 'MMICs, Modelling and CAD - where do we go from here?' 16th European Microwave Conference, p. 61.

Jansen, R.H. 'LINMIC: A CAD package for the layout-oriented design of single and multi-layer MICs/MMICs up to mm-wave frequencies' Microwave Journal, Feb. 1986, pg.151.

Lamnabi, M. 'Functional analysis of non-linear circuits: a generating power series approach' IEE MAP, Vol. 133, Oct. 1986, pg.375.

Rizzoli, V., Cechetti, C., Lipparini, A. and Neri, A. 'User-oriented software package for the analysis and optimisation of non-linear microwave circuits' IEE-MAP, Oct. 1986, pg.385.

Sobhy, M.I., Jastrzebski, A.K. 'Computer-aided design of non-linear microwave integrated circuits' Conference Procs. of 14th European Microwave Conf., pp.705-710, 1984.

Rizzoli, V., Lipparini, A., and Marazzi, E. 'A general-purpose program for nonlinear microwave circuit design'. IEEE Trans. on Microwave Theory and Techniques, Vol. MTT-31, pp.762-70, Sept. 1983.

Herte, D. and Jansen, R.H. 'Frequency domain continuation method for the analysis and stability investigation of nonlinear microwave circuits' IEE Proc. Vol. 133, Pt.H, No. 5, Oct. 1986, 351-362.

Brazil, T., et al. 'Large-signal FET simulation using time domain and harmonic balance methods', IEE Proc. Vol. 13, Pt.H, No. 5, October 1986, 363-367.

Curtice, W.R. 'A MESFET model for use in design of GaAs integrated circuits' IEEE Trans. MTT-28, pp.448-456, 1980.

Curtice, W.R. and Ettenberg, M. 'A nonlinear GaAs FET model for use in the design of output circuits for power amplifiers' IEEE Trans. MTT-33, pp.1383-1394, 1985.

Statz, H., Newman, P., Smith, I., Pucel, R. and Haus, H. 'GaAs FET device and circuit simulation in SPICE' IEEE Trans., ED-34, pp.160-169, 1987.

Materka, A. and Kacprzak, T. 'Computer calculations of large-signal GaAs FET amplifier characteristics' IEEE Trans. MTT-33, pp.129-135, 1985.

Hartmann, K. and Strutt, M.J.O. 'Changes of the four noise parameters due to changes of linear two-port circuits' IEEE Trans. ED-20, pp.874-877, 1973.

Hillbrand, H. and Russer, P.H. 'An efficient method for computer aided noise analysis of linear amplifier networks' IEEE Trans. CAS-23, pp.235-238, 1976.

Rohde, U.L., Pavio, A.M. and Pucel, R.A. 'Accurate noise simulation of microwave amplifiers using CAD' Microwave Journal, pp.130-141, Dec. 1988.

Podell, A.F. 'A functional GaAs FET noise model' IEEE Trans. on Electron Devices, Vol. ED-28, No. 5, May 1981, pp.511-517.

Cappy, A. 'Noise modelling and measurement techniques', IEEE Trans. MTT-36, pp.1-10, 1988.

Industrial Relevance of Device Modelling

Anthony J. Holden
GEC-Marconi Materials Technology Ltd.

INTRODUCTION

Right First Time

In the industrial environment the need to get things "right first time" has never been greater than it is today. A single mistake in the design of of a circuit mask set, or the incorrect specification of an epitaxial growth sequence can cost tens of thousands of pounds in process and engineering costs and many months of delay. In the development phase of a process, each process run is very expensive and each change to a process may involve new mask designs, procurement of fresh epitaxial material and, in extreme cases, investment in expensive new processing equipment. Getting the design wrong in manufacture and having to develop processes by multiple trials alone adds up to high costs and unacceptable delays. Device modelling offers the opportunity to cut out some process iterations, understand what the critical design areas are and reduce development times and costs.

Inputs and Outputs

Device modelling is most relevant to industry when it works closely with the practical activities, taking its inputs from measurements on the actual processes involved and making outputs in a form suitable for direct comparison with device and circuit measurements. Outputs should be at a level which enables designers to predict the performance of the optimum process and, most importantly, to pin-point the technology and design areas which are not performing to specification. The input data to any model is most critical but often the most difficult to determine accurately. Important parameters such as the precise doping profile or the depth of a gate recess can often only be inferred from relatively low resolution characterisation of test structures or by reverse engineering from the device characteristics. Such problems make device modelling less accurate and less predictive. On the other hand the existence of a reliable physical model of the device allows the sensitivity of the device output characteristics to variations in these key structural parameters to be determined and this can link performance variations to particular process steps.

Modelling from Top to Bottom

The ideal modelling environment begins with well characterised process models which translate specific process conditions into resulting device structures and properties. These structures translate directly into device parameter inputs to physical device models which can simulate the field and current distributions in the device and determine the terminal characteristics. Finally the terminal characteristics can be interpreted into CAD level models (usually non-linear equivalent circuits) which are used in circuit simulators to design and predict the performance of the finished analogue or digital modules. This route has been achieved in silicon device and circuit design as recent work in the ESPRIT "STORM" collaboration has demonstrated. In silicon well characterised processes and accurate process models are now available but in III-Vs this is not yet the case. This is due in part to the more complex physics involved in many of the III-V processes (we are dealing with a two component system) and in part to the less mature state of III-V process technology and device development compared to silicon.

Overview

We will discuss the use of device modelling in industry. We begin with the critical characterisation inputs from the process and then consider the nature and content of the basic physical models. The use of these models in the design of devices is then reviewed, followed by the incorporation of the models into a complete circuit design simulator. The paper concludes with a view of modelling as a manufacturing tool which leads into a discussion of the future role of modelling in industry.

PROCESS CHARACTERISTICS

Critical Inputs

This subject is arguably the most critical and crucial stage of device modelling and yet, at least in III-V materials, it is the most neglected. Physical device models require detailed information on doping profiles, epitaxial layer structures and device geometry. In various parts of the device they require the transport parameters (mobility, saturated drift velocity or the more fundamental scattering rates) as a function of doping concentration, temperature, field magnitudes, alloy concentration etc. In most devices it is not adequate to determine these parameters for bulk material but the effects of surfaces, junctions, dopant and implant interfaces and metal contacts may need to be taken into account. Furthermore the base material may be an implanted substrate or an epitaxial layer, it will certainly have undergone various heat treatments and "damage" processes (isolation implants, mesa etches etc) during the device and circuit process stages. All this adds up to the need to characterise the process you are attempting to model and not to assume that bulk GaAs is a universal well defined material.

Contacts and Surfaces

Where models can often depart from industrial reality is at surfaces and at metallic contacts. Depending on the technology a surface may have become damaged, have surface trace materials or have been deliberately passivated. Surfaces may also behave differently if they are on the edge of a mesa rather than the top, or if they have had an implant put through them. Contacts can be ohmic or insulating, they may be onto "p" material or "n" material, they may be diffused by an annealing process or "refractory" (having a barrier which prevents the contact metal entering the semiconductor). In MESFET style devices ohmic contacts can diffuse right through the structure leading to a contact "volume" rather than a surface. Conversely in a vertical structure like a HBT the emitter and base contacts go into thin layers and deep diffusion will lead to a shorting of the emitter-base or base-collector junctions.

Description of contacts and surfaces in a device model is not, therefore, a simple boundary condition problem and may in general involve modelling a complex two dimensional region. For this, accurate characterisation of these regions is necessary and thought must be given to the design of test structures and the use of electrical measurements microscopy and spectroscopy techniques to provide accurate profiles, bulk and sheet resistivities and contact resistances.

Materials Characterisation

Material parameters are only as good as the material they describe. This in turn is only as good as the characterisation which has been carried out on it. It should be born in mind that most profiling techniques such as SIMS and Auger are destructive and so can only be carried out on a test wafer grown at the same time or in the same batch run as the wafer used to process devices. Confidence is needed in the reproducibility of the material supplied and this must be built up by close liaison with the grower and a careful analysis to ensure that the growth rates and times set to achieve a particular specification do indeed achieve that specification on subsequent characterisation and analysis.

Some growth details which cannot be resolved using standard techniques can , nevertheless be crucial to device performance. For example, a 100Å movement of the tail of the p doping in the base of a HBT towards the emitter can have a catastrophic effect on the current gain (this moves the p-n junction into wider gap material) but such a movement is well below the resolution of SIMS. For such effects it is often necessary to use the transistor itself to detect variations, provided, that is, that this particular parameter is the only one that is varying! Another area where accurate device models are invaluable.

In MESFETS the base material is sometimes a GaAs substrate with one or more dopant implantations added. The precise profile produced depends in a complex way on dose, energy and anneal conditions with often a combination of deep high energy implantations with shallow, low energy implantations being used to create a reasonably uniform distribution. Since the gate is normally recessed the precise position of the channel may vary, making an accurate knowledge of the distribution of activated dopant critical to device performance (modelled and actual!). Models which can accurately describe the distribution of dopant following an implant and anneal sequence are not readily available for GaAs although some reasonable models do exist. It is important to fit the models to detailed characterisation of the actual implanted layers to ensure that correct data is input to the device models.

Experience shows that only in the case of epitaxial grown material can sufficient control be exercised over layer thicknesses and dopings to allow really accurate device models to be derived. Even here great care must be taken to control and characterise the grown layers.

Mask Layout and Design Rules

To complete the inputs to the device models the lateral dimensions must be determined. Here again care is needed since the specified metal widths and spaces may differ from those on the mask. For example, mesa edges tend to undercut the masks making the finished width up to $0.5\mu m$ narrower at each edge. Tolerances must also be considered especially where dimensions are close to the design rule minima.

Mesa Etching and Metal Deposition

The shape of the mesa recess can be very important. This is crystal orientation dependent (acute in one direction, obtuse in the orthogonal direction) and the precise shape and depth of a recess can have a profound effect on such things as lateral depletion region pinning in MESFETs and the corresponding gate-drain capacitance.

Models may need to take into account the shape and coverage of metals deposited into a recess. This includes the thickness variation and the effect of poor step coverage etc. For very small feature sizes the cross-sectional shape of the metal may also need to be taken into account. For example, a $0.2\mu m$ gate may be widened at the top to reduce resistance, this may lead to additional parasitic capacitance if the metal is close to or touching a passivation layer for example.

Models which predict the result of dry or wet etching and metal deposition by sputtering can be used to establish accurate mesa and metal shapes for input to device models.

PHYSICAL MODELS

Why Physics?

Much of engineering design is done using equivalent circuit models which have a limited topology and are fitted to measured s-parameters, usually at one bias point. What physics there is is buried in the equivalent circuit topology. This type of model provides a tractable representation of the device for use in circuit simulators, its use as a true design and investigative analysis tool is dangerously limited. To understand what makes a device tick (and more often than not why this particular device does not tick), you need a physical model. It is this aspect of proper device modelling which is probably of most relevance to industry.

Types of Physical Model

Physical models can be developed at many levels. These range from the simplest one dimensional Poisson solver to a full 3D finite element or Monte Carlo simulator. The model must contain enough of the right physics to determine the basic action of the device and the complexity of the model varies with the requirement. A one dimensional Poisson solver is extremely useful in determining the appropriate layer structure for a HEMT device. A coupled one dimensional Poisson solver and drift-diffusion transport model is adequate for understanding and optimising the alloy grading and doping profiles in a HBT. However, to understand the effect of recess shape on the capacitance of a MESFET or the surface and interface recombination currents in the emitter-base region of a HBT a more sophisticated 2D model with appropriate surface traps and geometrical structure is essential.

For power devices and high speed integrated circuits the effects of temperature may also need to be taken into account since the devices are likely to run at well above ambient temperature. Self consistent solutions considering the temperature distribution in the device due to dissipated electrical power and its effect on the device electrical parameters will be required in these circumstances. Significant degradation in gain and efficiency can occur in HBTs for example if adequate heat sinking is not supplied.

Relevance of Physical Models to Industry

Having a physical model in industry is important in at least three ways. First we can use a physical model to take us from our fully characterised device structure, described in the previous section through to device terminal characteristics, without actually making the device. In this way we can vary the device design to achieve the required performance specification without having to make repeated costly process runs. Second, when we have a batch of working devices we can use the model to determine in which areas of the technology the design is deficient or the process is insufficiently in control. This helps with trouble shooting, fault finding and enables us to monitor the process from batch to batch and ensure that all the stages continue to meet the design targets. Third we can use the model to explore the sensitivity of the device performance to changes in the design parameters. This enables us to make a more robust, yieldable device which is reliable.

DEVICE AND CIRCUIT DESIGN

Device modelling can be very relevant when a totally new device concept is being developed or when an existing device is to be developed for a new task. Recent examples would be the development of the pseudomorphic HEMT (PM-HEMT) to replace the conventional HEMT, the development of the power mm-wave PM-HEMT from the standard discrete J band device and the implementation of the discrete PM-HEMT into a 40GHz MMIC process. Modelling is also important in the development of novel analogue circuit functions in existing device types. For example the development of a logarithmic amplifier using HBTs, a low noise oscillator using HBTs or a 35 GHz mixer using PM-HEMTs. There are also important examples in the digital area such as the development of a K-band 4-bit counter for frequency detection, using HBTs.

Pseudomorphic HEMTs

In developing a new device, modelling is needed to understand the performance advantages to be gained, the optimum structure and layout, the potential risks to reliability of, and the sensitivity to, process variations. In the case of the PM-HEMT, rather like its predecessor the conventional HEMT, the right device was built for the wrong reasons. The original argument ran that the introduction of indium into the channel of a HEMT would increase the mobility and hence enhance the performance. When the devices were made they did indeed show enhanced performance but careful modelling showed that there was in fact no detectable increase in the channel mobility (the intrinsic increase in mobility observed in unstrained InGaAs is cancelled out by the change in effective mass brought about by the strain induced by pseudomorphic growth onto GaAs). Modelling shows that the real performance enhancement is due, largely, to the increased conduction band offset.

Power mm-wave PM-HEMTs

Here the device is required to move up in frequency and handle more power. A proper physical model will handle the large signal operation of the device giving indications of what compromises in design will be necessary to delay the onset of compression. An investigation of field distributions will indicate how design modifications can push the onset of breakdown to higher voltages. Finally a study of the frequency dependence will determine what gate dimensions and device layer specifications will be necessary to achieve the target frequency.

PM-HEMTs for MMICs

The role of device modelling becomes critical when integration is considered. Not only must the individual devices perform according to specification, they must be designed to perform within the environment of other active and passive components. Parameters which may not be very sensitive in a discrete device with external matching may become critical if they affect the interaction with other parts of the circuit. An MMIC leaves no flexibility for tuning the match between components after manufacture. In an MMIC the number of possible variables multiplies alarmingly and the cost of repeated iteration of the circuit process becomes unacceptable. The only viable alternative is accurate circuit modelling based on well characterised physical models.

Analogue Circuits

Limiting and logarithmic amplifiers, oscillators and mixers are very valuable components which need a small number of transistors in a compact circuit. The critical aspect here is non-linearity. The traditional approach of using single bias point equivalent circuits fitted to s-parameters will not work here. Non linear circuits are required with components which vary with bias across the full extremes of device operation. Again characterisation of the devices is essential with a full DC analysis and s-parameters measured at multiple bias points in the whole space. However to provide a non-linear model robust enough to give accurate answers in the latest non-linear simulators such as LIBRA^{TM} or HARMONICA^{TM} it is essential that a physically correct model is used as the basis.

Digital Circuits

As with the analogue circuits the non-linear performance of the devices is tested to the limit in a digital circuit and the DC model becomes an integral part of the problem (for level setting etc). Often digital circuits are simulated in time domain "SPICE" like simulators rather than the harmonic balance techniques referred to under analogue circuits. Physical models again allow you to optimise the device design within its circuit environment and link the circuit performance back through the device structure to the process steps which made it.

COMPUTER AIDED DESIGN

Integrated Design

The trend within industry has been towards a more integrated design capability. Computer aided design is now expected to allow designers to assemble libraries of active and passive units into working circuits, predict and then optimise the circuit performance and then lay out the circuit. At each stage the appropriate design rules have to be implemented and the final output is a full mask set which is dumped to tape ready for manufacture.

The outputs from device simulation must be compatible with this integrated approach. Traditionally devices are described by equivalent circuit models which are electrical analogues of the device and respond to inputs in the same way as the device but over a limited range of operation. Limitations include frequency range, bias condition and amplitude of input and output signals. Most equivalent circuits are 'small signal' representations at a fixed bias. Different bias conditions can be simulated by changing the circuit elements but large signal operation requires bias dependent circuit elements described either by an analytical function or a "look up table" of values. The current CAD simulators have been largely small signal in the analogue microwave field although digital circuits have required large signal descriptions from the outset. However, the need to simulate large signal analogue circuits such as power amplifiers, oscillators, mixers, limiting amplifiers etc has led to the introduction of non-linear simulators into many of the CAD packages.

Relevance of Device Modelling to CAD

A crucial factor in making useful CAD design tools is the provision of active and passive device models which describe the actual devices in the process which is to be used to fabricate the circuit. It is here that device modelling should be making a direct impact. To date that has not been the case! Small signal equivalent circuits are normally fitted by optimisation routines to measured s-parameter data taken directly from the processed devices. This has the important advantage that the inevitably imperfect devices with their associated parasitics are used in the design without the need to understand the source of such imperfections. The disadvantage is that you must have a well controlled, high yield, high uniformity process in place and running before the CAD can begin. The CAD cannot be predictive, indicating the potential performance of the next generation process and , perhaps even more importantly, the CAD cannot respond to variations in the process without a complete re-characterisation of the devices. Such variations may be intentional, due to a change of equipment or source material for example, or unintentional due to a variety of factors including non-uniform base material, poor control of a particular process step etc.

To extend the capability of CAD and make it robust enough to cope with rapidly changing processes, device simulation must make inputs to the device models. In particular it is necessary to provide some physical basis to the equivalent circuits which are used, especially when full non-linear dependence is required. Blind fitting of fixed equivalent circuits to measured data can be very slow and may lead to unphysical values for some parameters. This may be because the basic topology of the circuit is inadequate to describe the device (this is particularly relevant for new novel devices which may not behave in the same way as more traditional devices) or it may be that the physical correlations between the intrinsic parameters have not been properly allowed for. In a MESFET parameters such as C_{GS} and g_m are correlated to physical properties of the device such as gate geometry, a simple fit to s-parameters will vary them independently. Conversely, variations in the physical properties of the device will affect combinations of equivalent circuit parameters. Variations in the gate recess depth will cause variations in C_{GS}, g_m, C_{GD} etc. These effects are accentuated when the devices are operated in large signal.

CAD needs to be capable of accurate design on well characterised processes, but it also needs to be predictive and responsive to developments and variations. To achieve this a careful combination of device modelling through physical models and detailed characterisation and fitting to experiment must be achieved in the creation of device models. This requires the equivalent circuits (or more sophisticated analogues if such can be devised) to be rooted in the physical reality of the device AND the device simulations to be rooted in the physical realities of the process (tolerances, parasitics, complex geometries etc).

SUPPORT FOR MANUFACTURE

In addition to basic design, device simulation has a role to play in the manufacture of devices. Much is made of the use of simulation to "optimise" device or circuit performance. To optimise what? Usually what is meant is to make it go faster or with more power with smaller feature sizes. In manufacturing it is necessary to step back from this and, not an "optimum" device but rather an "appropriate" device is required.

In manufacture, a circuit has to be produced time and time again. To do this it must be robust and very tolerant to the inevitable variations in the process from run to run and batch to batch. Device simulation can help here by showing where tolerances can be relaxed without detracting from the required performance and where a particular parameter is crucial to meeting the specification. Such an analysis is required, not just for the device, but for the complete circuit. A parameter which makes little difference to the individual device performance may be more critical to the overall circuit performance, because of its effect on inter-stage matching for example.

An accurate device model used within a circuit simulator can be used to investigate the effects of statistical variations in the process. These can include layer thickness, doping density or profile, recess depth or shape, gate length etc. Very quickly a picture can be built up of where the tolerances can be relaxed and where significant and perhaps expensive changes in the process have to be made to ensure high yield of circuits.

Without this modelling support the process engineers have to design test experiments to vary as many of the parameters as they can in a controlled way. This is not only very expensive and time consuming but the experiments themselves are subject to the kind of statistical variations which we are seeking to investigate and this may mask the results of the "controlled" experiment.

THE FUTURE RELEVANCE OF DEVICE MODELLING

In the medium term device modelling has to find its way into the main stream of device design and manufacture. CAD models still rely largely on measured data fitted to simple equivalent circuits and process control is established by laborious experiment. Modelling will only break into this field when it has established cast iron links with the processes it seeks to simulate. This means exhaustive process characterisation linked to accurate model development. To be predictive it must also include accurate process modelling making direct inputs to the device models.

The advent of new device and circuit technologies will put new demands on modelling. There are still many fundamental problems in material properties especially in the region of heterojunctions which need to be addressed and as devices become smaller and layers

280

thinner new physics related to fluctuations in doping and interface morphology will need to be taken into account. Quantum effects in the so called "Mesoscopic" devices are now a real prospect for completely new circuit concepts and device models which merge the classical with the quantum will need to be developed.

The move to higher frequencies and the need for higher packing densities to reduce cost will demand a more integrated approach to modelling. Interactions between devices and circuit elements will become important and the microwave discontinuities at device inputs and outputs will have to be taken into account as whole circuits become "active waveguides".

Overall the emphasis will be on fully integrated systems allowing designers to call upon fundamental device models which are fully characterised against a specified process and insert them into full circuit layouts. Process variations can then be fed in and a full statistical analysis performed. The designer will then be able to develop appropriate manufacturable designs which meet the required specification.

Subject Index